Lecture Notes in Computer Science 1156

Edited by G. Goos, J. Hartmanis and J. van Leeuwen

Advisory Board: W. Brauer   D. Gries   J. Stoer

Springer
Berlin
Heidelberg
New York
Barcelona
Budapest
Hong Kong
London
Milan
Paris
Santa Clara
Singapore
Tokyo

Arndt Bode   Jack Dongarra
Thomas Ludwig   Vaidy Sunderam   (Eds.)

Parallel Virtual Machine – EuroPVM '96

Third European PVM Conference
Munich, Germany, October 7-9, 1996
Proceedings

Springer

Series Editors

Gerhard Goos, Karlsruhe University, Germany

Juris Hartmanis, Cornell University, NY, USA

Jan van Leeuwen, Utrecht University, The Netherlands

Volume Editors

Arndt Bode
Thomas Ludwig
Technische Universität München, Institut für Informatik
D-80290 München, Germany
E-mail: (bode/ludwig)@informatik.tu-muenchen.de

Jack Dongarra
University of Tennessee, Department of Computer Science
Knoxville, TN 37996-1301, USA
E-mail: dongarra@cs.utk.edu

Vaidy Sunderam
Emory University, Department of Mathematics and Computer Science
Atlanta, GA 30322, USA
E-mail: vss@mathcs.emory.edu

Cataloging-in-Publication data applied for
Die Deutsche Bibliothek - CIP-Einheitsaufnahme

Parallel virtual machine : proceedings / EuroPVM '96, Third
European PVM Conference, München, Germany, 7 - 9, 1996.
Arndt Bode ... (ed.). - Berlin ; Heidelberg ; New York ;
Barcelona ; Budapest ; Hong Kong ; London ; Milan ; Paris ;
Santa Clara ; Singapore ; Tokyo : Springer, 1996
  (Lecture notes in computer science ; Vol. 1156)
  ISBN 3-540-61779-5
NE: Bode, Arndt [Hrsg.]; EuroPVM <3, 1996, München>; GT
CR Subject Classification (1991): D.1.3, D.3.2, F1.2, G.1.0, B.2.1, C.1.2,
C.2.4, C.4

ISSN 0302-9743
ISBN 3-540-61779-5 Springer-Verlag Berlin Heidelberg New York

© Springer-Verlag Berlin Heidelberg 1996
Printed in Germany

Typesetting: Camera-ready by author
SPIN 10525573   06/3142 – 5 4 3 2 1 0   Printed on acid-free paper

Preface

High performance computing and networking is one of the key technologies for the development of future products. This is reflected by the large number of funding measures for this field in the U.S., Japan, and the European Union. High performance computing can be delivered by various parallel and distributed computer architectures. Because of the availability of a large number of machines interconnected by high speed networks, the idea of distributed computing, i.e., using a number of potentially heterogeneous machines as one single computer for high performance applications, is quite natural.

To support distributed computing, the parallel virtual machine PVM was developed at the University of Tennessee and Oak Ridge National Laboratory in collaboration with Emory University and Carnegie Mellon University. This software package permits heterogeneous collections of UNIX and Windows/NT computers hooked together by a network to be used as a single large parallel computer. PVM uses the paradigm of message passing for communication and synchronization. This form of explicit parallelism has been adopted to solve parallel scientific computing applications as well as general purpose applications.

The first version of PVM was available in 1991. Since that time the number of PVM users all over the world has grown considerably and PVM has been further developed regarding functionality and reliability.

This is the third in a series of European PVM users' group meetings. It is held at Technische Universität München October 7–9, 1996 after Lyon 1995 and Rome 1994. It covers PVM in its version 3.4, applications, programming environments, tools, fast networks, and enhancements. Tutorials on programming in PVM and tools for PVM are also covered, as well as full scientific papers, posters, and vendor sessions. Invited lectures cover the relation between PVM and MPI as well as the newest developments in PVM version 3.4. The meeting provides an opportunity for European and international users of PVM and other programming environments to meet each other, share ideas and experiences, and meet some of the members of the PVM team.

July 1996

Arndt Bode
Program Chair
EuroPVM'96

Organization

EuroPVM'96 is organized by the *Lehrstuhl für Rechnertechnik und Rechnerorganisation / Parallelrechner* (LRR-TUM) at *Institut für Informatik, Technische Universität München.*

Executive Commitee

Program Chair: Arndt Bode (LRR-TUM, Germany)

                                     Jack Dongarra (Univ. of Tennessee / ORNL, USA)

                                     Thomas Ludwig (LRR-TUM, Germany)

                                     Vaidy Sunderam (Emory Univ., Atlanta, USA)

Organizing Chair: Roland Wismüller (LRR-TUM, Germany)

Organization Committee: Rolf Borgeest, Bernward Dimke, Christian Röder (LRR-TUM, Germany)

Program Committee: A. Bode (LRR-TUM, Germany)

                                     R. Barrett, (Los Alamos National Lab, USA)

                                     A. Beguelin (CMU, USA)

                                     M. Bubak (Inst. of Computer Science, AGH, PL)

                                     H. Burkhart (Univ. of Basel, CH)

                                     J. Cook (Parallab, Univ. of Bergen, Norway)

                                     M. Cosnard (LIP ENS, Lyon, F)

                                     K. Decker (CSCS/SCSC, CH)

                                     F. Desprez (LIP ENS, Lyon, F)

                                     J. Dongarra (Univ. of Tennessee / ORNL, USA)

                                     G. Fagg (Univ. of Reading, UK)

                                     A. Geist (Oak Ridge National Labs, USA)

                                     M. Gengler (LIP ENS, Lyon, F)

                                     W. Gentzsch (Genias, Germany)

                                     B. Hirsbrunner (Univ. of Fribourg, CH)

                                     O. Krone (Univ. of Fribourg, CH)

                                     T. Ludwig (LRR-TUM, Germany)

                                     B. J. Overeinder (Univ. of Amsterdam, NL)

                                     G. Scheuerer (ASC, Germany)

                                     W. Schreiner (RISC, Univ. of Linz, Austria)

                                   P. Sguazzero (Italy)

                                   G. Stellner (LRR-TUM, Germany)

                                   V. Sunderam (Emory Univ., Atlanta/USA)

                                   O. Sykora (Slovak Academy of Sciences, Slovakia)

                                   B. Tourancheau (LIP ENS, Lyon, F)

                                   P. Tvrdik (Czech Technical Univ., Czech Republic)

                                   M. Valero (UPC, Barcelona, Spain)

                                   V. Van Dongen (CRIM, Canada)

                                   R. Wismüller (LRR-TUM, Germany)

Sponsoring Institutions

TU München, SFB 342
ASC GmbH
Cray Research GmbH
Harms Supercomputing
Stiftungsfonds IBM Deutschland
Intel GmbH
Siemens Nixdorf Informationssysteme AG
SUN Microsystems GmbH

Contents

Tutorials

Invited Talk

Session F1: Evaluation of PVM

Session F2: Applications: CFD Solvers

Session F6: Applications: Solvers

Vendor Sessions

Session F7: Extensions to PVM (2)

Session F8: Applications: Miscellaneous

Session F9: Implementation Issues

Session F10: Programming Environments

Session F11: Load Distribution/Load Balancing

Session F12: Implementation Issues (2)

Posters

Advanced Programming in PVM

G. A. Geist *

Oak Ridge National Laboratory, USA

Abstract. This paper describes many of the advanced features available in PVM (Parallel Virtual Machine) and how to take advantage of these features to create sophisticated distributed computing applications. The reader is expected to already have an understanding of parallel and distributed computing and a knowledge of the basic features provided by the PVM software. Topics covered in this paper include improving application performance, adding fault tolerance, building interactive applications, debugging, developing PVM "plug-ins", and a brief overview of the expanding scope of PVM with the creation of packages like CU-MULVS, JavaPVM and TkPVM.

1 PVM Programming Model

PVM's programming model begins with the concept of a *Virtual Machine*. This machine is composed of one or more *hosts* that can communicate with each other using standard TCP/IP network protocols. The hosts in turn can be any heterogeneous collection of workstations, supercomputers, PCs, and even laptops. ¿From the programmer's view, the virtual machine is a single large distributed memory parallel computer. In the PVM programming model there can be an unlimited number of hosts in the virtual machine (although the actual implementation limit is 4096 hosts each of which can be multiprocessors with up to 2048 nodes). The programming model allows hosts to be added and deleted dynamically from a virtual machine.

In the model the hosts are assumed to have sufficient memory to run a given application. This means it is up to the user not to exceed the virtual memory execution limits on any host.

Processes running on the virtual machine are called *tasks*. The model assumes that many tasks can run on each host and that any task can communicate with any other task. In the cases where the hosts can not multitask, it is up to the user to assign tasks to hosts in an appropriate way. The PVM programming model supports a dynamic task pool where tasks can be added or killed inside a larger running application.

The model also supports dynamic *groups* – sets of tasks that logically work together. Communication between groups of tasks such as broadcast, use the

---

* This work was supported in part by the Applied Mathematical Sciences subprogram of the Office of Energy Research, U.S. Department of Energy, under Contract DE-AC05-96OR22464 with Lockheed Martin Energy Research Corporation

group model. Tasks can join and leave groups without synchronizing with any other group members, and using group names, it is possible to for independently started PVM tasks to communicate with each other.

The PVM software is designed to manifest the above model. When a host is added to the virtual machine, a daemon process called pvmd is started on it. Even in the case of a multiprocessor host there is only one pvmd per host. The pvmds perform the distributed control and message routing for the virtual machine. The second part of the PVM design is libpvm, which is the library the users link to their applications in order to access the PVM functions. The third part of the PVM design is the task identifier TID. All tasks inside a virtual machine including the pvmd, group server, and user tasks all have unique TIDs across the entire virtual machine.

Communication between PVM tasks inside a multiprocessor uses the vendor's native functions to move data. For example, on an SGI, PVM uses shared-memory functions to move messages between tasks. On a Paragon, PVM uses NX to transfer messages between tasks, and on the SP-2, PVM uses IBM's MPI to move messages. Between hosts, PVM uses UDP to send data between the pvmds and uses TCP to send data directly between tasks. Unix domain sockets are used to move data between tasks and the local pvmd. All the above choices are made transparently to the application. In all cases the application simply sends data between two tasks in the virtual machine.

In the past year the Message Passing Interface (MPI) specification has been released and several implementations exist. People developing parallel applications often wonder whether to use MPI or PVM. Each package has its strengths and weaknesses. MPI was designed to provide a standard message passing API for parallel computers. Vendors are expected to provide and support high performance implementations for their systems. Given this, MPI is expected to give better performance for applications that are run predominately within a single multiprocessor. MPI programs are portable in the sense that the application should be able to run on another vendor's multiprocessor with little or no source changes.

PVM is a better choice in cases where the application is run across a heterogeneous network of computers. None of the vendor's MPI implementations can communicate with each other or even outside their own multiprocessors. Thus it is not possible for for a task using SGI's MPI to send a message across the network to a SP2 task using IBM's MPI. MPI also does not include any concept of a virtual machine – it is strictly a message passing API. PVM is a better choice when an application requires fault tolerance. The PVM model supports the dynamic adding and deleting of both hosts and tasks and supplies notification functions to determine when a fault occurs.

The flexibility of PVM model comes at the cost of peak communication performance. In the next section we describe ways to tune PVM applications to achieve higher performance.

2   Improving PVM Application Performance

PVM provides several options for sending messages, allowing the programmer to tune his application's communication. The standard three step method to send a message in PVM involves initializing a buffer, packing data into the buffer, and sending the data to the destination task.

The packing options are set on a per message basis in the pvm_initsend() call. They are PvmDataDefault, PvmDataRaw, and PvmDataInplace If the virtual machine contains a heterogeneous mix of data formats, then PvmDataDefault must be used. The default method converts all data into XDR format needed for heterogeneity, but which slows down packing and unpacking of messages. If the data formats between sender and receiver are the same, then PvmDataRaw can be used which avoids the conversion overhead. The raw method still has the cost of copying the data into the send buffer. This overhead can be avoided by using PvmDataInPlace. With PvmDataInPlace set, only pointers to the data are placed in the send buffer. At send time the data is copied directly from user's memory to the network. There are three restrictions to using PvmDataInPlace. First, only contiguous blocks of data can be packed. Second, sender and receiver must have the same data format. Third, if data is changed after packing, the changed values are sent.

The send options are set in the pvm_setopt() routine. PvmRouteDefault is the default setting. This method is scalable and should work in all cases. In the default method messages are routed and buffered through the pvmds. A potential 4X bandwidth increase can be made by setting PvmRouteDirect. The direct method sets up a TCP socket directly between the two tasks bypassing the pvmd. In general the direct method should be used, but be aware that setting up the TCP socket has a one-time very large latency. It only makes sense if many messages or a large volumes of data are going to pass between these two tasks. Also be aware that Unix operating systems only allow a fixed number of TCP sockets be set up. The number varies from system to system and can be as low as 32 and as high as 64,000. When no more are available, PVM automatically uses the default method.

To get performance equal to vendor's MPI implementations on MPP, applications should use pvm_psend()/pvm_precv() pair. Psend combines the three sending steps into one and matches very closely with vendors' native functions for moving data. On the SP-2 it maps directly to mpi_send(). When run on a network of hosts, psend is mapped to PvmDataInPlace and pvm_send with PvmRouteDirect.

¿From an application design standpoint consider the following two observations. First, limit the size and number of outstanding messages. These messages have to be buffered inside the PVM system and this increases overhead within PVM. Second, numerous tests have been conducted were PVM was connected to an ATM network. This boosted message bandwidth by 5X over the best ethernet speeds. All real applications run on this virtual machine showed little or no improvement in total execution time despite the higher bandwidth. Why? Because the execution time is much more sensitive to load balancing the application than improving the communication bandwidth.

The bottom line is work on improving the load balance of an application first, then spend time improving the communication performance.

3 Adding Fault Tolerance

There are two types of fault tolerance in PVM. There is virtual machine fault tolerance and application fault tolerance.

Virtual machine fault tolerance is analogous to having a single large multiprocessor continue to operate even if a few nodes fail. And being able to hot-swap in new nodes. In PVM, the pvmds form a distributed control system that watch for and detect the loss of a host or network connections. When this happens, PVM automatically reconfigures the virtual machine around the loses and continues running. New hosts can be added on the fly to replace lost resources, but this is not automatic. It is left up to the user.

Application fault tolerance refers to the ability of an application to detect and recover from the loss of one or more tasks. These tasks may have been lost when a host crashed, or they may have been killed by the owner of a workstation. PVM does not guarantee that an application can survive a fault, but PVM provides functions required to allow an application to detect and recover from a failure.

If a task calls **pvm_notify()**, then if the specified fault occurs, this task will be sent a message with a special message tag. Depending on the type of fault, the task may use **pvm_spawn()** to start up replacement tasks and/or **pvm_addhost()** to replace hosts.

4 Building Interactive Applications

PVM tasks are usually either computing or blocked in receive. But there are distributed applications that would like to multiplex PVM communication with other I/O sources. If the application waits in pvm_recv(), then other input will starve. If the application waits on other input, then PVM messages will starve.

PVM provides two methods to multiplex communication. The simplest method is to poll both sources, using **pvm_trecv()**. Trecv is a blocking receive with a timeout. a drawback to the polling method is that it wastes some CPU cycles. The second method is to wait for messages on PVM sockets. The function pvm_getfds() returns the file descriptors of sockets used by PVM. The application can add these FDs to its own I/O FDs in a **select()** . Before using getfds there are a few things to be aware of: It is only available on Unix systems. Getfds is simple to use with PvmRouteDefault, but is messy with PvmRouteDirect. Pvmd is always **fds[0]**. And finally, Socket ready doesn't mean message received. It may be just a fragment, or there may be more than one message waiting.

Another useful technique for interactive applications is to hide PVM from users. Normally, the virtual machine is started before the application. But there are cases where an application might want to start PVM, add hosts to the virtual machine, do some work, and then shut down PVM. The two routines designed for this are pvm_start_pvmd and pvm_halt. The advantage of using these two functions is that application users don't have to know anything about PVM. The PVM console and XPVM make use of these functions to startup PVM for the user.

5   Debugging Tips

Debugging distributed applications can be much more difficult than debugging serial programs because there are several programs running, several different hosts and potentially compilers involved. Subtle synchronization errors can be caused by race conditions between programs that disappear during debugging.

One of the quickest ways to check for problems in a PVM application is to start it inside the XPVM graphical user interface for PVM. Even though the primary function of XPVM is to visualize what is going on inside a PVM application, the interface actually performs four separate functions. First, XPVM is a graphical console allowing a user to start PVM, add hosts, spawn tasks, reset the virtual machine, and stop PVM all by clicking buttons. Second, XPVM is a real-time performance monitor. While an application is running, XPVM displays a space-time diagram of the parallel tasks showing when they are computing, communicating, or idle. The display also animates the message traffic between tasks. Third, XPVM is a "call level" debugger. The user can see the last PVM call made by each task as the application is running. Fourth, XPVM is a post-mortem analysis tool. XPVM writes out a trace file that can be replayed, stopped, stepped, and rewound in order to study the behavior of a completed parallel run.

If the extent of debugging is adding print statements to the different PVM tasks, then the easiest way to be sure these print statements are seen is to use the PVM console. Start the application from the console with the command

   pvm> spawn -> appname

All prints generated within any of the PVM tasks will then appear in the console window. The user should be sure to flush the print statements so that they don't get buffered within the runtime environment.

Several groups around the world have developed parallel debugging tools that work with PVM. A list of these packages is available on the PVM Web page (http://www.epm.ornl.gov/pvm).

6   Customizing PVM - "Plug-ins"

There are three plug-in interfaces defined in PVM. These allow developers to extend the PVM environment and add additional functionality. These interfaces were first requested by developers of distributed resource managers and parallel debuggers who wanted to allow PVM applications to be used with their software.

The interfaces are designed so a user written task can take over some of the functionality of the pvmd. If a task calls **pvm_reg_rm()**, then PVM defers all of its resource management decisions to this task. Examples would include deciding what hosts to add or delete from a virtual machine, where to place new tasks on a virtual machine, and when to migrate tasks. If a task calls **pvm_reg_hoster()**, then PVM defers all of the responsibility for accessing and starting new pvmds on the selected hosts. This can be the same task as the resource manager or it can be a separate task. The ability to create a hoster allows tasks to be started in environments that do not support standard PVM (Unix) startup, such as Windows NT and managed workstation farms. If a task calls **pvm_reg_tasker()**, then PVM defers all of requests to start new tasks on this host to the calling task. This ability is important for debuggers which need to start a process in order to have control to stop and breakpoint the process.

There can be only one tasker per host, and only one hoster per virtual machine. Resource management can be performed by an arbitrary collection of processes. All protocol definitions related to the plug-in interfaces are defined in pvm3/src/pvmsdpro.h.

7 Users expand PVM's scope

In addition to the ability to customize PVM, many groups around the world have developed PVM ports to other languages, and architectures. These ports greatly expand the scope of problems that can be attacked with PVM and heterogeneous distributed computing in general. Links to these packages can be found on the PVM Web page. Details about these packages is being covered in a separate tutorial at this conference. Below are listed some of the more interesting and recent software packages related to PVM.

PVM ported to WIN'95 and NT
CUMULVS - adding visualization and steering to PVM applications
PVMPI - allows applications to use features of both MPI and PVM
JavaPVM - allows Java applications to use PVM
Perl-PVM - Pearl extension to PVM
TkPVM - allows tcl/tk programs to use PVM
LispPVM - allows Lisp programs to use PVM

Tools for Monitoring, Debugging, and Programming in PVM

Adam Beguelin[1] and Vaidy Sunderam[2]

[1] Carnegie Mellon University, Pittsburgh, PA 15213, USA
[2] Emory University, Atlanta, GA 30322, USA

Abstract. In this paper we briefly describe several tools related to programming in PVM. The monitoring and debugging systems covered are XPVM, PGPVM, and PVaniM. XPVM is a graphical console for executing and tracing PVM programs. It uses PVM's built in trace facility. PGPVM is also a system for tracing PVM programs, but uses its own buffered tracing techniques and generates tracefiles for use with the ParaGraph visualization tool. PVaniM provides online and postmortem visualization but utilizes sampling rather than tracing. This sampling approach allows PVaniM to also provide steering. Systems which augment PVM are also presented. These include PIOUS, PVMRPC, tkPVM, JavaPVM, and JPVM. PIOUS is a system that supports parallel I/O from PVM programs. PVMRPC supports the remote procedure style of programming using PVM for the underlying messaging and process control. The tkPVM and JavaPVM systems are interfaces to PVM from the TCL and Java languages. JPVM is an implementation of PVM written in Java.

1   XPVM — A Graphical PVM Console

It is often useful and always reassuring to be able to see the present configuration of the virtual machine and the status of the hosts. It would be even more useful if the user could also see what his program is doing — what tasks are running, where messages are being sent, etc. The PVM GUI called XPVM was developed to display this information, and more.

XPVM combines the capabilities of the PVM console, a performance monitor, and a call-level debugger into a single, easy-to-use X-Windows interface. XPVM is available from netlib in the directory pvm3/xpvm. It is distributed as precompiled, ready-to-run executables for many architectures. The XPVM source is also available for compiling on other machines.

XPVM is written entirely in C using the TCL/TK toolkit and runs just like another PVM task. If a user wishes to build XPVM from the source, he must first obtain and install the TCL/TK software on his system. TCL and TK were developed by John Ousterhout at Berkeley and can be obtained by anonymous ftp to sprite.berkeley.edu The TCL and XPVM source distributions each contain a README file that describes the most up-to-date installation procedure for each package respectively.

Like the PVM console, XPVM will start PVM if PVM is not already running, or will attach to the local pvmd if it is. The console can take an optional hostfile argument whereas XPVM always reads $HOME/.xpvm_hosts as its hostfile. If this file does not exist, then XPVM just starts PVM on the local host (or attaches to the existing PVM). In typical use, the hostfile .xpvm_hosts contains a list of hosts prepended with an &. These hostnames then get added to the Hosts menu for addition and deletion from the virtual machine by clicking on them.

XPVM uses PVM's built in tracing facility to trace PVM programs that are spawned from the console. Trace messages are sent from the program back to XPVM. These messages are visualized by XPVM using a number of graphical views.

Post mortem visualization of PVM programs is also supported by XPVM. Traces are saved to a tracefile using Reed's "self defining data format" (SDDF). Thus the analysis of PVM traces can be carried out on any of a number of systems such as Pablo.

2 Profiling PVM Programs with PGPVM

PGPVM is an enhancement package for PVM 3.3 that produces trace files for use with standard ParaGraph. PGPVM attempts to give an accurate portrayal of applications by minimizing the perturbation inherent with this type of monitoring. PGPVM does not utilize standard PVM tracing but instead its own buffered tracing techniques to provide more accurate monitoring information. Further, PGPVM provides a shell script that performs some post-processing and produces a **<file>.trf**, i.e. a standard ParaGraph trace event file. Tachyon removal and clock synchronization are performed during post-processing when necessary.

Using PGPVM requires only two minor modifications to a PVM application. First, the application must include a new header file, "pgpvm.h". This should go directly under the standard "pvm3.h" header file. The new header file provides macros that replace normal PVM routines with calls to the PGPVM library. Calls to the PVM library in the application source code need not be modified. The source does however need to be recompiled. The only other modification is the addition of the **pg_tids (int *tid, int nprocs)** library routine to the source code. The array of tid's (nprocs denotes the number of elements in the array) informs PGPVM of which processes should produce tracing information. All processes that wish to participate in tracing must call **pg_tids()** and all must pass it the same tids array.

Once the application has been modified as described and recompiled; upon execution PGPVM produces a single tracefile named **pgfile.<uid>** in the **/tmp** directory. This tracefile can be found on the host machine of the zero'th element of the tids array passed into the **pg_tids()** routine. This file should be moved to the pgpvm/$ARCH directory. From here the user types, **PGSORT pgfile.<uid>**. This shell script converts the file to standard ParaGraph trace file format and the trace file will appear in the directory as **pgfile.<uid>.trf**. An advanced feature allows users to dictate the pathname and filename of where the tracefile is created, and many find this more convenient than retrieving the file from `/tmp`. Further details and the software itself can be obtained from [4].

3 PVaniM 2.0 — Online and Postmortem Visualization

The PVaniM 2.0 system provides online and postmortem visualization support as well as rudimentary I/O for long running, communication-intensive PVM applications. PVaniM 2.0 provides these features while using several techniques to keep system perturbation to a minimum. The online graphical views provided by PVaniM 2.0 provide insight into message communication patterns, the amount of messages and bytes sent, host utilization, memory utilization, and host load information.

For online visualization analysis (referred to as *pvanimOL*), PVaniM 2.0 utilizes sampling to gather data regarding interesting aspects of the application as well as interesting aspects of the cluster environment. With sampling, necessary statistics are collected and sent to the monitor intermittently. The rate at which the application data is sent to the monitor is a parameter that may be set by the user as well as "steered" from the PVaniM 2.0 interface. The user may wish to have the statistics sent to the monitor every 5 seconds, every 30 seconds, once a minute, etc. With lower sampling rates, the application will experience less perturbation, but the graphical views will not be updated as frequently. Currently, PVaniM 2.0 uses 5 seconds as a default sampling rate. Users are encouraged to try different sampling rates and find one that is most suitable to their personal tastes and the needs of their application.

For off-line visualization analysis, PVaniM 2.0 utilizes buffered tracing to provide support for fine-grain visualization systems. It is our philosophy that fine-grain views provided by such systems as ParaGraph or the original PVaniM are best suited for postmortem analysis. Typically these views need to be examined repeatedly to absorb the vast amount of information provided in them. By supporting the fine-grain visualization systems with buffered tracing (and hence not attempting to support them in real-time), we are able to provide these views with minimal perturbation. The techniques we use are similar to those used by our previous systems, PVaniM and PGPVM. PVaniM 2.0 produces buffered traces for use with the original prototype PVaniM system, but a trace file converter is provided to allow the user to also use the popular ParaGraph system.

Although minor modifications need to be made to PVM applications (2 lines modified), these modifications do not have to be repeatedly added or removed. They perform correctly whether or not the monitor is being used. The use of the monitor to spawn the application determines whether or not visualization is utilized. Full details as well as the PVaniM software can be obtained at [5].

4 Parallel I/O for PVM — PIOUS

PIOUS, the Parallel Input/OUtput System, implements a virtual parallel file system for applications executing in a metacomputing environment. PIOUS supports parallel applications by providing coordinated access to file objects with guaranteed consistency semantics. For performance, PIOUS declusters file data to exploit the combined file I/O and buffer cache capacities of networked computer systems. The PIOUS software architecture implements a virtual parallel file system within the PVM environment that consists of a service coordinator (PSC), a set of data servers (PDS), and library routines linked with client processes (PVM tasks). PVM provides both the process management facilities required to initiate the components of the PIOUS architecture, and the transport facilities required to carry messages between client processes and the PIOUS virtual file system.

A single PIOUS Service Coordinator initiates major activities within the virtual file system. For example, when a process opens a file the PSC is contacted to obtain file meta-data and to ensure that the requested file access semantics are consistent with those of other processes in a logical process group. A PIOUS Data Server resides on each machine over which a file is declustered. Data servers access disk storage in response to client requests via the native file system interface. Though ideally each PDS accesses a file system local to the machine on which it resides, two or more PDS may share a file system via a network file service.

Finally, each client process is linked with library functions that define the PIOUS file model and interface (API). Essentially, library functions translate file operations into PSC/PDS service requests. The PIOUS architecture emphasizes asynchronous operation and scalable performance by employing parallel independent data servers for declustering file data. Coordinated file access is provided via a data server protocol based on volatile transactions [6]. Further design and implementation details can be found in [7, 8].

5 Client-server tools — PVMRPC

PVM, which is based on the message passing model, has become a de-facto standard and widely used system. However, providing only a message-passing interface has restricted PVM application categories significantly. Furthermore, the message-passing model requires a number of potentially complex tasks to be programmed explicitly by the user, including process identification and table maintenance; message preparation, transmission/reception, ordering, and discrimination; and task synchronization. In contrast, traditional distributed computing is based on the "client-server" model. In this paradigm, entities known as servers provide services that may be invoked by clients. Implicit in this abstract definition is the potential for varying levels of implementation semantics. To support client-server applications in a more natural manner, we have developed an RPC system for the PVM framework [9]. PVM-RPC is loosely based on the RPC model in which references to remote services mimic a procedure call. The service provider designs and implements a server which may export one or more services. These services can then be accessed by a client via an invocation that looks and behaves like a procedure call. PVM-RPC provides two types of remote procedure calls, synchronous and asynchronous.

PVM-RPC has four major components, the pvm daemon (PVMD), the service broker (SB), the servers, and the clients. The remainder of this section describes the function and interaction of these components. The PVMD is an integrated part of the native PVM system, and is responsible for identifying and managing global resources (such as available hosts), message routing, and authentication. Before any PVM program can run, the master PVMD must be started. After the master PVMD starts and processes the list of available hosts, local PVMDs start on each available host. Once a local PVMD is running on a host, that host is available for use by the PVM system. The SB maps service names, as known to clients, to a tuple consisting of a server and service id recognizable to the server. In order to ensure that PVM-RPC is at least as reliable as PVM, with respect to hardware failure, the SB runs on the same machine as the master PVMD. While there are various ways that the SB can start, any server that is trying to register a service will start the SB if it is not already running.

A complete scenario of SB interaction with the rest of the PVM-PRC system is now illustrated: when a server starts, it first locates the SB from the PVMD using the table facility. If the SB is not running, then the server starts an SB using *pvm_spawn* and, again, attempts to locate the SB from the PVMD (the second lookup of the SB is done to prevent race conditions between two servers). The server finally registers its services with the SB and no further communication between the server and the SB is required. Similarly, when a client wishes to use a service, it first locates the SB from the PVMD and then gets a list of available services from the SB. In the client's case, a request for a list of registered services results in a complete list of all known services and addresses, which the client caches locally to eliminate unnecessary communication between the client and SB. While the list of servers and services may change during the client's execution, this is not an issue unless the client requires a service whose address is not cached, or there is a problem with a known provider of a service (due either to server failure or congestion). In either of these cases, the client can update its list of known services. If more than one server is available for a particular service, the client will, by default, interleave requests among available servers.

6 The tkPVM Interface to PVM

While the standard distribution of PVM supports programming in C, C++, and Fortran, special interfaces must be built for other languages to use PVM. The tkPVM system allows TCL/TK programmers to use PVM. TCL is an interpreted scripting language that does not require compilation. This allows the TCL programmer to quickly create scripts that interact with PVM and PVM tasks. TkPVM programs can send and receive PVM messages from other programs whether or not those programs were written in C, C++, Fortran, or TCL. The TCL/TK package supports quick prototyping of user interfaces. By combing PVM and TCL/TK, tkPVM provides a natural mechanism for adding user interfaces to PVM programs.

7   The JavaPVM Interface to PVM

Just as tkPVM provides PVM functionality to TCL programmers, the JavaPVM [3] system allows Java programmers to write programs using PVM. One of the main purposes of Java is to provide a safe and portable language. Java programs are compiled to bytecodes which are interpreted at runtime by the Java Virtual Machine (JVM). This means that Java bytecodes will run on any machine to which the JVM has been ported. The implementation of the JVM insures that the Java program does not access restricted resources on the host where it is executing. Safety of the Java application is assured as long as all the code being executed is Java code. However, in some cases it may be useful to allow the Java program to access libraries which are not written in Java. Thus, Java provides a mechanism for calling native functions. It is this facility that is used by JavaPVM for creating a Java interface to PVM. Of course, Java's normal safety guarantees no longer hold when native functions are introduced into a Java program.

8   JPVM — A Java Implementation of PVM

The JPVM [1] system is somewhat different from JavaPVM and tkPVM in that it is not a PVM interface for another language, but rather another implementation of PVM itself. In fact, JPVM is not actually a full implementation of PVM but actually a *PVM-like* library. The main disadvantage of this approach is that JPVM programs can not interact with PVM programs written using standard PVM. The advantage of the system is that it is written entirely in Java and thus has some hope of retaining the portability and safety promised by the Java language.

9   Summary

In this paper we briefly present several PVM related tools. These tools can be broadly categorized as either monitoring tools or programming extensions to PVM. The monitoring tools aid in the debugging and performance tuning of PVM programs. The programming extensions add new functionality to PVM either through supporting new programming abstractions such as RPC or by providing an interface to PVM from other languages such as TCL or Java. In order to better understand the contributions of these systems, it is important to further analyze them not only in terms of functionality, but also in terms performance and programmability.

References

1. A. Ferrari, "JPVM — The Java Parallel Virtual Machine", http://www.cs. virginia.edu/~ajf2j/jpvm.html, 1996.
2. J. Kohl and G. A. Geist, "The PVM 3.4 Tracing Facility and XPVM 1.1", Oak Ridge National Laboratory, 1996.

3. D. Thurman, "JavaPVM — The Java to PVM Interface", http://homer.isye.gatech.edu/chmsr/JavaPVM/, 1996.

4. B. Topol, V. Sunderam, A. Alund, http://www.cc.gatech.edu/gvu/people/Phd/Brad.Topol/pgpvm.html, 1995.

5. B. Topol, V. Sunderam, J. Stasko, http://www.cc.gatech.edu/gvu/softviz/parviz/pvanimOL/pvanimOL.html, 1996.

6. S. Moyer and V. S. Sunderam, "Characterizing Concurrency Control Performance for the PIOUS Parallel File System", *Journal of Parallel and Distributed Computing*, to appear, 1996.

7. S. Moyer and V. S. Sunderam, "Parallel I/O as a Parallel Application", *International Journal of Supercomputer Applications*, Vol. 9, No. 2, pp. 95-107, summer 1995.

8. S. Moyer and V. S. Sunderam, http://www.mathcs.emory.edu/pious/, 1995.

9. A. Zadroga, A. Krantz, S. Chodrow, V. Sunderam, "An RPC Facility for PVM", *Proceedings – High-Performance Computing and Networking '96*, Brussels, Belgium, Springer-Verlag, pp. 798-805, April 1996.

The Status of the MPI Message-Passing Standard and Its Relation to PVM

Rolf Hempel[1]

Computations and Communications Research Laboratories, NEC Europe Ltd.,
Rathausallee 10, 53757 Sankt Augustin, Germany

Abstract. In a similar way to its sister initiative HPFF, the Message–Passing Interface Forum (MPIF) has established a de–facto standard for parallel computing, in this case using the message–passing paradigm. MPI builds upon experience gained with portability interfaces such as PARMACS or PVM, and tries to combine the strongest features of all those existing packages. With MPI now available on any parallel platform, including several versions in the public domain, other interfaces have to re–define their role.

Following an overview of the established standard, MPI-1, we give an introduction to the ongoing MPI-2 initiative. Finally, we compare MPI with PVM, and discuss where MPI learned from the successes and weak points of PVM.

1 Programming Interfaces Using Message–Passing

Message–passing is the most popular programming model for MPP systems. The basis is a collection of parallel processes with private address spaces, which communicate by some kind of messages. While this model requires the applications programmer to explicitly code the data exchange between processes, it has two important advantages:

1. With the exception of the message–passing extensions which are usually handled by some subroutine library, the processes can be programmed in a traditional language, such as Fortran 77 or C. Thus, there is no need for special parallel compiler technology.
2. Since the programmer has full control of the content of the messages, he only sends the data which is really required at the destination, and can optimize the time when it is sent. In contrast, in more sophisticated programming models, such as virtual shared memory, a large overhead of superfluous data is often sent along with the required items, just because it happens to be in the same page or cache line.

For more than ten years, programmers have gained experience with a variety of message–passing interfaces. With the advent of parallel computers with local memory in the mid–eighties, every hardware vendor defined their own interface and distributed it with their products. However, the variety of different interfaces resulted in non–portability of user programs and incompatibility of libraries and tools.

Some national laboratories and universities developed portable message–passing interfaces which they implemented on a broad range of different platforms. Some of those activities were later taken over by commercial companies, such as Parasoft in California with EXPRESS [3] and PALLAS GmbH in Germany with PARMACS [1]. Whereas the commercial support offered and the guaranteed long–term compatibility were important, especially for industrial users, the price to pay for the software limited the distribution in academia. The most widely distributed interface is PVM [4], which up to the present day is available in the public domain.

Portable interfaces make the user program independent of a particular machine, but they do not resolve the incompatibility problem if software based on different interfaces is to be merged. Thus, in 1992 a consensus developed among hardware vendors, tool writers and users that message–passing should be standardized. A first workshop at Williamsburg, Virginia was held to assess the available interfaces and interested parties; this led to the decision that a standardization process should be started. It turned out that no interface in existence at that time was complete and efficient enough to justify its becoming the standard. Therefore, it was decided to define a new interface, combining the strongest features of its predecessors, and avoiding the bad ones.

At Supercomputing '92 a very preliminary draft was presented [2] and the formal rules for an open discussion forum, called MPI (for Message Passing Interface), were set up. After more than a year of intensive discussions on email and at regular meetings in Dallas, the MPI–1 standard was finally published in May, 1994 [7].

2 MPI–1: the Existing Standard

The aim of the MPI–1 project was the formalization of current practice in message–passing and to finalize the standard in a short time. Two decisions helped to make this possible:

1. MPI deliberately chose not to act within the framework of an official standards body, such as ANSI or ISO. Experience with similar initiatives shows that this can slow down the process considerably, sometimes even to the point where the results are useless when the standard is published. MPI, as a de–facto standard, gets its importance from the fact that most interested parties, and in particular virtually all hardware vendors, participated in setting it up.

2. Parallel programming with message–passing includes many aspects, ranging from simple point–to–point communication to parallel I/O. The short time–frame for the first phase made a concentration on the most important topics necessary. They include: point–to–point, collective operations, process groups and contexts, and process topologies. Other important areas were left out because they either seemed too immature for standardization (e.g., dynamic process management), or because other projects were already working on a standard (parallel I/O).

2.1 Key features of MPI–1

The fact that MPI–1 standardized current practice at that time is reflected in the similarity of its key functions with their counterparts in other interfaces. The point–to–point communication chapter defines the basic communication mechanism between two MPI processes. In order to facilitate porting from other interfaces, MPI supports a variety of communication protocols:

- MPI_SEND: standard send; does not wait until message is received at its destination, but may block the sending process because of lack of system buffering.
- MPI_BSEND: buffered alternative to MPI_SEND; avoids deadlock situations by using user–supplied memory area for system buffering.
- MPI_SSEND: synchronous communication; sender is blocked until receive operation is finished at destination.
- MPI_RSEND: forced communication; saves some protocol overhead if receive is already posted when the send is issued. An unrecoverable error may occur in other situations, so this feature requires a good understanding of the time behavior of the application.

For all these send protocols there are blocking and non–blocking versions. Non-blocking functions only start the transaction and immediately return control while the communication proceeds in the background. Special functions test the status of a transaction or wait for its completion, thus enabling the programmer to overlap the communication with other useful work if the hardware supports that.

MPI–1 provides a wide range of collective operations, such as broadcast or gather/scatter. Their scope is defined by a process group concept, similar to that in PVM 3.

In a modular program structure it is most desirable to be able to encapsulate the communication within a module. If, for example, a user program calls a linear algebra library routine, care must be taken that a message sent by the library routine is not mistakenly intercepted by the user code, or vice versa. Traditionally this is achieved by assigning tags to messages, and carefully reserving exclusive tag ranges to different modules, but this strategy completely relies on the discipline of the programmers. In Zipcode [9], and now in MPI–1, all communication functions are relative to a context, and a message sent in one context is not visible in any other context. By using different contexts, communication in different modules can be kept separate.

The concept of process topologies was first realized in EXPRESS and PAR-MACS. The topology functions in MPI–1 further generalize their PARMACS counterparts. Many data–parallel applications have a geometric background which results from the partitioning of a global data structure onto parallel processes. Knowledge of the resulting process graph, which can either be a Cartesian structure or a general graph, can be exploited in two ways:

1. It provides a more intuitive process naming. If the processes are laid out in a grid, it is more natural to send a message "to the right neighbor", than to

the process which happens to have the group rank 4711 (for example). Other functions, like shift operations, only become meaningful in the presence of a process topology.

2. In most data parallel applications, a process does not communicate with an equal frequency with all other processes, and in many cases most messages are exchanged with neighbors in the process topology. Knowledge of this topology can therefore enable the MPI library to optimize the placement of processes to hardware processors, so that communication is kept as local as possible. This optimization gains importance with the hardware trend towards clustered architectures.

A variety of information is available on MPI. The world wide web pages http://www.mcs.anl.gov/mpi and http://www.erc.msstate.edu/mpi provide up–to–date info on the MPI process, and the full MPI–1 standard document. Programmers looking for an introductory text, however, should not try to learn MPI from that document: few people would learn Fortran 77 by reading the ANSI standard! Perhaps the best introduction is found in [5], which has been written by some of the original MPI–1 authors. For programmers who want to learn about more sophisticated details, the reference book [8] contains the full standard plus many examples and comments.

2.2 Available MPI–1 Implementations

There are two categories of MPI–1 implementations: public–domain packages for virtually any parallel system, including workstation networks, and vendor–optimized versions for specific hardware platforms. The most important public-domain versions are:

- MPICH [6]: developed by Argonne National Laboratory and Mississippi State University. MPICH plays a special role, because it evolved in parallel with the standard itself, and thus provided early feedback on implementation issues related to proposed interface functions. To some extent, MPICH has been regarded as *the* incarnation of the standard. There are MPICH versions for most parallel systems. MPICH can be obtained by anonymous ftp from info.mcs.anl.gov/pub/mpi.
- LAM: a product of the Ohio Supercomputing Center, LAM is available for UNIX workstation clusters (anonymous ftp from tbag.osc.edu/pub/lam).

Most vendors of parallel hardware meanwhile provide optimized MPI implementations for their systems. They are usually as efficient as the proprietary interfaces, so that MPI today becomes more and more the interface of choice. Some hardware vendors, including NEC, did not even define proprietary message–passing libraries, and promote MPI as the standard interface.

3 The Continuation: MPI–2

One year after the finalization of MPI–1, the MPI forum initiated a second series of regular meetings, this time convened by Argonne National Laboratory in Chicago. Experience with the initial version of the standard led to a number of clarifications and a few minor modifications, which were included in MPI version 1.1. Of course, with the growing number of MPI users, extreme care was taken to modify the standard only where absolutely necessary, and most of the changes only affected very esoteric MPI features which most likely had not yet been used by any programmer.

The main task for the second MPI phase, however, was to add chapters to the standard on topics which had been left out of MPI-1. They include

- dynamic process management,
- single-sided communication,
- parallel I/O,
- extended collective operations, and
- new language bindings.

At the time of writing, all these chapters are still in a very preliminary state, but their general structure will probably not change too much. Whereas any MPI-1 implementation must support the full MPI-1 standard, this may not be feasible for MPI-2. Rather, the expectation is that it will define standard additional "packages" (corresponding to the new chapters) which may not be available on all platforms. Another observation is that the MPI-2 activities tend to move away from standardizing current practice, as many features are not available yet in any of the well-established programming interfaces. The following section outlines some of the topics covered by MPI-2.

3.1 Main MPI-2 Additions

Early user feedback on the MPI-1 standard often contained criticism of the static process concept and missing parallel I/O. Both topics are important for applications programmers, and they are provided by several existing programming interfaces. Therefore, their non-availability is a major obstacle in porting applications to MPI.

In addition to those essential extensions, MPI-2 will also contain chapters on single-sided communication, additional collective functions, and a C++ binding.

Dynamic process management: In MPI-1 it is assumed that the parallel processes of an application are started by some implementation-dependent mechanism which is not defined by the standard. All processes are started at the same time, and they are represented by the group associated with the MPI_COMM_WORLD communicator. It is not possible to start additional processes later. The MPI-2 function MPI_SPAWN and its derivatives provide this functionality. The new processes are arranged in groups which can either be connected with the spawning processes via an MPI intercommunicator, or they can constitute a separate communication domain with no connection with the parents.

A difficult issue encountered by client-server applications is establishing communication between independently created applications (independent jobs), as, for example, between a new client application and an already existing set of server processes. In MPI–2 this contact can either be realized by explicitly specifying port names or by using a name server mechanism.

Parallel I/O: In MPI–1 this subject was left out for several reasons:

- the complex topic would have delayed the finalization of the standard considerably;
- the MPI–IO project of IBM and NASA Ames already started work on a portable, MPI–based interface, and the MPI forum did not want to duplicate their effort;
- there was no general consensus at that time of how to standardize I/O.

Meanwhile, the MPI–IO project has produced a complete proposal and gained some experience with implementations. In June 1996 the MPI forum created a new I/O working group. It took the MPI–IO document as a first draft, and the final version will become a chapter in MPI–2.

Single–sided communication: In message–passing using send/receive, both the sender and receiver process must actively participate in the communication. This paradigm does not fit to applications in which some process has to put or get data to/from another process' memory without the target being aware of it. The single–sided communication functions MPI_PUT and MPI_GET support this alternative communication paradigm. They act on specially created communicators which associate each participating process with a *remote memory access* (RMA) window. Put and get operations always point to locations in RMA windows. The most difficult part in defining these operations consists in providing the appropriate mechanisms for process synchronization and cache consistency. This area is particularly sensitive because put and get operations are expected to realize a lightweight protocol with as little synchronization overhead as possible. The importance of single–sided communication is growing with the hardware trend towards SMP architectures, where remote put/get operations are realized by accessing shared memory segments.

Extended collective operations: This chapter contains some extensions to the existing functions in MPI–1. While collective functions in MPI–1 operate on intra–communicators only, they will be generalized in MPI–2 for inter–communicators. Since this class of operations tend to be expensive, the non-blocking variants proposed for MPI–2 could save execution time by avoiding processor idle times and by overlapping arithmetic operations with the ongoing communication.

New language bindings: With the growing popularity of the new languages Fortran 90 and C++, there is obvious demand for appropriate MPI bindings. MPI–2 will introduce a C++ binding, but most likely there will be no Fortran 90 binding. The reason is that MPI implementations for Fortran 90 require direct compiler support which make implementations very much more difficult.

4 MPI versus PVM

PVM is still by far the most popular message–passing interface on workstation networks and other heterogeneous, loosely–coupled processor configurations, and some vendors of MPP systems even chose it as their primary interface, including Cray Research for their T3D machine. The main reasons for PVM's success seem to have been

1. its public domain status,
2. its simplicity,
3. the portability of its application programs,
4. its emulation of a complete virtual parallel machine, including resource management, on networks of machines which were not originally set up for parallel programming.

PVM has shown the advantages of portable parallel programming to many applications programmers, and has thus been very helpful for the development of standards in general.

On the other hand, PVM was never designed for tightly–coupled parallel systems. Already at the Williamsburg MPI kick–off workshop there was general agreement that MPI could not just copy PVM if it was to provide near–optimal performance on every parallel system and the functionality for supporting a wide enough range of parallel applications. These, however, were among the aims of the MPI standardization process, and experience has shown that they have been met by MPI implementations.

As explained in the previous sections, MPI–1 is based on a static process concept without the possibility of spawning processes at run–time. Until MPI–2 implementations will become generally available, this will remain an advantage of PVM, in particular on workstation networks. Some users ask for dynamic process management because they got used to the PVM function pvm_spawn, whereas their applications do not really require dynamic process creation. However, other applications are less elegant or efficient if limited by the static concept in MPI. Of course, the experience with process spawning in PVM has helped in defining the dynamic process functions in MPI–2.

Looking at the strong points of PVM listed above, how does MPI compare? There are mature MPI implementations in the public domain, most notably MPICH of ANL and LAM of the Ohio Supercomputing Center. MPICH, in particular, is available for virtually every parallel system, and most hardware vendors of MPP systems meanwhile provide optimized versions. So, MPI obviously fulfills items 1 and 3. An often repeated criticism against MPI is its complexity, which results from its aim to be as generally applicable and as efficient as possible. Novice programmers, however, do not have to use all the available functions, and a simple MPI program is no more difficult to write than its PVM counterpart. Introductory textbooks help the beginner in learning the basic functions quickly. In this sense, MPI also matches point 2.

Since the MPI–1 standard does not contain process management, providing a satisfactory programming environment is left to the implementations. The mpirun start–up script of MPICH became so popular, that it is currently being discussed to be included in the MPI–2 standard. Thus, apart from the dynamic process management, point 4 is also fulfilled by available MPI implementations.

Since the publication of the first standard, MPI has gained enough momentum that there is hardly any doubt that it is establishing itself as *the* message–passing standard. This will decrease the popularity of other interfaces. Some issues for further research, however, are still open. At present, for example, there are no MPI implementations which couple the optimized implementations on MPP systems of different vendors. The reason is that MPI defines the programming interface, but leaves the definition of communication protocols, data structures, etc. open. This and other issues will keep implementors of message–passing interfaces busy for several more years.

5 Acknowledgements

I gratefully acknowledge the organizers of the "Third Euro PVM Users Group Meeting" for inviting me to present the MPI view of message–passing. In addition, I would like to thank my colleague Guy Lonsdale for checking the manuscript and correcting my English.

References

1. Calkin, R., Hempel, R., Hoppe, H.-C., Wypior, P.: Portable programming with the PARMACS message–passing library. Parallel Computing **20** (1994) 615–632
2. Dongarra, J., Hempel, R., Hey, A.J.G., Walker, D.W.: A proposal for a user–level, message passing interface in a distributed memory environment. Technical Report TM–12231, Oak Ridge National Laboratory, February, 1993
3. Flower, J., Kolawa, A.: Express is not just a message passing system: current and future directions in Express. Parallel Computing **20** (1994) 597–614
4. Geist, A., Beguelin, A., Dongarra, J., Jiang, W., Manchek, R., Sunderam, V.: PVM 3 user's guide and reference manual. ORNL/TM–12187, May, 1994
5. Gropp, W., Lusk, E., Skjellum, A.: Using MPI. MIT Press, 1994, ISBN 0–262–57104–8.
6. Gropp, W., Doss, N., Lusk, E., Skjellum, A.: A high–performance, portable implementation of the MPI message passing interface standard. Parallel Computing, to appear
7. Message Passing Interface Forum: MPI: a message–passing interface standard. International Journal of Supercomputer Applications, **8(3/4)** (1994), special issue on MPI
8. Snir, M., Otto, S.W., Huss–Lederman, S., Walker, D.W., Dongarra, J.: MPI: the complete reference. MIT Press, 1996, ISBN 0–262–69184–1.
9. Skjellum, A., Smith, S.G., Doss, N.E., Leung, A.P., Morari, M.: The design and evolution of Zipcode. Parallel Computing **20** (1994) 565–596

A Comparison of Message-Passing Parallelization to Shared-Memory Parallelization

Stefan Pokorny [1]

[1] German Aerospace Research Establishment DLR, D-51140 Cologne, Germany

Abstract. In this paper we shall present a comparison of the performance of a unsteady, interactive flow solver implemented via PVM to the performance of the same solver implemented via multithreading. Based on these results a programming method used to implement such CFD-codes for distributed memory machines with symmetric multi-processing nodes is constructed.

1 Introduction

To investigate new propulsion concepts at the DLR a parallel program to simulate the unsteady flow within turbomachines has been developed. It covers the unsteady effects produced by the relative motion and the interaction of its components (Fig. 1). The unsteady simulation poses two major problems on the computational environment. One is the large numerical load produced by the explicit, second order accurate upwind scheme used, to provide sufficient numerical accuracy. The second problem is the large amount of data produced. In unsteady simulations every time step yields a three dimensional flow field and therefore increases the amount of data by several orders in magnitude as compared to steady simulations.

To deal with these problems, the program must be designed to run interactively on a parallel distributed memory machine. Parallel computers constitute the most promising platform for the highest possible performance. The flow solver TRACE [1, 2, 3], developed at the DLR, is designed originally to run on a parallel distributed memory machine.

The interactive program combined with the ability to visualize the results on-line is used to reduce the amount of data produced by the simulation system. Instead of specifying the entire simulation process beforehand and running the program as a batch process, the user may solve his application problem in a similar way as in a real test facility. While the program is running, parameters such as the exit pressure or the speed of the stage may be changed and immediately analyzed. The simulation program may be used to reproduce data instead of storing hundreds of gigabytes to some medium and analyzing the entire data set for one set of parameters only.

Fig. 1. New propulsion concepts based on a counter rotating propfan require detailed knowledge of the flow field within such configurations.

In the light of upcoming parallel computers based on symmetric multi-processing (SMP) nodes, the question arises wether the message passing model is still suitable to address the parallelization on these nodes. The efficiency of pure message passing programs greatly depend on the bandwidth of the slowest communication link and on the granularity, the ration between computation and communication, of the problem. Thus, it is not straight forward to exploit the benefit of a communication system via shared memory on a mixed architecture. Fine grain parallelism requires to transfer small messages across the tasks. The efficiency of message passing will decrease for small messages. Thus, we expect this to significantly effect the overall efficiency of a program.

To investigate this problem, we started a first experiment on what difference in performance we might expect when changing from message passing parallelization to shared memory multithreading parallelization. We shall further outline what gain may be reached by mixing the two programming models, the message passing and the shared memory parallelization.

The flow solver TRACE, to be analyzed here, is used to simulate the unsteady flow in turbomachines. It uses an explicit TVD method combined with Runge-Kutta time integration to solve the three dimensional Navier-Stokes equations. The solver is accelerated by implicit residual damping for the steady initialization phase and by a simple two level multigrid technique when switching to time accurate solutions. It is parallelized and implemented using the portable parallelization library Opium [7]. All parallel tasks within this parallelization library may be executed either as separate threads, exchanging data via shared global memory or by explicitly spawning tasks with their own local address space, exchanging data by calling massage passing routines of a message passing library. The code itself has been ported to a variety of parallel and scalar platforms including the IBM SP2, SGI PowerChallenge, and workstation clusters, as well as single CPU workstations.

2 Implementation

The design philosophy of the parallelization library is based on the premise that programming complexity of numerical codes and applications should be greatly reduced as compared to pure message passing implementations. Instead of directly implementing parallel numerical algorithms by explicitly spawning tasks and exchanging messages between them, the notion of a numerical model has been introduced [7]. A numerical model represents a parallelized numerical algorithm which may be used as a building block for the implementation of applications. From the application point of view, numerical schemes consist of parallelized resources, the numerical models, but are not parallelized by themselve. This decouples the implementation of the numerical algorithm from its parallelization.

The parallelization library is responsible for partitioning the computational domain, setting up the necessary communication links, and mapping the entire system onto the nodes of the parallel computer. It also contains methods to perform global IO, where all data is collected or distributed according to the current distribution or partitioning of the solver.

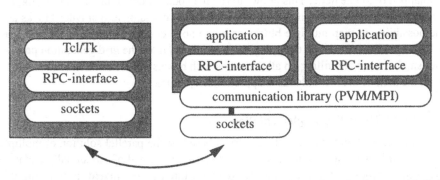

Fig. 2. The left side shows the implementation of the user interface. Its front end is Tcl/Tk based. On the right two numerical blocks (application) or tasks coupled by a communication library are shown.

Besides the parallelization environment, the library also allows the interactive control of the entire application. Its user interface is based an Tcl/Tk [6] which provides the necessary scripting language. Tcl is extended by a special remote procedure call (RPC) mechanism to invoke routines on the remote parallel computer, where the application is running (Fig. 2). This allows to describe the entire initialization and simulation process interactively via user defined scripts.

The application is integrated into the server-side of this library by special hooks which automatically provide all necessary synchronization for the parallel environment. Its main routine waits for RPC-requests within a loop. The application is responsible for executing the procedure. Thus, the application provides its data as a service to the user interface.

A batch version of the program is obtained by replacing the socket binding from the parallel applications to the user interface by a file containing all commands. This allows to quickly calculate flow-fields for a fixed set of parameters without the necessity to interactively specify all actions.

The basic design of the parallelization library stems from the originally targeted distributed memory architectures. Every block of the simulation domain uses its own distinct address space. Pointers must not be passed between blocks. This restriction implies, that the programming model cannot take full advantage of the shared memory system where all threads operate on the same address space. Thus, instead of simply accessing remote data in main memory all data from other blocks must be explicitly copied.

2.1 Message-Passing Implementation

Within the current version of this library, the computational domain of the solver is decomposed into independent subdomains, which are subsequently mapped onto the nodes of a parallel computer. The number of subdomains must be equal to or greater than the number of nodes which are to be used for the calculation. Multiple blocks may be mapped onto one node. This allows to run the code for development and debugging purposes on single processor workstations. Splitting domains into smaller blocks and the possibility to map multiple blocks onto one node enables a simple means for load balancing. This is based on chopping the large blocks from the grid generation process into smaller ones and filling up all the nodes with blocks until their total load resulting from the sum of the local blocks is approximately equal.

2.2 Multi-Threading Implementation

The multi-threading version of the library is based on the parallel solution of multiple blocks within the single-node version of the program. Instead of sequentially updating every local block of the computational domain, threads are created, performing this task in parallel. The number of threads to be started may vary from the number of nodes available to the number of locally available blocks. All the data remain in main memory and are shared among all started threads. Synchronization is achieved by explicitly allocating semaphores and using mutex-locks when accessing shared data.

Due to the structure of the program, all block data for each block remain local to the thread. Coupling between the blocks is achieved by explicitly copying the data from one location in memory to another. Although data is shared among all tasks the overall message passing design of the program requires explicit data movement.

3   Results

The first measurements show the communication bandwidth for PVM in relation to the message sizes. We compare three different implementations of PVM (PVM version 3.3.11 socket based/shared memory based and PVMe on the IBM-SP2) and two communication methods pvm_send/recv and pvm_psend/precv (Fig. 3).

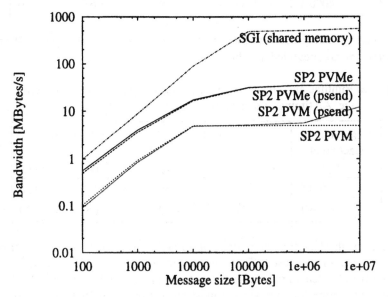

Fig. 3. Bandwidth of message passing with PVM, PVMe on an IBM-SP2, and shared memory based PVM on an SGI-PowerChallenge

The shared memory implementation on the SGI-PowerChallenge yields the highest possible bandwidth which is approximately 80% of the memory bandwidth. It outperforms the fastest communication method on the SP2 by more than one order in magnitude. Second best is the performance of PVMe on the IBM-SP2. Its performance reaches approximately 85% of the peak communication bandwidth. Another order in magnitude below this bandwidth we find the socket based versions of PVM.

The decrease in bandwidth for all types of communication results from the increasing portion of start-up time when transferring small chunks of data. This overhead reduces the effective transfer rate by almost two orders in magnitude.

Table 1 shows the runtime per iteration and the corresponding efficiency of the solver for three different problem sizes on the IBM-SP2 under PVMe and PVM. All timing results are the same except for the smallest load per node. The effect of load imbalances due to boundary conditions and the double calculations on the boundary lower the efficiency of the code.

Table 1. Runtime and efficiency for different parallelizations on the SP2. All timings are measured in seconds.

local problem size/ no of procs	SP2/PVMe		SP2/PVM	
	runtime [s]	efficiency	runtime [s]	efficiency
100/40	0.097	0.75	0.101	0.72
500/8	0.53	0.86	0.53	0.86
4000/1	3.65	1.00	3.65	1.00

The corresponding numbers for the SGI PowerChallenge based on PVM and on multithreading are presented in Table 2.

Table 2. Runtime and efficiency for different parallelizations on the SGI PowerChallenge. All timings are measured in seconds.

local problem size/ no of procs	SGI/PVM		SGI/threads	
	runtime [s]	efficiency	runtime [s]	efficiency
500/8	0.59	0.86	0.59	0.86
4000/1	4.05	1.00	4.05	1.00

These results show that the runtime of this solver is almost independent from the communication method used. The communication time for this numerical method must be several orders in magnitude below the computational time. Thus the decrease in efficiency for the 50 node version must come from a different source. In fact it largely results from double calculations done at the boundaries of the local domains. The current design of the program does not allow to simply replace these double calculations by multiple data communications.

4 Conclusions

Shared memory parallelization based on threads may be used in two different alternatives within our type of application. One is the ability to accelerate the computation by increasing communication speeds and thus increasing overall efficiency. The entire code remains a message-passing program. In distributed memory machines with shared memory multiprocessor nodes this programming model results in a two level communication hierachy. The fast shared memory communication within the nodes and the slower communication between the nodes.

The second alternative is to fully exploit the shared memory parallelization. This requires redesigning the message-passing code to avoid local communication completely by accessing local data directly. From this point of view a node of a distributed memory machine with shared memory multiprocessor nodes is a powerful node within the message-passing environment. This programming model means a mixture of models which will increase code complexity as long as compiler technology is not able to take over this task. However, this will lead to a coarse grain parallelism with respect to the message passing model.

The goal of this paper is to show which model can be used for efficiently using mixed shared an distributed memory machines. The results presented here showed that the efficiency depends on the granularity of the problem. Simply switching to a new environment with a very efficient communication system, as the shared memory communication, does not necessarily change the efficiency of a program. To exploit the benefits of an architecture to its best it is necessary to adopt the appropriate programming model. Thus, we suggest to use a shared memory programming model to be used on shared memory nodes. Every SMP-node within distributed memory machines must be seen as one node.

To address the programming complexity resulting from the mixing models, we start with our experiences from optimizations on superscalar architectures. All programs on superscalar processor architectures must be designed to work on a local set of data only, which must fit into the local data cache. Performance decreases dramatically if coding does not follow this rule.

This suggests a programming model which is structured according to these constraints. All numerical computations are based on local data only. On initialization of the computation either global data is read from main memory or data is communicated from a different node within the network. This concept yields an homogeneous approach to numerical solvers within a distributed environment. Local computational threads may be started at the beginning of the calculation according to the locally available resources.

By structuring the global data according to the physical memory structure of the machine, we should be able to fully exploit the shared memory facilities. Especially for fine grain parallelism the efficiency is greatly influenced by double calculations. A simple estimate shows that efficiency could be increased by another 20% when restructuring the code. Shifting all communication requests to a seperate thread enables entirely asynchronous communication for more communication limited methods in the future. This method allows the most flexible approach to multi-style programming methods.

Acknowledgments

The entire system is part of a collaboration between the Institute for Propulsion Technology and the Department for High Performance Computing at DLR. Dr. Karl Engel, Frank Eulitz, and Dr. Herman Gebing from the Institute for Propulsion Technology did most of the work in developing, designing, and implementing the numerical solver TRACE. The infrastructure for the parallel environment was developed by the Dr. Michael Faden and the author from the Department for High Performance Computing.

References

[1]   Engel, K.; Eulitz, F.; Pokorny, S.; Faden, M.; *3-D Navier-Stokes Solver for the Simulation of the unsteady Turbomachinery Flow on a Massively Parallel Hardware Architecture*, Flow Simulation with High-Performance Computers II, Notes on Numerical Fluid Mechanics, Vieweg Verlag, 1995

[2]   Engel, K.; Eulitz, F.; Faden, M.; Pokorny, S.; *Numerical Investigation of the Shock Induced Interaction in a Transonic Compressor Stage*, ASME Symposium, 6-11 November, Chicago, USA, 1994

[3]   Eulitz, F.; Engel, K.; Pokorny, S.; *Numerical Investigation of Inviscid and Viscous Interaction in a Transonic Compressor*, Proceedings of the 85th AGARD-PEP meeting, 8-12 May, Derby, UK, 1995

[4]   Ferrari, A.; Sunderam, V.S.; *TPVM: Distributed Concurrent Computing with Lightweight Processes*, Dept. of Math. and Computer Science, Emory University, 1995

[5]   Message Passing Interface Forum; *MPI: A Message-Passing Interface Standard*, University of Tenessee, Knoxville, Tenessee, May 1994

[6]   Ousterhout; J. K.; *Tcl and the Tk Toolkit*, Addison-Wesley, ISBN 0-201-63337-X, 1994

[7]   Pokorny, S.; *Objektorientierte Modelle zur interaktiven Stroemungssimulation auf Parallelrechnern*, Dissertation Uni-GHS Essen, 1995

[8]   Sunderam; V. S.; *PVM: A framework for parallel distributed computing.* Concurrency: Practice & Experience, Vol 2, No 4, pp 315-339, Dec 1990

[9]   Zhou, H.; Geist, A.; *LPVM: A Step Towards Multithread PVM*, 1995

Practical Experiments to Improve PVM Algorithms

J. Roda, J. Herrera, J. González, C. Rodríguez, F. Almeida, D. González

Centro Superior de Informática
Dpto. E.I.O. y Computación
Universidad de La Laguna
Tenerife - Spain
casiano@ull.es

Abstract. This paper describes several experiments to measure different network parameters under PVM. Mapping of tasks over different
branches of the network and different broadcasting strategies based on
pvm_mcast and pvm_send are compared. Intensive communication experiments have been also measured. As an example of how to apply this
information, we broach the implementation of two parallel algorithms
using PVM to perform the communications on a LAN.

1 Introduction

Parallel and Distributed Computing is currently one of the areas of Computer
Science which has achieved a remarkable progress. With the technology reaching
the limits, it appears as a true alternative to reduce the running times of the
applications. To achieve parallel computation, a collection of processors has to be
available. Since these processors have to cooperate, some mechanism supporting
communication and synchronization is needed.

A Distributed System is a collection of independent computers linked by a
network with the necessary software to coordinate its functions and to share the
resources of the system. The principal keys of distributed systems are: concurrency, scalability, fault tolerance, transparency and support for resource sharing.
Among a variety of approaches, these systems can be implemented over clusters
of workstations connected by a Local Area Networks.

The emergence of networks of high performance workstations in companies
and research institutions has introduced a significative advance in the popularity
of distributed computing. These computers are idle for long periods of time [7]
(editing, mailing, ...) and distributed computing can use this potential power to
solve large computational problems. A lot of effort is being developed to exploit
the computing power of this environments: PVM [4], Linda [2], P4 [1], EXPRESS
[3], MPI [9].

On the other side, the increase of network speed will benefit many aspects of
these systems. Low latency and high bandwidth networks make distributed computing efficient. The recent advances in network technology have made feasible
to build gigabit Local Area Networks with more than 100 times actual bandwidths [5]. New applications will emerge and a new perspective of computing

will appear. But not only network technology is involved, also protocols technologies, interaction between networks, computer architecture, operating systems and software applications. Until this gigabit networks becomes a common situation, we should take all the advantages of our 10 M bits Local Area Networks.

In this work, we describe our experiences measuring network parameters on an ethernet local area network. The values of Latency and Data Transfer Rate parameters have been obtained for different message passing functions different communication routing, different data packing functions and different network system loads. Several experiments are presented to measure the actual Total System Bandwidth for a local area network with several branches connected through a high speed switch router. Two different broadcasting strategies based on pvm_mcast() and pvm_send() are compared.

As an example of how to apply this information, we broach the implementation of two parallel algorithms using PVM to perform the communications on a LAN. The Parallel Sorting Algorithm by Regular Sampling [6] and a Dynamic Programming Algorithm, the Integer Knapsack Problem [8] are used to prove that an appropriate selection of packing functions, broadcasting strategies and mapping of tasks leads to a significant improvement of the execution time.

Our experiments were performed in two different ethernet networks. In the first case, we used 8 Sun workstations sharing the same communication media. The second, have 140 heterogeneous workstations distributed over different branches. The results obtained in both networks were similar although light user interactions were perceived in the second. Each experiment was accomplished during the night and was executed 100 times. The variation in the results obtained from one network to the other for all the parameters and all the experiences were irrelevant.

2 Performance measurements for networks of workstations

2.1 Latency and Data Transfer Rate

Several factors have a strong influence in the performance of network cluster based environments. Among the parameters depending on the subcommunication system, latency, data transfer rate and total system bandwidth have a primary interest in this work.

We can define latency as the time required to transfer an empty message between the relevant computers. This time includes not only the delays incurred in obtaining access to the network but also the time spent within the network itself. We measured latency as half of round-trip time for a data size of 4 bytes. When we communicate small messages, latency must be the parameter to consider. The fixed overhead dominate transfer times. Table 1 shows the latency values obtained for pvm_send() function expressed in milliseconds.

Data transfer rate denotes the speed, in bytes per second, at which data can be transferred between two computers once the transmission has begun.

	DataDefault	DataRaw	DataInPlace
TCP	1,98	1,77	1,95
UDP	3,87	3,81	4,65

Table 1. Latency (ms.)

Size	64	256	1024	4096	16384	65536	262144	1048576
DataDefault	14669	48521	155542	244235	315791	345610	351425	349498
DataRaw	16016	58127	176906	280012	414047	463658	484767	468821
DataInPlace	7167	27968	87191	218616	440259	511582	519549	524770

Table 2. TCP routing for pvm_send()

Obviously, the time of transferring a message of length l between two directly connected computers is given by the linear function:

$$Message\_transfer\_time = latency + l/Data\_transfer\_rate$$

For the more general case of two non adjacent computers, the message transfer time depends of the number h of hops to traverse:

$$Message\_transfer\_time = latency + l/Data\_transfer\_rate$$
$$+(h-1)\ delay\ per\ hop$$

The extended use of worm-hole routing in most current multicomputers weakes the impact of the third term due to the pipeline effect that arises when the length of the message is large enough. For this reason, the data transfer rate is also called asymptotic bandwidth. This assumption is especially accurate when the diameter of the network is small, since in this case the number of hops to traverse tends to be negligible.

Data Transfer Rate can be deduced from the time in seconds to send and receive the information for different packet sizes and is expressed in bytes per second. Table 2 and 3 corresponds to data transfer rate values for pvm_send() function for UDP and TCP with different data packing functions.

We can conclude that TCP routing communication and PvmDataInPlace packing encoding reports the best values. When the circuit is set, the data transfer rate reports 50% better results than for UDP for many cases. The PvmDataInPlace encoding reduces the number of copies of data we want to transmit.

Size	64	256	1024	4096	16384	65536	262144	1048576
PvmDataDefault	8040	30415	92965	151988	197198	215676	220964	221187
PvmDataRaw	8311	31596	98447	174652	228845	253420	263579	269682
PvmDataInPlace	7608	27479	95713	170550	236171	259930	264517	257667

Table 3. UDP routing for pvm_send()

Size	64	256	1024	4096	16384	65536	262144	1048576
TCP	7568	27226	70828	170348	146387	201097	205255	218058
UDP	6933	21232	68576	114891	159757	188624	191414	178411

Table 4. Network load over 30% for TCP and PvmDataInPlace

Some considerations must be done at this point. First, the limited number of connection oriented TCP sockets available in the systems. Second, the loose of exchange data between network elements, and third, the possible change of the data to send because of the pointer to data used instead of its copy to a new buffer.

The impact of network load was also considered. Table 4 shows the values obtained when network load was over 30%.

A design consideration when solving a problem between several computers could be leaving in the master computer a subproblem of greater size than the subproblems sent to the other slaves computers, so the time to solve the subproblem in the master is equal to the time that takes to send the subproblems to the slaves plus the time to solve it. This size can be calculated from the tables, but is strongly dependent on the computation time of the application.

2.2 Total System Bandwidth

Usually, the Total System Bandwidth is defined as the total number of bytes that can be transferred across the network in a given time. For a simple bus ethernet local area network, once the transmission has begun the message holds the full channel capacity for itself and, therefore, the total system bandwidth coincides with the data transfer rate. However, when several communication channels are available, several messages can be transmitted simultaneously and the bandwidth overcomes the data transfer rate.

The variations of the system bandwidth are studied by mapping pairs of tasks (sender and receiver processes) on the different branches of the network according to different allocation patterns. The branches are interconnected through a 1,6 G bits router.

Size	64	256	1024	4096	16384	65536	262144	1048576
Pattern A	7394	26543	77514	127098	151245	167661	174901	180014
Pattern B	9641	36347	117732	204153	274522	294108	306836	310143

Table 5. Total system bandwidth: Pattern A and Pattern B

Size	256	1024	4096	16384	65536	262144	1048576
2	-25,61	-20,51	-7,95	-0,45	8,26	8,82	10,54
4	-16,27	-18,77	-11,76	-3,36	6,52	6,37	8,04
8	-15,31	-17,97	-28,21	-2,71	6,20	4,89	6,40

Table 6. Comparing pvm_send() and pvm_mcast() functions

Pattern A is obtained by allocating senders and receivers on different branches. All the messages have to pass through the router. For pattern B, the sender and the receiver are always allocated in the same branch while the number of communicating processes is fairly distributed among the branches.

These two patterns correspond to extreme situations. The first one stresses the router to the limit and the router capacity becomes the bottle neck of the system. The second one looks for a balance that minimizes the load of both resources. Table 5 presents the values for a configuration with four segments and four pairs of processes communicating on each segment. The first column contains the packet sizes and second and third columns show the experimental average values for both patterns, expressed in bytes per second.

From a simple glance to Table 5 that communication between sender and receiver on different segments is handicapped by the router. Better performance results from having both processes in the same segment.

2.3 Comparing pvm_send() and pvm_mcast() functions

One of the most common communication patterns is broadcasting, and more in general, multicasting. i.e. the sending of the same message to a group of processors. Two different communicating strategies based on pvm_mcast() and pvm_send() are compared according to the number of processors and the packet size. These experiments can help to decide which of these function is more suitable for the requirements of a given application making use of multicasting.

The size M of the messages varies from 256 to 1 Mbytes and the number of receiving processors P varies from 2 to 8. All the experiments were done with the UDP protocol and PvmDataRaw data encoding.

The master sends with pvm_mcast() routine a packet of size M to the slaves and they return to the master packets of size M/P. This broadcasting strategy is

Size	64	256	1024	4096	16384	65536	262144	1048576
2	7989	28095	98757	173489	225524	246400	260865	264917
4	12680	46616	162735	192448	196769	209925	206813	209029
8	7214	35510	52700	51717	53824	55361	54871	55110

Table 7. ALL TO ALL one branch and 2, 4, 8 processors

Size	64	256	1024	4096	16384	65536	262144	1048576
1 branch	7214	35510	52700	51717	53824	55361	54871	55110
2 branches	8919	39430	69971	73171	84426	95525	102531	106447
4 branches	6870	30831	129979	135555	154576	210109	224532	217770

Table 8. ALL TO ALL 8 processors in 1, 2 and 4 branches

compared with this other: the master sends using pvm_send() routine P different packets of size M/P and the slaves return packets of size M/P. For all the range of sizes and number of processors pvm_send() reported higher results.

Table 6 shows the values when the master sends with pvm_mcast() routine a packet of size M and the slaves return to the master packets of size M. This is compared with this other: the master sends with pvm_send() routine P different packets of size M and the slaves return packets of size M. In this case, for packet size up to 16 Kbytes pvm_send() reported higher results and over this size pvm_mcast() presents the best values. Negative values show the case for pvm_send() and positive for pvm_mcast(). The quantity is expressed in percentage, i.e. for -25.61, indicates that pvm_send() reported 25.61 % better values than pvm_mcast().

2.4 Intensive broadcast communications

Many applications need to communicate intensively. Pattern communication where all the computers send and receive messages from other elements of the virtual machine is a normal scheme that we must consider in our experiments.

We choose two interesting patterns for our experiments. Table 7 presents the results when we allocate 2, 4 and 8 processors on one branch. The increasing of processors stresses the bus and we can observe how decreases the performance. Table 8 represents the values obtained for 8 processors allocated in 1, 2 and 4 different branches. All values are in bytes per second.

Processors	2	4	8
UPD-DataDefault	14,34	14,69	18,33
TCP-DataInPlace	10,81	11,15	12,33

Table 9. PSRS: Comparing UDP-DataDefault with TCP-DataInPlace

Distribution	8 processors
1 branch	12,33
2 branches	12,04
4 branches	10,22

Table 10. PSRS: Tasks on different branches

3   Practical Improve of PVM Applications

3.1   Parallel Sorting by Regular Sampling (PSRS)

The Parallel Sorting Algorithm by Regular Sampling [6] is used to prove that an appropriate selection of packing functions and mapping of tasks leads to a improvement of the execution time.

In Table 9 we present the execution time in seconds of the algorithm using UDP-PvmDataDefault and TCP-PvmDataInPlace. We can improve our results if we choose the correct functions and data packing coding.

Table 10 compare two different ways of mapping tasks to differents branches. An appropriate distribution of the tasks over different branches leads to a significant reduction of the times.

3.2   Integer Knapsack Problem (IKP)

The Integer Knapsack Problem [8] has intensive communication requirements. In Table 11 we compare the execution time of this algorithm using UDP-PvmData-Default and TCP-PvmDataInPlace. In this table the significative improvement leads us to select the correct functions.

The mapping of IKP tasks over different branches of the network decreases the execution times. An appropriate distribution of the tasks over the different branches improve the results. See Table 12.

4   Conclusions

We broach the implementation of differents experiences to prove that an appropriate selection of data packing functions, broadcasting strategies and mapping

Processors	2	4	8
UPD-DataDefault	13,60	7,15	4,35
TCP-DataInPlace	9,73	4,99	2,76

Table 11. IKP: Comparing UDP-DataDefault with TCP-DataInPlace

Distribution	8 processors
1 branch	4,34
2 branches	4,09
4 branches	2,51

Table 12. IKP: Tasks on different branches

of tasks leads to a significant improvement of the execution times. The knowledge of this information has been successfully proved in two different algorithms.

Acknowledgments We thank to the Instituto de Astrofísica de Canarias for allowing us the access to their LAN.

References

1. Butler R. and Lusk E. User's Guide to the p4 Parallel Programming System. University of North Florida, Argonne National Laboratory. 1992.
2. Carriero N. and Gelernter D.. Applications experience with Linda. Proceedings ACM Symposium on Parallel Programming. 1988.
3. Flower J. and Kolawa A.. The Express Programming Environment. Parasoft Corporation Report. 1990.
4. Geist A., Beguelin A., Dongarra J., Jiang W., Mancheck R., Sunderam V.. PVM: Parallel Virtual Machine - A Users Guide and Tutorial for Network Parallel Computing. MIT Press. 1994.
5. Kung H.T. Gigabit local Area Networks: A Systems Perspective. IEEE Communications Magazine. 1992.
6. Li X., Lu P., Schaeffer J., Shillington J., Sze Wong P. and Shi H. On the Versatility of Parallel Sorting by Regular Sampling. Parallel Computing, North Holland. 1993.
7. Mutka M.W. and Livny M.. Profiling Workstations' Available Capacity For Remote Execution. Performance'87. Elsevier Science Publishers B.V.. North Holland. 1988.
8. Morales D., Roda J., Almeida F., Rodríguez C. , F. García. Integral Knapsack Problems: Parallel Algorithms and their Implementations on Distributed Systems. Proceedings of ICS'95. ACM Press. 1995
9. Snir M., Otto S., Huss-Lederman S., Walker D., Dongarra J.. MPI: The Complete Reference. The MIT Press, Cambridge, Massachusetts. London, England. 1996.

PVM, Computational Geometry, and Parallel Computing Course

Michal Šoch, Jan Trdlička and Pavel Tvrdík*

Department of Computer Science and Eng.,
Czech Technical University,
Karlovo nám. 13, 121 35 Prague,
Czech Republic
{soch,trdlicka,tvrdik}@sun.felk.cvut.cz

Abstract. In the last two years, PVM has been used in our SUN workstation lab as a programming tool to teach our upper-level undergraduate students parallel programming of distributed-memory machines. More than 160 students have implemented in PVM non-trivial parallel algorithms for solving problems in computational geometry as their term projects. This paper summarizes our experience with using PVM in the parallel computing course and evaluates the results of student PVM projects.

Keywords: PVM, computational geometry, divide&conquer algorithms, SUN workstations, SP-2

1 Introduction

Many universities have recently included courses on parallel computing in their undergraduate curricula. At the Department of Computer Science and Engineering (DCSE), Czech Technical University (CTU), Prague, such a course has started in 1994. The lectures have provided students with theoretical background of designing efficient parallel and communication algorithms for distributed-memory machines. The aim of the lab for this course has been to gain practical experience with parallel programming. Students have worked on term projects to design and implement nontrivial parallel algorithms in PVM on a LAN of 10 workstations SUN of various configurations.

Our experience with using PVM as a programming tool in a parallel computing course has been encouraging and positive so far. It turned out that such an imperfect parallel computing environment, as a LAN of older SUN workstations, forced the students to learn a lot about the parallel programming techniques and to realize the tradeoffs of programming parallel distributed-memory computers.

In this paper, we describe the basic philosophy of the course, summarize the feedbacks from the students on using PVM for parallel programming, and evaluate the performance of selected PVM programs.

---

* Corresponding author

2 Parallel computing course and PVM term projects

In 1994, a course on parallel computing for upper-level undergraduate students in Computer Science at DCSE CTU in Prague, entitled *Parallel Systems and Computing*, has started, and since then, it has been taken by more than 160 students. The course of lectures has been focused on architectures of parallel distributed-memory machines, issues of interconnection networks, routing and collective communication algorithms, and about 50% of the lectures have been devoted to the design of fundamental parallel algorithms and sorting and matrix algorithms [3].

The lab of this course has been oriented to practical parallel programming. Students formed teams of two. Each team could choose one of a collection of classical problems from 2D computational geometry, dealing with unordered sets of simple objects (points, line segments, etc.) in 2D plane. The collection included: convex hull, all nearest pairs of points, the closest pair of points, all intersections or detection of an intersection of line segments or isothetic rectangles, visibility pairs of vertical line segments, minimum circle cover. We believe that these problems are ideal for the lab of such a course, since

1. they are easy to understand and code sequentially,
2. parallel algorithms can be designed using divide&conquer approach and require parallel sorting and nontrivial interprocess communication,
3. using Brent's rule, the students can experiment with scalability, granularity, speedup, and work-optimization.

For each problem, the students were given a detailed description of the problem with hints on how to parallelize it. In addition, they had to choose an interconnection topology to be implemented in PVM. The options were the hypercube, 2D mesh or torus, binary tree or mesh-of-trees, pyramidal mesh, and de Bruijn graph.

Unfortunately, DCSE has currently no access to a massively parallel computing platform able to provide resources for about 50 two-student teams for a period of 3 months. The only choice was a LAN of 10 older Sun workstations (SUN4, Sun Sparc ELC, and Opus machines) with SunOS and connected with a standard 10Mbit Ethernet.

In the course of lectures, the students were given a short (10 minutes) introduction into the philosophy and architecture of PVM 3.3.x. We used PVM 3.3.4 in 1994/5 and PVM 3.3.10 in 1995/6. Then they have been given detailed instructions on setting up their programming environment for implementing PVM programs. They were given no special lecture on using PVM for parallel programming and were only asked to study examples in the PVM installation and use on-line man pages.

3 Evaluation of PVM student projects

Our experience with the ability of students to master PVM 3.3.x under these conditions is exceptionally good. Having studied the examples, most of the teams understood the PVM programming techniques very quickly. In 2–3 weeks, they started to design and code in PVM their project programs. All students used C or C++. Most of them claimed that it was relatively easy to understand the PVM message-passing and process-management philosophy. They appreciated the simple structure and organization of the PVM library, including the self-explanatory names of functions and richness of useful parameters, mainly easy message-passing functions (both blocking and non blocking primitives have been used), message compression, and availability of broadcasting and multicasting.

In 1994/95, the students had no debugging tool at hand. This caused the biggest difficulties and they had to develop their own tracing methods. A common method was to collect tracing information into files. Each PVM process was assigned a unique output trace file. Post-mortem analysis of tracing profiles was then needed to locate and correct the malfunctioning parts.

In 1995/96, we have installed xPVM 1.1.1 and the situation has improved dramatically. The students mastered xPVM in several hours and have been using it intensively throughout the development period. The xPVM's ability to visualize interprocess communication, to inspect message buffers, and mainly to redirect easily the standard input and output, solved the main troubles. The integration of full PVM functionality into an intelligent GUI plus a possibility of real-time and post-mortem analysis makes xPVM a useful tool fully satisfying the needs of such middle-scale parallel programming tasks. At the end of the last term, we got a one-month trial installation of TotalView (BBN Systems and Technologies). However, the students expressed little interest in it, claiming that xPVM was good enough for debugging and monitoring their programs.

The projects included performance evaluation of the PVM algorithms. The students had to measure the speedup and efficiency, to assess the scalability of the algorithm, and to experiment with granularity. Execution time can be measured very easily within xPVM, but due to large overhead in xPVM, the applications run considerably slower compared to console-controlled PVM tasks. Therefore, students used other methods to measure time: either the UNIX command time or the standard function gettimeofday in sys/time.h. Unfortunately, neither method can eliminate the OS overhead (task switching, swapping, etc.) in our LAN environment.

For that, a parallel machine where a user can allocate a partition of processors dedicated to his task would be needed. Our university has recently installed an 11-node IBM SP-2 machine. Unfortunately, the IBM's customized version of PVM, PVMe, has not been installed yet, so that we had to do all the measurements in the standard public-domain PVM, version 3.3.10. It was compiled for RS6K architecture and not for the SP2MPI one. Therefore, it cannot take advantage of the high-performance switch of SP-2 [2] and uses the standard Ethernet links for interprocessor communication. Consequently, the allocation of partitions dedicated to PVM users is not supported. Nevertheless, even these measurements have approved that the PVM programs developed on a LAN of workstations by novices in parallel programming and then ported to a state-of-the-art MPP machine, exhibit very good performance figures.

At the end of each term, our LAN of SUNs was fully saturated, students were running many PVM tasks simultaneously on all our workstations, some of them on large sets of data (e.g., 100K input elements). This resulted usually into system crashes: exceeding the maximal number of UNIX processes, lack of memory for message buffers, incorrect termination of PVM tasks. Version 3.3.10 seemed to be more stable than version 3.3.4 as to the frequency of crashes.

4 Performance evaluation of PVM programs

Within last two years, more than 80 student PVM projects have been completed. Average work per project has been about 50 man-hours, including performance measurements and comparisons. About 95% of the projects produced functional codes. Of course, we could not do detailed performance evaluations of all of them, but trial measurements indicate that about 50% of them give more or less expected results in terms of speedups or scalability (absolute performance figures differ due to relative performance of node processors) and about 15% of the programs have achieved nearly optimal complexity figures under the conditions described above. From these, we have chosen 3 programs representing three different problems and tested them on both the LAN of SUNs and on the SP-2 machine in detail.

All the time measurements were done using `gettimeofday()` (by taking snapshots at the beginning and end of the task runs). Each measurement was repeated 7 times on SP-2 and 5 times on SUNs and after deleting the lowest and highest values, we took the average value. The mapping of PVM virtual processors (= processes) to physical processors was always 1-to-1 if the number of physical processors was less than the number of PVM virtual processors, and many-to-1 using the PVM default mapping strategy otherwise.

For each HW platform and for each algorithm, we provide two graphs. The first one shows the parallel time as a function of input data set size, with the number of processors being a parameter. The second one shows the speedup of the algorithm for various input data sets.

4.1 2D convex hull on a hypercube

Results on Figures 1 and 2 are based on an algorithm implemented by students *T. Rylek* and *D. Shwe* in C++ and PVM 3.3.4. The source code size is about 70KB (widely commented). It is a standard convex hull problem for a set of randomly generated 2D points solved on a hypercube topology using the divide&conquer strategy. The data are evenly distributed among processors using binary splitting on a binomial spanning tree of the hypercube and globally sorted using Batcher's bitonic sort (the local sort is QuickSort). Then each processor constructs a local convex hull and using the binomial spanning tree reversely, the local convex hulls are successively merged into the final convex hull. Smaller hulls are merged into larger ones using the lower and upper tangent method [1].

4.2 The nearest pairs of points on a pyramidal mesh

Results on Figures 3 and 4 are based on an algorithm implemented by students *L. Sýkora* and *J. Tučan* in C++ and PVM 3.3.10. The source code size is about 40KB. Given a randomly generated set of 2D points, the task is to find for each point its closest neighbor. The parallel algorithm is again of divide&conquer type, based on the hierarchical structure of the pyramidal mesh. The pyramid is constructed

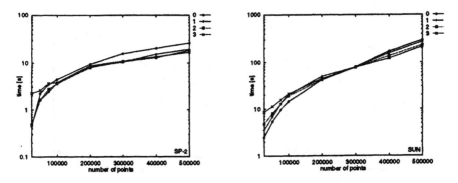

Fig. 1. Parallel time of the convex hull algorithm parameterized with the dimension of the hypercube

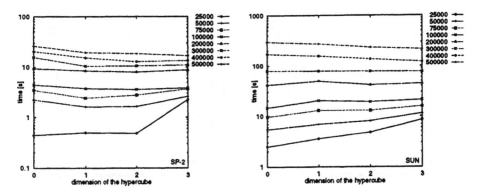

Fig. 2. Speedup for the convex hull algorithm parameterized with the input data size

recursively top-down. Each intermediate processor sorts its data, creates its four children, and splits its data evenly among them. Each leaf processor exchanges data with its parent to learn about the boundary coordinates, solves the all-nearest-pairs problem sequentially, selects a subset of local points whose nearest neighbors might belong to other processors, and sends them to its parent for further processing. This scheme continues recursively up to the root of the pyramid.

4.3 All intersections of line segments on a binary tree

Results on Figures 5 and 6 are based on algorithm implemented by students *P. Farda* and *L. Janík* in C and PVM 3.3.10. The source code size is about 32KB. The algorithm is based on K-d trees. 2D plane filled with line segments is successively split into rectangular regions using alternatively directions X and Y so that the number of line segments in both parts is approximately the same and the line segments intersecting the boundary between the parts are kept in the parent process. This splitting generates a complete binary tree of processes. The leaf processes calculate the intersections sequentially by testing all pairs exhaustively, and similarly, all local

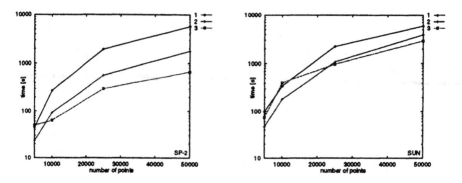

Fig. 3. Parallel time of the algorithm for the nearest pairs of points problem parameterized with height of the pyramidal mesh

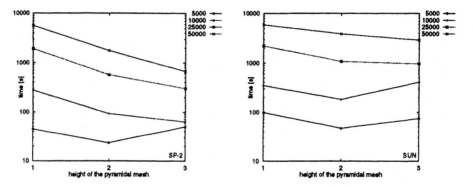

Fig. 4. Speedup of the algorithm for the nearest pairs of points problem parameterized with the input data size

line segments have to be tested with the line segments kept in the processes on the path to the root of the tree.

5 Conclusions

PVM 3.3.x and xPVM 1.1.1 turned to be very useful and simple tools for students to design and code efficient nontrivial coarse-grained message-passing parallel algorithms. The feedback from more than 160 students has proved that PVM philosophy and architecture is simple for upper-level undergraduate students to comprehend. xPVM helped reduce the problems with debugging to a reasonable level. Although the sample programs were developed on a slow and unreliable LAN of workstations, they performed well on an IBM SP-2 system. Measurements for sequential cases of described algorithms have shown that the performance of a SP-2 node is in average 2.5 times higher than that of a SUN workstation, which explains the significant difference in performance between the SUN LAN and the SP-2 machine. This has supported our hypothesis that the divide&conquer class of parallel algorithms implemented by novice parallel programmers on a low-bandwidth LAN of workstations

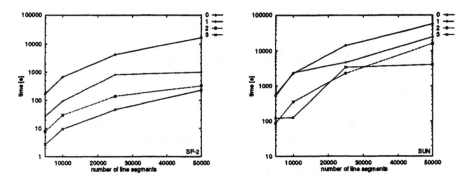

Fig. 5. Parallel time of the algorithm for the all intersections of line segments problem parameterized with the height of the binary tree

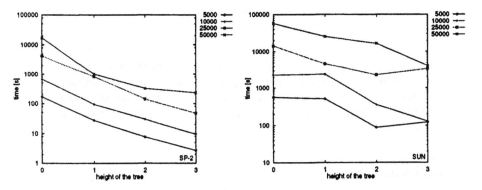

Fig. 6. Speedup of the algorithm for all intersections of line segments parameterized with the input data size

in PVM can be ported to a MPP (e.g., SP2) while keeping good performance figures. Our aim is to conduct further tests of these and similar PVM programs on our SP-2 machine when PVMe is installed.

Acknowledgments

We thank the students for providing their source codes for this study. The IBM SP-2 machine used for performance evaluation of the PVM algorithms is installed in the Joint Computing Center of CTU, Prague Institute of Chemical Technology, and IBM, Prague 6, Zikova 4.

References

1. S. G. Akl and K. Lyons. *Parallel computational geometry*. Prentice-Hall, 1993.
2. Special issue on SP-2 architecture. *IBM Systems Journal*, 34(2):144–322, 1995.
3. P. Tvrdík. Theoretical background of parallel computing for undergraduates. In *Proc. of Conf. on Par. Comp. for Undergraduates*, Hamilton, N.Y., USA, 1994. Colgate Un.

Parallelization of a State-of-the-Art Industrial CFD Package for Execution on Networks of Workstations and Massively Parallel Processors[*]

P. Luksch[1], U. Maier[1], S. Rathmayer[1], F. Unger[2], M. Weidmann[1]

[1] Lehrstuhl für Rechnertechnik und Rechnerorganisation (LRR-TUM)
Institut für Informatik der Technischen Universität München
D-80290 München
e-mail: luksch, maier, maiers, weidmann@informatik.tu-muenchen.de

[2] Advanced Scientific Computing GmbH (ASC)
D-83607 Holzkirchen
e-mail: fu@ascg.ISAR.de

Abstract. Parallelizing TfC, a state-of-the-art industrial 3d CFD code for execution on both networks of workstations (NOWs) and massively parallel processors (MPPs) is one of the main objectives in the interdisciplinary research project SEMPA[3]. In this paper, the concept for parallelizing TfC is presented as well as results obtained with a first prototype.

1 Motivation

Computational Fluid Dynamics (CFD) is one of the grand challenge applications that have been and still are the driving force behind high performance computing. Wide spread use of simulation is mainly limited by the amount of time to compute a solution. Parallel processing therefore provides an important contribution in the effort to replace experiments by simulation. Especially for small and medium size companies, parallel processing on networks of workstations therefore has become an attractive option.

This situation has been the motivation for the interdisciplinary project SEMPA. The central objective of SEMPA is to develop methods for the efficient design and implementation of portable parallel and distributed software in scientific computing. The parallelization of an industrial CFD simulation package (TfC [TASCflow for CAD]) is the major case study from which methods are to be derived that can be generalized to a certain class of applications. Making efficient use of NOWs for parallel production runs is another objective of the project. A resource manager is being developed that executes parallel programs in batch mode and dynamically schedules them on hosts that

---

[*] The work presented here is being funded by the German Federal Department of Education, Science, Research and Technology, **BMBF** (*Bundesministerium für Bildung, Wissenschaft, Forschung und Technologie*) within the project SEMPA.

[3] SEMPA = *Software Engineering Methoden für parallele Anwendungen aus dem wissenschaftlich-technischen Bereich*, Software Engineering Methods for Parallel and Distributed Applications in Scientific Computing

institution	background
LRR-TUM	research in parallel and distributed computing (computer architecture, programming environments and tools, applications)
ASC	software company, CFD package TfC
ICA III	University Stuttgart, numerical analysis, especially adaptive multigrid methods
GENIAS	software company, batch queueing system for NOWs: CODINE

Table 1. SEMPA: project partners

currently are not claimed for interactive use. Batch jobs are to be executed in parallel to interactive operation without interfering interactive users. The resource manager will make use of PVM's resource manager and task interfaces.

In this paper we focus on the PVM based parallelization of TfC. Section 2 gives an introduction to the software package. The concept for its parallelization is outlined in section 3. First results obtained with a partially parallel prototype are given in section 4. The software engineering aspects are considered in more detail in [LMRW96].

2 The CFD package TASCflow for CAD (TfC)

TfC is a widely applicable CFD package that is used to predict laminar and turbulent viscous flows in complex three dimensional geometries. Flows can either be steady or transient, isothermal or convective, incompressible or compressible (subsonic, transonic and supersonic). Application fields include pump design, turbomachines, fans, hydraulic turbines, building ventilation, pollutant transport, combustors, nuclear reactors, heat exchangers, automotive, aerodynamics, pneumatics, ballistics, projectiles, rocket motors and many more.

TfC solves the Navier-Stokes equations and scalar transport equations in 3d space on unstructured grids. It implements an element-based finite volume discretization. The system of linear equations that is assembled in the discretization phase, is solved by an algebraic multigrid method (AMG). The basic control flow in TfC is outlined in Fig 1. For each time step a number of iterations is performed, where a system of linear equation is assembled (discretization) and solved (AMG solver). In the discretization phase, the influence between nodes is computed from approximations of the nodal values ϕ_i (ϕ stands for the physical quantity to be computed, e.g. velocity U, pressure p, etc). It is sufficient for the solver to reduce the error by one or two orders of magnitude. Equation assembly in the next iteration will then compute more accurate coefficients for the next run of the solver.

The (unstructured) grid is generated in a pre-processing step by filling the solution domain with elements of predefined topologies (see Fig. 2). The node connectivity defines elements (which have nodes as their corners). Control volumes (CV) are defined around a grid node as illustrated in Fig. 3 for the 2d case. The surface of a control volume is subdivided into faces. For each face, an integration point is defined. The

---

CFD Simulation
<u>**begin**</u>
    int *Time* = *StartTime*
    bool *converged* = FALSE
    <u>**while**</u> *Time* < *EndTime* <u>**do**</u>   *time step loop*
        converged = FALSE
        <u>**while**</u> **not** *converged* <u>**do**</u>   *coefficient loop*
            discretization (equation assembly)   *computes new coefficients*
            solution of the linear equation   *updates ϕ values*
            converged = **CheckResidual**()
        <u>**od**</u>
    <u>**od**</u>
<u>**end**</u>

Fig. 1. Control flow in TfC

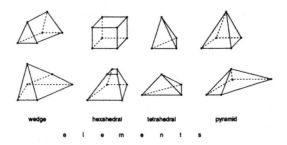

wedge   hexahedral   tetrahedral   pyramid

e l e m e n t s

Fig. 2. Element Topologies in TfC

flow through a face is approximated by the flow at the integration point multiplied with the faces' area. For each CV, a balance equation is obtained by summing up the flows through its faces. This equation states the physical principle of conservation: the total ϕ-flow out of the CV must be equal to the total flow into the CV plus the generation of ϕ inside the CV. This is stated in the general transport equation (where ρ is density, U_j velocity in x_j direction, and J_j the x_j component of the diffusion vector):

$$\underbrace{\frac{\partial(\rho\phi)}{\partial t}}_{\text{time derivative}} + \underbrace{\frac{\partial(U_j\rho\phi)}{\partial x_j}}_{\text{convection}} = \underbrace{\frac{\partial J_j}{\partial x_j}}_{\text{diffusion}} + \underbrace{S_\phi}_{\text{source term}} \tag{1}$$

The flows through the faces of the CVs are computed in an element-based way: For each element, TfC computes the flow through all its faces. The flow through the faces determines the coefficients in the system of linear equations that is being set up. A coefficient $a_{i,j}$ expresses the influence that node j has on node i. From each element

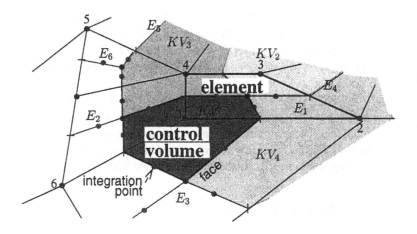

Fig. 3. Finite volume discretization in TfC

evaluation, local coefficients are obtained that contribute to the global coefficients. $a_{i,j}$ is the sum of the contributions from all elements that have nodes i and j.

The system of linear equations is solved by an algebraic multigrid method. We will not discuss general concepts of multigrid here (see [Bri87] for a tutorial introduction). The definition of coarse grids typically is based on geometry only. *Algebraic* Multigrid (AMG), however, also takes into account the values of the coefficients. The AMG concepts implemented in TfC are explained in [Raw94]. As coefficients are updated in the discretization phase, the coarse grid hierarchy must be determined in each iteration of the coefficient loop. AMG has proved to be a very fast and robust solver in TfC.

3 The design of ParTfC

3.1 Requirements

The parallel version of TfC, ParTfC, has to meet a number of requirements. Most important is portability. ParTfC has to run efficiently on a wide range of hardware platforms from NOWs to high performance MPPs. Having one software package that covers a wide range of platforms not only reduces the software vendor's development and maintenance cost but also saves users' investments in designing models for the CFD package. In order to keep the software package consistent for monoprocessor and multiprocessor usage, there should be only one version of the code for sequential and parallel execution.

The parallel computation should be scalable. Therefore, no instance of the parallel program should have to hold global information (such as a list of *all* elements) that scales up in size proportional to the problem size. ParTfC must fit into the present environment, i.e. the interface to pre- and postprocessing tools has to be maintained.[4]

---

[4] Pre- and postprocessing are left unchanged for the time being. Therefore, they provide a poten-

For NOW users it is important to be able to execute parallel production runs in batch mode. A resource management system that allows for parallel batch jobs to be executed concurrently to interactive use is being developed in **SEMPA**. The resource manager and tasker interfaces of PVM are needed to obtain the necessary control over the virtual machines.

From the user's point of view, performance certainly is the most important objective to be followed with **ParTfC**. Efficiency is most critical on NOWs because LANs typically have a much higher latency than MPP networks and because multiuser operation often causes tasks to progress unevenly. In order to minimize these effects, the number of synchronization points is kept as small as possible in **ParTfC**.

3.2 The Concept for parallelizing TfC

SPMD[5] is the prevailing paradigm for parallelizing applications in the domain of scientific computing since it scales well with the problem size and the number of available processors. SPMD requires the problem to be partitioned. Data distribution essentially is a graph partitioning problem that can be solved by using one of the available software packages (In **ParTfC**, we use MeTiS [MET95]). The decision to be made is which type of objects from the application is to be mapped to the nodes of the partitioning graph and which relation between objects is to be mapped to its edges.

For TfC it proved to be most convenient to take the grid nodes as nodes and their connectivity as the edges of this graph, i.e. the partitioning graph is identical to the grid. Alternatively, one could think about taking the elements as nodes and draw an edge between nodes that represent adjacent elements in the grid.[6].

Partitioning based on grid nodes was chosen because it enables us to compute all the coefficients needed by the solver in the equation assembly phase by introducing an overlap region around the core partition which is determined by the graph partitioning tool. To compute the coefficients $a_{i,j}$ in the equation for ϕ_i we need to evaluate all elements that have node i as one of their corners. Some of these elements will have nodes that are not assigned to "our" partition. They, together with their nodes define the overlap region as illustrated in Fig. 4.

By computing all coefficients needed in the solver during equation assembly we save one synchronization point in the coefficient loop body at the cost of some redundant computation. Cut elements are evaluated in multiple partitions. For the sake of keeping the code modular the element evaluation routines were left unchanged although local contributions to coefficients for nodes in the overlap region need not be computed.[7] The main control loop of **ParTfC** is illustrated in Fig. 5.

---

tial bottleneck because only problems that can be pre- and postprocessed on a monoprocessor can be dealt with. Parallel pre-/post processing, however, is beyond the scope of **SEMPA** but will possibly be addressed in future projects.

[5] Single Program Multiple Data

[6] This alternative was discussed but rejected because it would require more communication and administrative overhead than the solution we adopted

[7] Distinguishing core nodes from overlap nodes would certainly need more time than than simply computing local coefficients for all nodes.

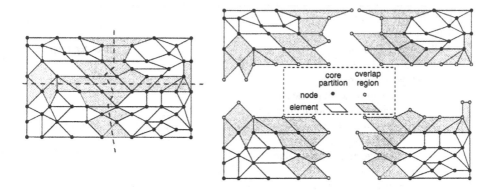

Fig. 4. Data distribution in ParTfC

The problem description consists of geometry information and nodal values, ϕ. Geometry information consists of the list of nodes and the list of elements. For each node, the list contains its (global) number (0 ... NumNodes), its Cartesian coordinates and, for the parallel version, its partition number. An entry in the element list consists of the (global) element number (0 ... NumElems), the element type (cf. Fig 2) and the list of nodes. This representation is identical to the one used in the sequential program, except for the partition numbers. Geometry information for the partitions is prepared by the preprocessor which does the partitioning.

The nodal values define the initial values for velocity, pressure, etc. at the start of the computation. They may either be estimates or results from a previous computation. Nodal values are stored in a single file which has the same structure as for the sequential program. Each process reads the values for all nodes in its partition.

Now as geometry is known and initial values for ϕ_i have been set, each process has to find out which nodal values it has to sent to (or receive from) other processes in the following computation. Note, that the list of values that has to be sent from P_k to P_j, $S_{k \to j}$ is identical to the list of values that has to be received by P_j from P_k, $R_{j \leftarrow k}$.

P_i can build its lists $R_{i \leftarrow j}, 1 \leq j \leq$ NumPartitions, concurrently to the initialization of its partition. P_i makes an entry n in $R_{i \leftarrow j}$ whenever it encounters a node n with partition number j in the node list of an element that belongs to its partition (i.e. it has a node in partition p_i). Having completely initialized the local partition, the $R_{i \leftarrow j}$ lists are sorted by global node numbers to guarantee a globally unique order of nodal values in messages. Thus, (global) node number can be omitted in messages that communicate nodal values since the assignment of values to nodes is clear from the order in which they appear in the message. Before parallel computation can start, each process P_i sends its $R_{i \leftarrow j}$ list to P_j for all $j \neq i$, where it is stored as $S_{j \to i}$.

While equation assembly itself is only marginally affected by parallelization, parallelizing the AMG solver requires some modifications to the existing code. In the "local" system of equations for parition i, $A^{[i]} x^{[i]} = b^{[i]}$, nodal values x_k appear for nodes in

---

ParTfC Main Control Loop()
Time = StartTime
while Time < EndTime **do**   *time step loop*
    Converged = `false`
    while ¬ Converged **do**   *coefficient loop*
        compute nodal gradients
        communicate nodal gradients
        EquationAssembly()   *Finite Volume discretization*
        AMG Solver()
        communicate nodal values
    od
    Time = Time + TimeStep
od

Fig. 5. Control flow in ParTfC

partitions other than p_i. For them, new equations are introduced. They have the form $x_k = b_k$. b_k is updated with the x_k values received from the process that computes it. In AMG, additional communication is required in the computation of the coarse grid hierarchy. Coarse grid blocks are determined independently by each node; blocks are not allowed to contain fine grid nodes from different partitions. Having determined its coarse grid block, process P_i knows to which fine grid nodes in a remote partition its coarse grid blocks are connected. It must "ask" its neighbors to which coarse grid block they have mapped their fine grid nodes to complete its connectivity data structures for the coarse grid.

4 A first Prototype of ParTfC

A first prototype of ParTfC has been presented at the TASCflow Users' Group Conference in May, 1996 [Ung96]. It implements parallel equation assembly. The solver has been extended by adding the additional equations for the nodes in the overlap region as explained in the previous section. Nodal values are, however, communicated only once per coefficient loop iteration, i.e. the b_k's are not updated within the solver itself. This results in a Block-Jacobi approach, which, as may be expected, slows down convergence.

    Some profiling has been done for a test case, where the flow around a cylinder is computed (grid: 22100 nodes). Due to the Block-Jacobi iteration implemented in the prototype, the number of iterations that were needed to obtain a given accuracy increased by a factor of more than two compared to the sequential program. The CPU time per iteration increased by about 20%, which can be attributed mainly to the redundant computation caused by the overlap region. With four partitions, 17% percent of the nodes and 9% of the elements are replicated. The partition sizes were about 6500 nodes and 5200 elements for our test case, which is rather small compared to production runs. As the percentage of overlap will decrease as partition size increases, redundant computation is not expected to degrade efficiency.

5 Conclusion and Future Work

First results for the (yet incomplete) parallel prototype seem quite promising. Next, parallel AMG has to be completed. Up to now, the optimization goals in partitioning the problem description are to generate partitions of equal size and to minimize the number of cut edges. For heterogeneous networks, however, it is desirable to be able to assign weighting factors to partitions that reflect the relative speed of the host on which the partition is to be processed. Instead of minimizing the *number* of cut edges it would be much more useful in CFD computations to take into account the flow direction in the partitioning phase if flow information is available from a previous run. We will investigate how these modified objectives can be implemented in general purpose partitioning software.

ParTfC is designed to be portable not only with respect to hardware platforms but also with respect to the message passing library. Like most applications in scientific computing, ParTfC uses only a small subset of the functionality implemented in PVM, which can be found in most other libraries, too. Communication is encapsulated in a library so that a port to another message passing system is very easy. For production runs especially on NOWs, support for heterogeneity as provided by PVM is an important feature. For the design of a resource manager for parallel batch jobs, the resource manager and tasker interfaces of PVM are of vital importance since they provide the necessary control over the application and enable us to implement the resource management functionality that is specific to *parallel* applications as an encapsulated module whose interactions with the message passing system and the batch queueing system are implemented by precisely defined interfaces.

References

[LMRW96] Peter Luksch, Ursula Maier, Sabine Rathmayer, and Matthias Weidmann. SEMPA: Software Engineering Methods for Parallel Scientific Applications. In *International Software Engineering Week, First International Workshop on Software Engineering for Parallel and Distributed Systems*, Berlin, March 1996.
http://wwwbode.informatik.tu-muenchen.de/archiv/artikel/pdse96/PDSE96.ps.gz.

[Bri87] William L. Briggs. *A Multigrid Tutorial*. SIAM, Philadelphia, PA, 1987.

[Raw94] Michael Raw. A Coupled Algebraic Multigrid Method for the 3D Navier Stokes Equations, 1994.

[MET95] *METIS: Unstructured Graph Partitioning and Sparse Matrix Ordering System.* George Karypis and Vipin Kumar, University of Minnesota, 1995.

[Ung96] Friedemann Unger. SEMPA Software Engineering Methods for Parallel Applications, Project Status. In Georg Scheuerer, editor, *4th TASCflow User Conference*. ASC, May 1996.

A Parallel Multigrid Method for the Prediction of Incompressible Flows on Workstation Clusters

K. Wechsler, M. Breuer and F. Durst

Lehrstuhl für Strömungsmechanik, Universität Erlangen-Nürnberg,
Cauerstr. 4, D-91058 Erlangen, Germany

Abstract. A parallel multigrid method for the prediction of incompressible flows in complex geometries is investigated with respect to the performance on workstation clusters. The parallel implementation is based on grid partitioning and follows the message passing concept, ensuring a high degree of portability. A high numerical efficiency is obtained by a nonlinear multigrid method with a pressure-correction scheme as smoother.

Within this investigation, up to 64 workstations located at the TU Munich connected by an ethernet switch are used as well as two workstations located at the University of Erlangen. The workstations in Erlangen and Munich are connected by a high-speed ATM-network. Global and local communication which is required within the solver is examined for different configurations of workstations. These results are used to understand the performance of the flow solver. Moreover, it is shown that speedups can be achieved for the computation of complex flow problems on workstation clusters, although these speedups are smaller compared to other parallel architectures.

1 Introduction

In recent years intensive research has been undertaken to improve the performance of flow computations in order to extend their applicability to a cost-effective solution of practically relevant flow problems. It is especially interesting to being able to perform optimizations of fluid flow problems based on predictions of numerical computations and verification by experimental work. However, practically relevant flows are almost always confined by complex geometries and are governed by complex physical and chemical processes. Accurate prediction of these flows requires reliable physical and chemical models as well as fine grids and small time steps in order to achieve a reasonable accuracy. Recently, significant improvements have been achieved both on the development of efficient solution algorithms and the acceleration of computations by more efficient computer hardware such as high-performance parallel computers. However, only the combination of efficient algorithms and high-performance computers will provide the possibility to enable fast computations of practically relevant flows.

The parallel multigrid method used within this paper, which is an example of an efficient implementation of highly-advanced numerical methods on parallel computers, has been described in detail in [1, 2]. It is suitable for the prediction of laminar and turbulent incompressible flows in complex geometries. The geometrical complexity is handled by a block structuring technique, which also serves as the base for the parallelization by grid partitioning. The high numerical efficiency is obtained by a global nonlinear multigrid method with a pressure-correction smoother also ensuring only slight deterioration of the convergence rate with increasing processor numbers. A similar procedure for flows in simpler geometries which can be described by one geometrical block is given in [3].

The method has already been examined with respect to its numerical and parallel efficiency on different parallel platforms [4],[5]. Efficiencies have also been given for small numbers of powerful workstations connected by a FDDI network and clustered to be exclusively used for the computations without being deteriorated by interactive work [5]. It has been shown in that work, that these twelve HP735 workstations clustered to the CONVEX META system give the possibility to reach nearly the performance of the early vector supercomputer CRAY-YMP. Within this paper, the performance of the described parallel multigrid method using a larger number of less powerful workstations, which are connected to provide a pool for educational and research purposes, is investigated. For this, up to 64 workstations connected in sub-networks which in turn are connected by an ethernet switch at the Technical University of Munich and two workstations located at the University of Erlangen-Nürnberg are applied. The workstations in Erlangen and Munich are connected by an ATM-network.

In order to discuss the performance of a parallel method it is helpful to measure or estimate the time required for computational effort and communication. After a brief description of the basic numerical algorithm, the multigrid technique and the parallelization strategy, some simplified benchmarks measuring the necessary communication processes for the applied method will be explained. These communication primitives can be used to estimate the parallel performance of the underlying method.

2 Numerical Method

The numerical solution method is based on a fully conservative finite volume discretization on blockstructured non-orthogonal boundary-fitted grids with a non-staggered arrangement of variables. In order to ensure the correct coupling of the solution between momentum and pressure, the well known interpolation technique of Rhie and Chow [6] is applied. Second order discretization is used for all terms (central differences, linear interpolation) together with a deferred correction approach for the convective fluxes. The continuity equation is used to obtain a pressure-correction equation according to the SIMPLE algorithm of Patankar and Spalding [7]. The linearized equations for the velocity components, pressure-correction and other scalar variables are assembled and relaxed sequentially, where the ILU approach of Stone [8] is employed as a linear system solver. Outer iterations are performed to take into account the non-linearity, the coupling of the variables, and the effects of grid non-orthogonality, which are treated explicitly in all equations.

For increasing the rate of convergence a nonlinear multigrid method (full approximation scheme), in which the pressure-correction scheme acts as the smoother, is employed. V-cycles are used for the movement through the grid levels, and nested iterations are employed for improving the initial guesses on the finer grid levels (full multigrid). The basic concepts of the multigrid technique are described in detail in [9].

The parallelization of the method is achieved by a grid partitioning technique based on the blockstructuring. The geometrically complex integration domains, which is applied to handle the geometrical complexity, is mapped to a parallel blockstructure such that the resulting subdomains are suitably assigned to the available processors with respect to a a good load balancing. For handling the coupling of the blockstructured grids auxiliary control volumes containing the corresponding boundary values of the neighbouring block are introduced along the block interfaces. The coupling of the blocks is ensured by communicating the boundary values during the iterative solution algorithm. More details about the parallel implementation of the ILU-method of Stone and the overall parallel solution can be found in [10, 1].

3   Benchmarking Communications

In order to get a rough estimate of the performance of the method on a parallel computer, one can start by investigating the iterative solver. Usually, approximately 50 percent of the computing time is used within this procedure, and the granularity of this procedure reflects the one of the whole program quite well. For each inner iteration, one has to solve for the solution, update the boundary data, and check for the residuals:

- Local communication is needed to exchange data at the boundaries of the partitioned grid. In the case of three-dimensional flow computations, the local communication corresponds to the exchange of the data at the interface between two neighbouring blocks of the parallel blockstructure. Typical message length can exceed up to 50 kByte. As it is desirable to communicate as much as possible at the same time in parallel, efficient local communication requires a high total bandwidth. In order to simulate this local exchange, a benchmark has been implemented which exchanges data between processors pair by pair and in parallel. Thus, the total bandwidth of the network can be examined.
- Global communication is necessary to update the residuals within the iterative solution procedure in order to check for convergence. In this case, only small messages of eight bytes are needed to sum up globally the residual. It has been shown that global communication is the limiting factor for large number of processors [3]. Therefore, a hierarchical procedure is necessary to avoid a linear increase in communication time with increasing number of processors. With respect to the global communication, the subroutine of summing up the residuals directly serves as the benchmark.

4 Results

The results of the benchmarks for the performance of the communication on workstation clusters are shown in table 1. In cases A1 through A6, only workstations located in Munich are used, whereas B1 through B6 correspond to configurations where workstations in Erlangen are involved.

Table 1. Comparison of the measured performance of the communication primitives for different configurations of workstations, which are signed by A1 through B6 for better reference in the text. The total number of workstations, those located in Munich and those in Erlangen are indicated. Performance is demonstrated by the time required for global communication T_{glob}, the time required for local communication T_{loc} and the resulting bandwidth R_{loc} for this local communication.

Case:	A1	A2	A3	A4	A5	A6	B1	B2	B3	B4	B5	B6
Total:	2	4	8	16	32	64	2	4	8	16	32	64
Munich:	2	4	8	16	32	64	1	2	6	14	30	62
Erlangen:	0	0	0	0	0	0	1	2	2	2	2	2
T_{glob} in ms:	5	10	14	22	27	36	88	100	108	132	161	236
T_{loc} in ms:	32	64	134	160	161	180	547	147	231	186	275	240
R_{loc} in MB/s:	0.48	0.48	0.46	0.78	1.54	2.75	0.03	0.21	0.27	0.67	0.91	2.08

It is obvious from the table, that the hierarchical implementation of the global communication works satisfactorily. This can be seen, because the necessary time is approximately direct proportional to the logarithm of the number of processors. Unfortunately, this performance is destroyed for B1 through B6 due to the apparently high latency of the ATM-network which connects the universities of Munich and Erlangen. The topology of the network in Munich can clearly be seen form the benchmark for local communication. A1, A2 and A3 involve workstations on one subnetwork leading to a constant total bandwidth and therefore to an increasing time necessary for the local communication. A4, A5, and A6 involve two, three, and four subnetworks resulting in an increasing total bandwidth, which can clearly be seen in the table. Again, the performance of the ATM-network is not good enough to reach the performance of cases A1 through A6. Due to the bad performance of cases B1 through B6, following investigations will concentrate on the the workstation cluster in Munich.

In order to show the performance of the multigrid technique on workstation clusters, the three-dimensional flow around a circular cylinder in a square channel at $Re = 20$ is considered. Investigations with this multigrid method for another parallel architecture have already been done in [2]. However, the results concerning the computing times are not directly comparable, because significant improvement concerning the communication within the code has been done in the meantime. Computing times for ten outer iterations for different numbers of CVs and processors are given in table 2. In the case of 6144 CVs, there is even a speeddown for increasing number of processors due to high effort for communication. Situation is better for the case of 49152 CVs. However, the advantages of parallel computing can clearly be seen from the computation of the flow for 393216 CVs. In this case, the computing time for one processor had to be estimated, because the necessary memory requirement of approximately 350 MB could not be fulfilled by a single workstation. Thus, parallel computing enabled the computation of a problem which was not possible by a single workstation. In addition to that, there is a speedup of approximately 3.2 which is really not very good but indeed there. Moreover, table 2 shows the computing times for one V-cycle for three different configurations. However, these numbers do not include the numerical efficiency of the single-grid and multigrid method. This is included in table 3. Obviously, multigrid technique does not work very good for the case of computing 49152CVs on eight processors. In this case, the effort of communication is to big compared to the pure computing time, especially on the coarser grids. Situation is different for the computation of 393216 CVs with 32 processors. In this case, a speedup of 2.85 can be achieved. This is lower compared with other parallel architectures [2], but it shows that the multigrid technique is able to speed up convergence even for workstation cluster which are characterized by poor performance in communication.

Table 2. Comparison of single-grid (10 outer iterations) and multigrid (one V-cycle) with different numbers of CVs and processors for a three-dimensional flow around a circular cylinder. The computing times are given in seconds.

	6144 CV			49152 CV			393216 CV	
	P=1	P=8	P=32	P=1	P=8	P=32	P=1	P=32
Single-grid	20	40	71	178	108	132	1260	391
Multigrid	13				96			338

Table 3. Computing times (seconds) and numbers of fine grid iterations (in parentheses) for single-grid and full multigrid methods with corresponding acceleration factors with respect to computing time for different numbers of processors and control volumes for the flow around a circular cylinder.

	6144 CV	49152 CV	393216 CV
	P=1	P=8	P=32
Single-grid	109(58)	2260(207)	26460(675)
Full multigrid	58(25)	3295(27)	9283(26)
Acceleration	1.85	0.68	2.85

5 Conclusions

A parallel multigrid method for the investigation of incompressible flows has been examined with respect to the performance on workstation clusters. In earlier work, it has been shown, that the multigrid method is clearly superior to the corresponding single grid method. This is especially true for the numerical efficiency which reflects the amount of additional operations which result from the parallel algorithm. In addition to that, high parallel efficiencies have been demonstrated to be achieveable on parallel hardware as long as communication and calculation performance are balanced.

This is clearly a problem for workstation clusters due to the fact that neither latency time nor communication bandwidth are competeable compared with closer coupled, parallel computers. Because of the high implicity of the code and the resulting close coupling of the processors which is necessary to achieve the high numerical efficiency, special problems arise due to the fact that the performance of the code is significantly deteriorated as soon as only one of the workstations is loaded with other tasks like interactive work, additional computation tasks or supervising and administrative tools.

Nevertheless, it has been shown, that it is possible to compute complex flow problems on workstation clusters which cannot be solved on one workstation due to memory constraints. Moreover, the multigrid method gives the possibility to speed up these computations compared with single processor computation and parallel single grid method mainly due to the high numerical efficiency as long the workstation cluster can be used exclusively for the computation and the number of processors is not too large. The resulting speedups, however, are smaller because the parallel efficiency is not very good compared to other parallel architectures.

References

1. F. Durst and M. Schäfer: A Parallel Blockstructured Multigrid Method for the Prediction of Incompressible Flows. Int. J. Num. Methods Fluids **22** (1996) 1–17.
2. F. Durst and M. Schäfer and K. Wechsler: Efficient Simulation of Incompressible Viscous Flows on Parallel Computers. NNFM Flow Simulation with High-Performance Computers II (1996), 87–101.
3. E. Schreck and M. Perić: Computation of Fluid flow with a Parallel Multigrid Solver. Int. J. Num. Methods Fluids **16** (1993) 303–327.
4. M. Schäfer and E. Schreck and K. Wechsler: An Efficient Parallel Solution Technique for the Incompressible Navier-Stokes Equations. NNFM **47** (1994) 228–238.
5. F. Durst and H.J. Leister and M. Schäfer and E. Schreck: Efficient 3-D Prediction on Parallel High-Performance Computers. NNFM **43** 1994
6. C.M. Rhie and W.L. Chow: Numerical Study of the Turbulent Flow past an Airfoil with Trailing Edge Separation. AIAA Journal **21** (1983) 1525–1532.
7. S.V. Patankar and D.B. Spalding: A Calculation Procedure for Heat, Mass and Momentum transfer in Three Dimensional Parabolic Flows. Int. J. Heat Mass Transfer **15** (1972) 1787–1806.
8. H. Stone: Iterative Solution of Implicit Approximations of Multi-dimensional Partial Differential Equations. SIAM J. Num. Analysis **5** (1968) 530–558.
9. M. Hortmann and M. Perić and G. Scheuerer: Finite Volume Multigrid Prediction of Laminar Natural Convection: Benchmark Solutions. Int. J. Num. Methods Fluids **11** (1990) 189–207
10. M. Schäfer and E. Schreck: ILU as a Solver in a Parallel Multi-grid Flow Prediction Code. NNFM **41** (1992)

An Abstract Data Type for Parallel Simulations Based on Sparse Grids*

Michael May[1] and Thomas Schiekofer[2]

[1] Technical University of Munich, Institute of Computer Science,
Department of Computer Architecture and Computer Organization
Arcisstr.21,
D - 80290 Munich, Germany
email: maym@informatik.tu-muenchen.de
[2] University of Bonn, Institute of Applied Mathematics,
Department of Scientific Computing and Numerical Simulation,
Wegelerstr. 6,
D - 53115 Bonn, Germany
email: schiekof@iam.uni-bonn.de

Abstract. In this paper we introduce an abstract data type for the distributed representation and efficient handling of sparse grids on parallel architectures. The new data layout, implemented by means of the PVM message passing library, provides partitioning and dynamic load balancing transparent to the application making use of it. This enables parallelization of sequential partial differential equations (PDE) solvers based on sparse grids with hardly any source code modifications.

1. Sparse Grids

The numerical simulation of problems occuring in engineering and natural sciences need quite fine grids for discretization if they should supply reasonable results. The concept of adaptivity takes care of smoothness of the treated function, provides a grid which is appropriate to the given problem and allows efficient discretization. Adaptive full grids require only few grid points where the solution is smooth and possess many grid points where regularity vanishes.

Sparse grids were introduced 1990 by [7] (see also [2]) and represent a given function (e.g. a solution of a PDE) with even less grid points than adaptive full grids do. Moreover, sparse grids allow the concept of adaptivity, too.
Sparse grids only contain $\mathcal{O}(h_l^{-1} \cdot (\mathrm{ld}(h_l^{-1}))^{d-1})$ grid points compared to $\mathcal{O}(h_l^{-d})$ for usual full grids, where d denotes the dimension of the problem and h_l the mesh width on the boundary. On the other hand the interpolation error in the L_2 norm only increases slightly from order $\mathcal{O}(h_l^2)$ to $\mathcal{O}(h_l^2 \cdot (\mathrm{ld}(h_l^{-1}))^{d-1})$. Sparse grid points are a subset of grid points of a full grid. An important characteristic

---

* This work is being funded by the ministry for education, science, research and technology BMBF within the project PAR-CVD (Parallel simulation of Chemical Vapor Deposition processes)

of sparse grid points is that they can be described by a level and an index information (see equ. 1) which can be exploited for hashing.

The just mentioned properties of sparse grids are sufficient to understand the scope of this paper beginning in section 1.2. For the curious reader however, we give a short introduction into the theory of sparse grids in the next section. For more detailed information we refer to [7], [1], [2], [3] and [4].

1.1 Theoretical background

The typical *nodal basis* on $\Omega = [0, 1]$ in one dimension is given by $B_l = \{\varphi_l^i(x) |\ \varphi_l^i(x_j) = \delta_{ij}, \varphi_l^i(x)$ piecewise linear, $0 \leq i, j \leq 2^l\}$, where $x_j = j \cdot 2^{-l}$ $(0 \leq j \leq 2^l)$ denote equidistant grid points of grid Ω_l of mesh width 2^{-l} with $x_0 = 0$ and $x_{2^l} = 1$. Sparse grids are based on hierarchical bases H_l which can be obtained as follows:

$$H_1 = B_1$$
$$H_l = H_{l-1} \cup \{\varphi_l^j \in B^l | x_j \in \Omega_l / \Omega_{l-1}\}$$

Figure 1 shows the hierarchical basis in the $1D$-case for $l = 3$. It can easily be verified that $span\{B_l\} = span\{H_l\}$ $(l \in \mathbf{N})$.

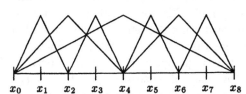

$$x_0 \quad x_1 \quad x_2 \quad x_3 \quad x_4 \quad x_5 \quad x_6 \quad x_7 \quad x_8$$
Figure 1: hierarchical basis in $1D$ for $l = 3$
(basis functions on the boundary are neglected)

The basis functions for the d-dimensional nodal basis on the hypercube $[0, 1]^d$ can be defined as the tensor product basis functions of the one-dimensional nodal basis functions, that is

$$\varphi_{l_1, \ldots, l_d}^{i_1, \ldots, i_d}(x^{(1)}, \ldots, x^{(d)}) = \varphi_{l_1}^{i_1}(x^{(1)}) \cdot \ldots \cdot \varphi_{l_d}^{i_d}(x^{(d)}). \tag{1}$$

Figure 2: The grids Ω_{l_1, l_2} in the case $d = 2$ for $1 \leq l_1, l_2 \leq 4$ and sparse grid $S_{4,4}^{(2)}$

We introduce the grids $\Omega_{l_1,\ldots,l_d}$ on $[0,1]^d$ with mesh width 2^{l_k} ($1 \le k \le d$) belonging to coordinate direction $x^{(k)}$. The space $S^{(d)}_{l_1,\ldots,l_d}$ denotes the space of piecewise d-linear functions on grid points of grid $\Omega_{l_1,\ldots,l_d}$. Additionally, we need spaces $T_{i_1,\ldots,i_d}$ which consist of piecewise d-linear functions on $\Omega_{i_1,\ldots,i_d}$ vanishing on all grid points of grids $\Omega_{k_1,\ldots,k_d}$ with $k_j \le i_j$ for $1 \le j \le d$ and $k_1 + \ldots + k_d < i_1 + \ldots + i_d$. Figure 2 shows the grids Ω_{i_1,i_2} induced by T_{i_1,i_2} in the 2D-case for $1 \le i_1, i_2 \le 4$ (grid points on the boundary are neglected).

Using $T_{i_1,\ldots,i_d}$ we can rewrite $S^{(d)}_{l_1,\ldots,l_d} = \oplus_{i_1=1}^{l_1} \cdots \oplus_{i_d=1}^{l_d} T_{i_1,\ldots,i_d}$. The interpolation of a d-dimensional continuous function $u(x^{(1)}, \ldots, x^{(d)})$ yields

$$u_I^{(d)}(x^{(1)}, \ldots, x^{(d)}) = \sum_{i_1=1}^{l_1} \cdots \sum_{i_d=1}^{l_d} u_{i_1,\ldots,i_d} \quad \text{with}$$

$$u_{i_1,\ldots,i_d} = \sum_{k_1=1}^{2^{i_1}-1} \cdots \sum_{k_d=1}^{2^{i_d}-1} c^{i_1,\ldots,i_d}_{k_1,\ldots,k_d} \varphi^{i_1,\ldots,i_d}_{k_1,\ldots,k_d}$$

with the hierarchical surplusses $c^{i_1,\ldots,i_d}_{l_1,\ldots,l_d}$ (cf. [1]). In the 1D-case the hierarchical surplus of a node x_k is $c_k^i = u(x_k) - 0.5 \cdot (u(x_{h_l}(x_k)) + u(x_{h_r}(x_k)))$ where $x_{h_l}(x_k)$ and $x_{h_r}(x_k)$ are the left and right hierarchical neighbor of x_k (for example in figure 1 x_0 and x_4 are the hierarchical neighbors of x_2) and $u(\cdot)$ is the function which is to be represented. The computation of a d-dimensional hierarchical surplus is straightforward using the tensor product notation.

Let us now consider the 2D-case for the introduction of sparse grids. The error contribution of a single space T_{i_1,i_2} with respect to the L_2-norm is of order $\mathcal{O}(4^{-i_1-i_2})$. Obviously, all spaces T_{i_1,i_2} with $i_1 + i_2 = c$ (c fixed) are of same dimension and give the same order of contribution to the error. Using a triangular scheme of T_{i_1,i_2} (as indicated in figure 2 by a line) instead of the quadratic scheme used by full grids yields the *sparse grids*.

Let us return to the d-dimensional case and denote the space of piecewise d-linear function on a sparse grid by $\tilde{S}^{(d)}_{l,\ldots,l}$. Using $T_{i_1,\ldots,i_d}$ again we can write $\tilde{S}^{(d)}_{l,\ldots,l} = \oplus_{i_1+\ldots i_d \le l+d-1} T_{i_1,\ldots,i_d}$. The interpolation of a continuous function $u(\cdot)$ in the sparse grid space yields now

$$\tilde{u}^{(d)}_{l,I} = \sum_{i_1+\ldots+i_d \le l+d-1} u_{i_1,\ldots,i_d}.$$

Evidently, sparse grids need much less grid points than full grids. They only involve $\mathcal{O}(h_l^{-1} \cdot (\mathrm{ld}(h_l^{-1}))^{d-1})$ grid points compared to $\mathcal{O}(h_l^{-d})$ for full grids where h_l denotes the mesh width on the boundary. error in the L_2-norm is only slightly worse than in the full grid case: it is of order $\mathcal{O}(h_l^2 \cdot (\mathrm{ld}(h_l^{-1}))^{d-1})$ instead of $\mathcal{O}(h_l^2)$. The same behavior can be obtained in the L_∞-norm. If we measure the error contribution of a space $T_{i_1,\ldots,i_d}$ with respect to the energy norm we can eliminate additional spaces $T_{i_1,\ldots,i_d}$ of our triangular scheme which was deduced by using the L_2-norm. In this case we get energy based sparse grids which consists only of $\mathcal{O}(h_l^{-1})$ grid points, but have same error order $\mathcal{O}(h_l)$ as the full grid.

1.2 Data Structures

In literature the only data structure considered useful to implement sparse grids are systems of binary trees. They represent the hierarchical structure and the low density of sparse grids very well, in contrast to inefficient and costly standard arrays. The interpretation of the individual binary trees is, for instance in a three-dimensional case, as follows:

A first binary tree, whose nodes represent *grid planes*, contains second order binary trees with nodes representing *grid lines*. On each of these second order nodes, third order binary trees eventually represent the actual *grid points* along the given lines. The problems arising from this data structure layout are obvious: Depending on the target function, the binary trees can become arbitrarily deep and hence degenerate into linked lists. This fact hinders both the efficient access and distribution of the data onto the available processors, which are only possible from balanced trees.

For use with sparse grid based PDE solvers we thus need efficient implementations of search operations for arbitrary grid points, while insertions and deletions during the building and destruction of the sparse grid may be slow.

Additionally pointered Binary Trees As a first step to avoid the problem of accessing neighboring points in the sparse grid, one could think of additional pointers to represent nearest neighbors and hierarchical son and father relationship. Still efficient access to the grid points in a possibly degenerated binary tree cannot be guaranteed.

Hash Tables A way to accelerate the access is to compute the address of the data instead of searching for it. This notion leads to hashing the grid points into a Hashing Table and resolving collisions in linked lists. In this layout no dimension of the grid is preferred and all access to neighboring grid points is equally fast and easy. However, the distribution of the Hashing Table onto the processors is hindered by the arbitrary scattering of data over the available table space, destroying the needed data *convexity*, i.e. adjacent data is *not* mapped on a neighboring space in the hash table.

Partitioned Hash Tables In order to avoid the scattering of the grid points, one could partition the sparse grid into convex areas, i.e. *d*-dimensional quads, and give each quad its own hash table, which then could be mapped onto the available processors. However, the partitioning of the hash tables is static, so any adapting of the sparse grid to the target function will lead to in-equal load of the processors.

In order to handle these limitations we need a representation of sparse grids that offers both efficient access and easy data partitioning for parallel processing. In the next section we present a module implementing an abstract data type Sparse_Grid for the parallel representation of such grids satisfying the demands just mentioned.

2 The distributed abstract Data Type Sparse_Grid

An abstract data type offers its functionality as a black box, i.e. its implementation and properties (variables) are encapsulated inside the module and can only be accessed by means of the exported module methods. This way the numerical algorithm using the data type `Sparse_Grid` does not and need not know anything about the internal representation and distribution of the grid points. Modifications to step from a sequential numerical program to a parallel one are thus minimal.

An incarnation of the data type is characterized by the size of the individual dimensions and the number of maximal grid points leading to the notion of a *virtual* array as the sparse grid points are internally stored in hash tables instead of arrays. Instances of `Sparse_Grid` can be initialized and destroyed by the methods

$$SG\_alloc(), SG\_dealloc()$$

where the first method needs the maximal number of grid points and the grid sizes as parameters.

The properties of each individual grid point are the level and index information (see equ. 1) and a special field to hold additional information like function values or the hierarchical surplus (see 1.1). In order to read and set these properties there are the methods

$$SG\_insertnode(), SG\_deletenode, SG\_findnode(), SG\_set()$$

which all need the following parameters: a sparse grid `SG_grid`, the level and the index of the grid point and the additional information. These parameters do not have all to be set, depending whether the access is read or write.

In order to initialize, synchronize and re-gather the data distribution and parallel calculation there are three more external methods,

$$SG\_distribute(), SG\_sync(), SG\_collect(),$$

which need an instance of `Sparse_Grid` and the number of processors as parameters.

Everything else, for instance the partitioning and load balancing schemes by means of processor *splits* and processor *merges* (see section 2.2), the internals of synchronization, the storing and accessing of grid points in processor local hash tables, is fully transparent to the numerical algorithm.

The data structures inside the sparse grid module are based on a multikey file structure for accessing disk storage presented by [6]. In the subsequent subsections we will give an overview of all hidden features of the abstract data type `Sparse_Grid` and discuss its suitability for partitioning and load balancing.

2.1 Process Arrays

For notational simplicity we only regard the two-dimensional case (d=2), as the adaption to the general case ($d > 2$) is straight-forward.

For a parallel computation the target area covered by a sparse grid should be divided into subregions of uniform point density. This is achieved by means of a so called *process array*, a two-dimensional array

$$\texttt{process\_array[0..nx-1][0..ny-1]},$$

defining the correspondence of convex 2-dimensional quads of the grid to associated processors or processes. Each cell of `process_array` holds the identification number of the processor or process responsible for the appropriate subregion. The grid points in these subregions are stored in (local) processor memory in (individual) hash tables.

The partition of sparse grid (that yields the process array) is determined by so called *scales*, vectors

$$\texttt{xscale[0..nx]} \; and \; \texttt{yscale[0..ny]}$$

describing the partition of each dimension of the grid respectively.

These data structures are replicated on each processor of the parallel machine.

The access (find, insert, delete) to a sparse grid point with level l and index i is as follows (see figure 3):

1. The level and index of the grid point are checked against the x-scale and y-scale to determine the appropriate row x and column y in the process array.
2. Next it is checked whether the pointer `process_array[x][y]` matches the processor of the computing process. If so, the grid point is accessed in the local processor memory by efficient hashing techniques. A write access can lead to subsequent processor splits or merges[3].
3. If `process_array[x][y]` does not match, the access to the grid point is delegated by passing a message to the corresponding processor.

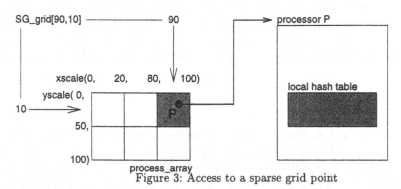

Figure 3: Access to a sparse grid point

---

[3] which triggers a broadcast update of the process_array and the scales to the remaining processors

2.2 Splitting and Merging

Partitioning and dynamic load balancing are achieved by means of processor *splits* and processor *merges*, if individual processor loads, i.e. the number of locally stored sparse grid points, reach a maximal (minimal) value by the repeated insertion (deletion) of grid points. Two parameters influence the dynamic correspondence of the processors to the sparse grid subregions, the *splitting policy* and the *merging policy*.

1. The **splitting policy** is characterized by three features:
   - ☐ The (maximal) *processor load*, of which any overflow leads to a split.
   - ☐ The *dimension* that is chosen for the next split, i.e. the axis to which the partitioning hyper-plane is orthogonal. For instance, a very simple splitting policy would choose the dimensions cyclically. More elaborate policies could favor some dimensions and thus lead to higher resolutions.
   - ☐ The *location* of the split point, i.e. the point at which the linear scale is partitioned

2. The **merging policy** is determined by the following decisions:
   - ☐ The (minimal) *processor load*, of which any underflow triggers a merge operation.
   - ☐ Which pair of adjacent subregions is a possible *candidate* for a merge. Examples for algorithms to choose possible merge partners are the well known buddy system and the more general neighbor system.
   - ☐ If there are several candidates, which one wins the *competition*.

These policies of the abstract data type enable an equal distribution of the virtual sparse grid array over the processors and re-balance the processor load after each split or merge implementing a dynamic partitioning.

As there is no a-priori information about the number of points contained in a sparse grid, a dynamic distribution using subsequent processor splits from the very beginning can lead to unequally balanced grid points and thus processor loads. In such cases it is more efficient to build up the full sparse grid on all processors, split the grid in subregions and then delete the grid points not needed for the local computation on each processor respectively. A further adaption of the sparse grid to the target function can be balanced again by processor splits.

2.3 Hash Tables

For memory and performance reasons the points of the virtual sparse grid are internally stored in hash tables. In the case of sparse grids it is evident from the introduction to sparse grids (equ. 1) that information of level and index is sufficient to identify a sparse grid point. Consequently, the position of a node is determined by the result of a hash function that computes an array-index of the hash table from the parameters level and index.

2.4 Synchronization with PVM

As already mentioned above, access to non-local grid points is delegated by passing a message to the processor responsible for the appropriate subregion. PVM messages in the abstract data type Sparse_Grid are used in a uniform manner:

1. The messages are tagged with a flag indicating the type of action to be performed by the receiving processor, and
2. contain all the data necessary for the indicated operation.

For example, in order to insert a new grid point in the grid region of another processor, a message containing the level, index and additional information is packed, tagged with SIG_INS and sent to the appropriate processor. At a given time, the appropriate function on the remote processor, here SG_insertnode(), is called with the parameters that were delivered in the message.
The message tags are

$$SIG\_INS, SIG\_DEL, SIG\_GET, SIG\_SPLIT, SIG\_MERGE$$

to signal a nonlocal insertion, deletion, find, split and merge operation respectively.

The messages are handled and buffered in a message queue by the PVM daemon until they are received by the appropriate PVM process. This message queue is worked off by a message handler inside the module, SG_sync(). Only by calling SG_sync() messages will be received and corresponding actions performed. By placing calls to the SG message handler on synchronization points in the numerical application, the user has complete control over the strictness and timing of synchronization among the processors.

For instance, in conventional scientific computing, a call to SG_sync() should be placed after every time step, i.e. at the end of the time-step loop. To support local time stepping inside the target domains, SG_sync() could be called only every, say 50 time steps. Eventually, calling SG_sync() after every write access to the sparse grid would implement a very strict synchronization.

3 Conclusion and Future Work

We have presented an abstract data type built on top of PVM to distribute and parallelize sparse grids nearly transparently to the numerical application. As a case study we are about to parallelize a sparse grid based Navier-Stokes-solver for use with Fastest 3D, a simulation package for chemical vapor deposition processes in the electronic and optoelectronic field.
As the data type Sparse_Grid is suitable for grids of arbitrary dimensions, it is obvious to implement simulations of d-dimensional problems as well i.e. to handle heat transport and chemical reaction equations.

References

1. Bungartz, H.-J.: *Dünne Gitter und deren Anwendung bei der adaptiven Lösung der dreidimensionalen Poisson-Gleichung,*
   Ph.D. Thesis, Technical University of Munich (1992)
2. Griebel, M.: *A parallelizable and vectorizable multi-level algorithm on sparse grids,*
   in Parallel Algorithms for partial differential equations, Notes on Numerical Fluid Mechanics, Volume 31, pp. 94-100, W. Hackbusch, ed., Vieweg, Braunschweig, 1991
3. Griebel, M.: *Parallel Multigrid Methods on Sparse Grids,*
   in Multigrid Methods III, Proceedings of the 3rd European Conference on Multigrid Methods, Bonn, Oktober 1990, International Series of Numerical Mathematics (Vol. 98), W. Hackbusch and U. Trottenberg, ed., Birkhäuser, Basel, 1991 (also: Technical University of Munich, SFB report no. 342/30/90 A)
4. Griebel, M.: *The combination technique for the sparse grid solution of PDE's on multiprocessor machines,*
   Technical University of Munich, SFB report no. 342/14/91 A, 1991
5. Griebel M., Schneider M. and Zenger C. *A combination technique for the solution of sparse grid problems,*
   in Iterative Methods in Linear Algebra, pp. 263-281, IMACS, P. de Groen und R. Beauwens, ed., Elsevier, North Holland, 1992, (also: Technical University of Munich, SFB report no. 342/19/90 A)
6. Nievergelt, J., Hinterberger, H., Sevcik, K. C.: *The Grid File: An Adaptable, Symmetric Multikey File Structure,*
   ACM Transactions on Database Systems, Vol. 9, No. 1 (1981)
7. Zenger, C.: *Sparse Grids in Parallel Algorithms for Partial Differential Equations,*
   Proceedings of the Sixth GAMM-Seminar, Kiel, 19.01.-21.01.1990, Notes on Numerical Fluid Mechanics Vol. 31, W. Hackbusch, ed., Vieweg, Braunschweig, 1991 (also: Technical University of Munich, SFB report no. 342/18/90 A)
8. Zenger, C.: *Hierarchische Datenstrukturen für glatte Funktionen mehrerer Veränderlicher,*
   Informatik und Mathematik, Broy, M., Springer, Berlin (1991)

The PVM System: Status, Trends, and Directions *

Vaidy Sunderam

Dept. of Math & CS, Emory University, Atlanta GA 30322, USA
email: vss@mathcs.emory.edu

Abstract. The Parallel Virtual Machine (PVM) system is a software framework that enables heterogeneous concurrent computing in heterogeneous environments. Since its conception in 1989, it has evolved into a very popular and widely used platform for cluster and network computing. The programming model supported by PVM, straightforward formulation of the API, robust portable implementations, and timeliness have all played an important role in PVM's impact on high performance computing. We discuss the overall design of PVM and describe its ongoing evolution, recently introduced features, and future directions.

1 Introduction

The PVM system for concurrent computing has evolved into a defacto standard and is in widespread use worldwide. PVoM (Parallel Virtual Machine) is a software framework that emulates a generalized distributed memory multiprocessor in heterogeneous networked environments. Such an approach, which obviates the need to possess a hardware multiprocessor, has proven to be a viable and cost-effective technology for concurrent computing in many application domains. For a number of reasons, including portability, support for heterogeneity, robustness, and a simple but complete programming interface, the PVM system has gained widespread acceptance in the high-performance concurrent computing community. Currently, over ten thousand users have obtained the software and documentation and it is estimated that a substantial fraction of them are actively using PVM on a regular basis. The PVM project is an ongoing research effort aimed at investigating novel concepts in distributed and concurrent computing. This research is carried out by the principal investigators of the project [1] as well as at many other institutions. Promising or useful ideas that demonstrate their viability through experimental research are incorporated into the release version of the software. In this paper, we first consolidate the model, features, and implementation of PVM, including some characteristic experiences, and then discuss ongoing enhancements and future trends.

---

* Research supported by the U. S. Department of Energy, under Grant No. DE-FG05-91ER25105, NASA under grant NAG 2-828, and the National Science Foundation, under Award Nos. ASC-9214149, ASC-9527186, and CCR-9523544

2 An Overview of the PVM System

The PVM software system is an integrated set of software tools and libraries that emulates a general purpose, flexible, heterogeneous concurrent computing framework on interconnected computers of varied architecture. The overall objective of the PVM system is to to enable such a collection of computers to be used cooperatively for a concurrent or parallel computation. Detailed descriptions and discussions of the concepts, logistics, and methodologies involved in this network-based computing process may be found in [2, 3, 4]; briefly, the principles upon which PVM is based include:

- Explicit message passing model: Concurrent and parallel PVM applications fall under the category of "message passing parallelism". Collections of computational entities, each performing a part of an applications workload using data- functional- or hybrid decomposition, cooperate by exchanging messages to accomplish the overall computational task.
- User configured host pool: The application computational entities execute on a set of machines that are selected by the user for a given run of the PVM program. Both single-CPU machines and hardware multiprocessors (including shared-memory and distributed-memory computers) may be part of the host pool, which may also be altered by adding and deleting machines during operation.
- Translucent access to hardware: Application programs may view the hardware environment either as an attribute-less collection of virtual processing elements or may choose to exploit the capabilities of specific machines in the host pool by positioning certain computational entities on the most appropriate computers.
- Process based computation: The unit of parallelism in PVM is a process, an independent sequential thread of control that alternates between communication and computation. No process to processor mapping is implied or enforced by PVM; in particular, multiple processes may execute on a single processor.
- Heterogeneity support: The PVM system supports heterogeneity in terms of machines, networks, and applications. With regard to message passing, PVM permits strongly typed heterogeneous messages (containing more than one type of data) to be exchanged between machines having different data representations.

2.1 The PVM Model

The PVM computing model is based on the notion that an application consists of several different sub-algorithms or "components". Components, depending on their functions, may execute in parallel with one another. Further, components that are parallelizable may be executed by replication, thereby reducing the processing time involved. Thus, PVM supports a two-level hierachy of parallelism – with the relationship between components based on functional decomposition, and intra-component parallelism typically achieved via data decomposition, commonly using the SPMD (single-program multiple-data) model of computing.

Under PVM, a user defined collection of serial, parallel, and vector computers emulates a large distributed-memory system, hereafter. termed the *virtual machine*. Multiple users may configure overlapping virtual machines, PVM supplies functions to initiate tasks on the virtual machine and allows these tasks to communicate and synchronize with each other. Applications, which may be written in Fortran77 or C, are parallelized based on the message-passing paradigm that is common to most distributed-memory computers. By sending and receiving messages, multiple tasks of an application cooperate to solve a problem in parallel. The model assumes that any task can send a message to any other PVM task, and that there is no limit to the size or number of such messages. The PVM communication model provides asynchronous blocking send, asynchronous blocking receive, and non-blocking receive functions. In addition to these point-to-point communication functions the model supports multicast to a set of tasks and broadcast to a user defined group of tasks. In addition, synchronization among tasks through the use of barriers, rendezvous, and signals is supported.

2.2 Using the PVM System

The PVM software system contains library routines that enable a user process to register with/leave a collection of cooperating processes, routines to add and delete hosts from the virtual machine, to initiate and terminate PVM tasks, to synchronize with and send signals to other PVM tasks, and routines to obtain information about the virtual machine configuration and about active PVM tasks. Routines are provided for packing and sending messages between tasks. The core communication routines include an asynchronous send to a single task, and a multicast to a list of tasks. PVM transmits messages over the underlying network using the fastest mechanism available e.g. either UDP, TCP on networks based on the Internet protocols, or other high-speed interconnects available between the communicating processors. Messages can be received by filtering on source or message tag (both of which may be specified as wildcards), with either blocking or non-blocking receive routines. A routine can be called to return information about received messages such as the source, tag, and size of the data. There are routines for creating and managing multiple send and receive buffers. This feature allows the user to write PVM math libraries and graphical interfaces that can be invoked by PVM applications without communication conflicts.

The PVM system is composed of two parts. The first is a daemon, called *pvmd3* (sometimes simply *pvmd*), that executes on all the computers making up the virtual machine. Pvmd3 is designed so any user with a valid login can install this daemon on a machine. A user wishing to use PVM first configures a virtual machine by specifying a host-pool list; the daemons are started on each, and cooperate to emulate a virtual machine. The PVM application can then be started from a shell command line prompt on any of these computers. The second part of the system is a library of PVM interface routines (libpvm3.a). This library contains user callable routines for message passing, spawning processes, coordinating tasks, and modifying the virtual machine. Application programs must be linked with this library to use PVM.

3   A Representative Performance Report

Cluster/network computing in general and PVM in particular are widely used for high-performance applications including scientific supercomputing. While the general purpose nature of the machines and networks involved preclude efficiencies attainable in dedicated MPP's, PVM performance can be quite satisfactory and frequently very good. To illustrate, we include a report on a recent performance exercise.

3.1   The NAS Parallel Benchmarks on PVM

The Numerical Aerodynamic Simulation (NAS) Program of the National Air and Space Administration (NASA) has devised and published a suite of benchmarks [5] for the performance analysis of highly parallel computers. These benchmarks are designed to substantially exercise the processor, memory, and communication systems of current generation parallel computers. They are specified only algorithmically; except for a few constraints, implementors are free to select optimal language constructs and implementation techniques.

The complete benchmark suite consists of eight applications, five of which are termed *kernels* because they form the core of many classes of aerodynamic applications, and the remaining three are simulated CFD applications. The five kernels, and their vital characteristics are listed in Table 1. NASA periodically publishes performance results obtained either from internal experiments or those conducted by third-party implementors on various supercomputers and parallel machines. The *de facto* yardstick used to compare these performance results is a single processor of the Cray Y-MP, executing a sequential version of the same application.

Table 1. NAS Parallel Benchmarks: Kernel Characteristics

Benchmark Code	Problem size	Memory (Mbytes)	Cray time (seconds)	Operation count
Embarassingly parallel	2^{28}	8	126	2.67×10^{10}
V-cycle multigrid	256^3	453.6	22	3.90×10^{09}
Conjugate Gradient	$2 * 10^6$	83.2	12	1.51×10^{09}
3-D FFT PDE	$256^2 * 128$	343.2	29	5.6×10^{09}
Integer sort	2^{23}	248.8	11	7.81×10^{08}

The five NPB kernels, with the exception of the embarassingly parallel application, are all highly communication intensive when parallelized for message passing systems. As such, they form a rigorous suite of quasi-real applications that heavily exercise system facilities and also provide insights into bottlenecks and hot-spots for specific distributed memory architectures. In order to investigate the viability of clusters and heterogeneous concurrent systems for such applications, the NPB kernels were ported to execute on the PVM system. Detailed discussions and analyses are presented in [6, 7]; here we describe our experiences with two representative kernels, using Ethernet and FDDI based clusters.

The V-cycle multigrid kernel involves the solution of a discrete Poisson problem $\nabla^2 u = v$ with periodic boundary conditions on a $256 \times 256 \times 256$ grid. v is 0 at all coordinates except for 10 specific points which are +1.0, and 10 specific points which are -1.0. The PVM version of this application was derived by substantially modifying an Intel hypercube version; data partitioning along one dimension of the grid, maintaining shadow boundaries, and performing nearest neighbor communications. Several optimizations were also incorporated, primarily to maximize utilization of network capacity, and to reduce some communication. Results for the multigrid kernel under PVM are shown in Table 2.

Table 2. V-cycle Multigrid: PVM timings

Platform	Time (secs)	Comm. volume	Comm. time (secs)	Bandwidth (KB/sec)
4*IBM RS6000/550 (Ethernet)	293	96MB	101	973
4*IBM RS6000/550 (FDDI)	185	96MB	18	5461
8*IBM RS6000/320 (FDDI)	110	192MB	32	6144
Cray Y-MP/1	54	—	—	—

¿From the table it can be seen that the PVM implementation performs at good to excellent levels, despite the large volume of communication which accounts for upto 35% of the overall execution time. It may also be observed that the communications bandwidths obtained, *at the application level*, are a significant percentage of the theoretical limit for both types of networks. Finally, the eight processor cluster achieves one-half the speed of a single processor of the Cray Y-MP, at an estimated one fourth of the cost.

The Conjugate Gradient kernel is an application that approximates the smallest eigenvalue of a symmetric positive definite sparse matrix. The critical portion of the code is a matrix-vector multiplication, requiring the exchange of sub-vectors in the partitioning scheme used. In this exercise also, the PVM version was implemented for optimal performance, with modifications once again focusing on reducing communication volume and interference. Results from our experiments are shown in Table 3.

This table also exhibits certain interesting characteristics. Like the multigrid application, the conjugate gradient kernel is able to obtain near theoretical communications bandwidth, particularly on the Ethernet, and a four processor cluster of high-performance workstations performs at one-fourth the speed of

Table 3. Conjugate Gradient: PVM timings

Platform	Time (secs)	Comm. volume	Comm. time (secs)	Bandwidth (KB/sec)
4*IBM RS6000/550 (Ethernet)	203	130	124	1074
4*IBM RS6000/550 (FDDI)	82	130MB	24	5433
16*Sun Sparc SS1+ (Ethernet)	620	370MB	360	1074
Cray Y-MP/1	22	—	—	—

a Cray Y-MP/1. Another notable observation is that with an increase in the number of processors, the communication volume increases, thereby resulting in lowered speedups. Our results from these two NPB kernels indicate both the power and the limitations of concurrent network-based computing with PVM — i.e. that with high-speed, high-capacity networks, PVM performance is competitive with that of supercomputers; that it is possible to harness nearly the full potential and capacity of processing elements and networks; but that scaling, load imbalance, and latency limitations are inevitable with the use of *general purpose* processors and networks that most cluster environments are built from.

4 Recent and New Features

Dynamic process groups are layered above the core PVM routines. A process may belong to multiple groups, and groups may change dynamically at any time during a computation. Routines are provided for tasks to join and leave a named group. Tasks can also query for information about other group members. Functions that logically deal with groups of tasks such as broadcast and barrier use the user's explicitly defined group names as arguments. The PVM system is designed so that native multiprocessor calls can be compiled into the source. This allows fast message-passing mechanisms of a particular system to be exploited by the PVM application. Messages between two nodes of a multiprocessor use its native message-passing routines, while messages destined for an external host are routed via the user's PVM daemon on the multiprocessor. On shared-memory systems the data movement can be implemented with a shared buffer pool and lock primitives. The MPP subsystem of PVM consists of a daemon that manages the allocation and deallocation of nodes on the multiprocessor. This daemon is implemented in terms of PVM core routines. The second part of the MPP port is a specialized library for this architecture that contains the fast routing calls between nodes of this machine.

5 Research Initiatives

The evolution of PVM is driven by numerous research projects at various institutions. Some of these initiatives are designed with the intention of incorporation into PVM as a feature, in a subsequent release. Such ideas tend to be straightforward, but nevertheless very valuable and lead to direct and immediate added value to the PVM system. Others are more experimental and/or novel in concept, and constitute research ideas that may or may not turn out to be viable or successful ideas. Even those that are demonstrably successful sometimes do not merge into PVM distributions for various reasons, though most good ideas are eventually manifested in PVM in one form or another. In this section we outline a number of such projects, systems, and features.

5.1 Parallel I/O with PIOUS

Parallel I/O is a much needed feature that will not only be required by scientific applications that manipulate large data sets or require out-of-core computations, but also by non-scientific applications, especially those in transaction processing, distributed data bases, and other high-performance commercial applications. Motivated by this reasoning, we have embarked on a project to provide parallel I/O facilities in network-based concurrent computing systems, with PVM as the first target environment. PIOUS [8] is an input/output system that provides process groups access to permanent storage within a heterogeneous network computing environment. PIOUS is a parallel distributed file server that achieves a high-level of performance by exploiting the combined file I/O and buffer cache capacities of multiple interconnected computer systems. Fault tolerance is achieved by exploiting the redundancy of storage media.

To better support parallel applications, PIOUS implements a parallel access file object called a *parafile* and provides varying levels of concurrency control for process group members. PIOUS is a file service built on top of existing file systems and accessed via library calls. PIOUS is itself implemented as a group of cooperating processes within (an enhanced version of) the PVM distributed computing framework. Because of its modular design and utilization of existing standards, PIOUS is easily ported to other distributed computing environments providing similar functionality. An alpha version has been implemented and preliminary results, in our evaluation of both functionality and performance, is very encouraging.

5.2 Threads-based concurrent computing

TPVM [9] is a collection of extensions and enhancements to the PVM computing model and system. In TPVM, computational entities are threads. TPVM threads are runtime manifestations of imperative subroutines or collections thereof that cooperate via a few simple extensions to the PVM programming interface. In the interest of straightforward transition and to avoid a large paradigm shift, one of the modes of use in TPVM is identical to the process-based model in PVM, except that threads are the computational units. TPVM also offers two other programming models — one based on data driven execution, and the other supporting remote memory access.

Architecturally, TPVM is a natural extension of the PVM model. In terms of the resource platform, TPVM also emulates a general-purpose heterogeneous concurrent computer on an interconnected collection of independent machines. However, since TPVM supports a threads-based model, it is potentially capable of exploiting the benefits of multithreaded operating systems as well as small-scale SMMP's — both of which are becoming increasingly prevalent in general purpose computing environments. Further, multiple computational entities may now be manifested within a single process. In combination, these aspects enable increased potential for optimizing interaction between computational units in a user-transparent manner. In other words, inter-machine communication may continue to use message passing, while intra-machine communication, including that within SMMP's, may be implemented using the available global address space. In addition, "latent" computational entities in the form of dormant threads may be instantiated either asynchronously or during initialization, at negligible or low cost. In many cases, this helps reduce the overhead of spawning new computational entities during application execution. Further, the concept of latent threads extends naturally to service-based computing paradigms that are more appropriate in general purpose, non-scientific, distributed and concurrent processing. TPVM is designed to be layered over the PVM system, and does not require any modifications or changes to PVM. User level primitives are supplied as a library against which application programs link; operational mechanisms are provided in the form of standalone PVM programs.

5.3 PVM RPC

To expand the scope and applicability of the PVM system, we have developed an alternate interface to PVM that supports client-server computing. The core of this interface is a remote procedure call (RPC) facility [10] that permits the specification and export of services which may be invoked by clients using the well-established RPC paradigm and mechanics. Previous experiences have established that this model is natural and effective for programming distributed applications, and that RPC can be implemented in message-passing systems like PVM with a minimum of overhead. This model also enables significant additional functionality such as user-transparent load balancing, failure resilience, and adaptive parallelism. A preliminary implementation of the RPC system has been completed and both functionality and performance have been found to be significantly improved and useful in preliminary experiments.

5.4 XPVM: A Graphical Console and Monitor for PVM

XPVM [13] provides a graphical interface to the PVM console commands and information, along with several animated views to monitor the execution of PVM programs. These views provide information about the interactions among tasks in a parallel PVM program, to assist in debugging and performance tuning. To analyze a program using XPVM, a user need only compile their program using the PVM library, version 3.3 or later, which has been instrumented to capture tracing information at run-time. Then, any task spawned from XPVM will return trace event information, for analysis in real time, or for post-mortem playback from saved trace files. XPVM provides "point-and-click" access to the PVM console commands. A pull-down menu allows users to add or delete hosts to configure the virtual machine. Tasks can be spawned using a dialog box that prompts for all spawn options, including the trace mask to determine which PVM routines to trace for XPVM. XPVM serves as a real time performance monitor for PVM tasks. Tasks spawned from XPVM automatically send back trace events that describe any desired PVM activity. User programs need not be recompiled or annotated for tracing, as the PVM 3.3 library has a built-in tracing facility. The monitor views scroll and zoom in unison and are time correlated, allowing the user to more easily compare different information about a particular occurrence in the program execution. A number of other views, including a Network view, a Space-Time vies, and a Utilization view are included; further, XPVM also has some very useful debugging facilities.

5.5 PVaniM

The PVaniM 2.0 system [11] provides online and postmortem visualization support as well as rudimentary I/O for long running, communication-intensive PVM applications. PVaniM 2.0 provides these features while using several techniques to keep system perturbation to a minimum. The online graphical views provided by PVaniM 2.0 provide insight into message communication patterns, the amount of messages and bytes sent, host utilization, memory utilization, and host load information. For online visualization analysis, PVaniM 2.0 utilizes sampling to gather data regarding interesting aspects of the application as well as interesting aspects of the cluster environment. With sampling, necessary statistics are collected and sent to the monitor intermittently. The rate at which the application data is sent to the monitor is a parameter that may be set by the user. With lower sampling rates, the application will experience less perturbation, but the graphical views will not be updated as frequently. For offline visualization analysis, PVaniM 2.0 utilizes buffered tracing to provide support for fine-grain visualization systems. Currently, PVaniM produces tracefiles for the prototype PVaniM system. A converter is provided that allows the user to also use the popular ParaGraph system using techniques similar to PGPVM.

5.6 Context

Contexts are a way to partition the communication space so that messages sent in one context can only be received by tasks in the same context. The use of context are particularly important in the design of parallel software libraries. For example, consider the case were task A in a parallel application calls a math library routine and this routine contains message-passing. If the math routine has been written by a third party, then the application developer does not know which message tags (or wildcards) are used inside it. There is no way he can guarantee that a message sent to task A by task B in the application will not be inadvertently intercepted by the math routine. Thus, causing both the math library and the application to fail.

Introducing a communication context is a solution to the above problem. Context provides a mechanism that allows tasks to choose arbitrary message tags and still have the guarantee that messages with these tags will not be received inadvertently by a task executing in a different thread or routine. One way to visualize context is to consider two tags associated with a message – a user tag and a context tag. The user tag is arbitrarily assigned by the programmer, while the context tag is assigned by the system. PVM views context [14] as merely another tag (without wildcarding). Primitives are provided for tasks to request a new system-wide unique context tag. The program may use and distribute this context in exactly the same way that MPI does. However, the programmer is not bound to the MPI communicator model. Distributing a new context among processes is simplified by adding a synchronous group formation operation that gets and distributes a new context to all members and is analogous to the MPI group formation. Our scheme insures that two different groups are each guaranteed a unique context, removing the need for separate intra and inter-communication constructs. As long as a process knows (or can find out) the context of a group, it can communicate with the group. Other advantages to this scheme are that new processes can assume old contexts and start communicating with an existing group. This is especially useful for fault-tolerant programs that want to replace failed tasks.

PVM 3.4 will introduce the concept of a base context but will not have a base communicator. All tasks will know the base context and will be able to use it for communication even in the event of task failure. To maintain compatibility with existing PVM codes, a context tag has not been explicitly added to the argument lists of pvm_send and pvm_recv. Instead, a task has an *operating* operating context that can be queried and set by the program. Sends and receives are carried out in the current operational context.

5.7   Name Service and Static Groups

PVM 3.4 will add some MPI-like communicator constructions [14]. The collective call pvm_staticgroup() takes a list of tids and returns a group communicator id. This allows PVM programs to arbitrarily construct groups of processes with context. Group send and receive functions will be added that take a group id and a rank as arguments. These functions are "syntactic sugars" that handle the mappings of group rank to tids and the proper management of the operating context. Collective operations such as broadcast, gather, scatter and reduce will also be supported on static groups.

It is often desirable for two programs to start independently and discover information about each other. One mechanism would be for each of the programs to key on a "well-known name" to look up information in a database. A program could be given essential information such as a task identifier (tid) and a context to use when communicating with the named task. Only one PVM task may insert a specific key into the database. When another task performs a lookup on the name, the message will be sent to the calling task. This sort of name service is a very general mechanism. The message could, for example, contain a list of task id's and an associated context, or just a task id, or simply initialization data. Programs may insert items into the database and tag them as "sticky." This means that the inserted item will remain in the database after the program has exited. The default is for the name server to eliminate information that has been inserted in the database when the inserting program terminates.

5.8   Windows PVM

WPVM is a PVM implementation for the "MS Windows" - Microsoft Windows Operating System. WPVM stands for "Windows Parallel Virtual Machine" and intends to exploit the potential computational power of the small and medium PC LANs that proliferate in corporations and universities. As the original PVM, WPVM creates the abstraction of a parallel machine from a cluster of computers connected by a network. This virtual machine appears as a single and manageable resource to the programmer. The difference is that the target machines for WPVM are personal computers with MS Windows instead of UNIX machines. However, this system is compatible with the original PVM, meaning that we can use virtual machines composed simultaneously of UNIX and MS Windows machines. WPVM is provided in two different flavors. One for Windows 3.x 16 bits and another for Windows 95 - 32 bits. The April 96 release offers WPVM libraries for the following compilers: Borland C++ 4.51; Ms VisualC++ 2.0 and Watcom 10.5. All the packages have examples (*.ide for Borland and *.mak for VisualC++ and Watcom). WPVM uses the Windows sockets standard v1.0 (WINSOCK).

5.9   Other Extensions and Enhancements

In addition to the auxiliary projects and enhancements to PVM outlined above, various other efforts are also in progress[12]. These include:

- Thread Safety - redesigning the PVM internals to greatly improve thread safety, based on research developing an experimental lightweight thread PVM.
- Buffered Tracing - to reduce the expense of large communication volume and frequency when the tracing option is turned on for support of tools such as XPVM.
- Secure Message Passing - The SecurePVM project is creating a compatible PVM version that securely establishes the virtual machine/pvmd structure and allows the user to specify in the pvm_initsend() call whether she wants the message to be (a) not encrypted; (b) authenticated that message is unmodified and sender verified; or (c) encrypted so that message is unreadable by outsiders.
- MPI port - The MPP versions of PVM are being updated to run over the native MPI message passing library that some MPP's like the SP-2 support. This enhances standardization and leads to improved performance.

– Faster Heterogeneous encodings - Researchers are investigating Receiver Makes Right data encodings, where transmitters send data in native form to be corrected by receivers when necessary.

6 Discussion

Our experiences with PVM and related systems have been very valuable. From the research perspective, the important issues in heterogeneous network computing have been highlighted, and the benefits and limitations of this paradigm have been demonstrated. The project has also provided a deeper insight into important research issues that warrant further investigation and has led to the evolution of novel techniques that are promising and should be pursued further. Like many experimental research projects, PVM has also had an important side-benefit; it has produced software systems that are pragmatic and robust enough for actual production quality use by applications in a variety of domains. From the point of view of applications developers, PVM has provided a cost-effective and technically viable solution for high performance concurrent computing, as indicated by the large number of installations adopting this system, often as their primary concurrent computing facility. Both these facets of PVM have encouraged us to continue research and development, both on aspects that enhance the effectiveness and efficiency of the systems software, as well as on new features and techniques.

References

1. V. S. Sunderam, G. A. Geist, J. J. Dongarra, and R. Manchek, "The PVM Concurrent Computing System: Evolution, Experiences, and Trends", *Journal of Parallel Computing*, **20**(4), pp. 531-546, March 1994.
2. V. S. Sunderam, "PVM : A Framework for Parallel Distributed Computing", *Journal of Concurrency: Practice and Experience*, **2**(4), pp. 315-339, December 1990.
3. A. Beguelin, J. Dongarra, G. Geist, R. Manchek, and V. Sunderam, "Solving Computational Grand Challenges Using a Network of Supercomputers", *Proceedings of the Fifth SIAM Conference on Parallel Processing*, D. Sorensen, ed., SIAM, Philadelphia, 1991.
4. G. A. Geist and V. S. Sunderam, "The Evolution of the PVM Concurrent Computing System", *Proceedings - 26th IEEE Compcon Symposium*, pp. 471-478, San Fransisco, February 1993.
5. D. H. Bailey, et. al., The NAS Parallel Benchmarks, International Journal of Supercomputer Applications, 5(3):63-73, Fall 1991.
6. S. M. White, Implementing the NAS Benchmarks on Virtual Parallel Machines, Emory University M. S. Thesis, April 1993.
7. S. M. White, A. Anders, and V. S. Sunderam, Performance Optimization of the NAS NPB Kernels under PVM, Proc. Distributed Computing for Aeroscience Applications, Moffett Field, October 1993.
8. S. Moyer and V. S. Sunderam, "Parallel I/O as a Parallel Application", *International Journal of Supercomputer Applications*, Vol. 9, No. 2, summer 1995.

9. A. Ferrari and V. S. Sunderam, "TPVM: Distributed Concurrent Computing with Lightweight Processes", *Proceedings – 4th High-Performance Distributed Computing Symposium*, Washington, DC, pp. 211-218, August, 1995.

10. A. Zadroga, A. Krantz, S. Chodrow, V. Sunderam, "An RPC Facility for PVM", *Proceedings – High-Performance Computing and Networking '96*, Brussels, Belgium, Springer-Verlag, pp. 798-805, April 1996.

11. B. Topol, V. Sunderam, J. Stasko, "Performance Visualization Support for PVM", preprint, Emory University, March 1996.
    `http://www.cc.gatech.edu/gvu/softviz/parviz/pvanim/pvanim.html`

12. `http://www.epm.ornl.gov/pvm/research.html`

13. J. A. Kohl, G. A. Geist, "The PVM 3.4 Tracing Facility and XPVM 1.1," Proceedings of the 29th Hawaii International Conference on System Sciences (HICSS-29), Heterogeneous Processing Minitrack in the Software Technology Track, Maui, Hawaii, January 3-6, 1996.

14. G.A. Geist, J.A. Kohl, and P.M. Papadopoulos, "PVM and MPI: A Comparison of Features", *Calculateurs Paralleles*, to appear, 1996.

OCM — An OMIS Compliant Monitoring System

Thomas Ludwig, Roland Wismüller, and Michael Oberhuber

Technische Universität München (TUM), Institut für Informatik
Lehrstuhl für Rechnertechnik und Rechnerorganisation (LRR)
Arcisstr. 21, D-80333 München
email: {ludwig|wismuell|oberhube}@informatik.tu-muenchen.de
http://wwwbode.informatik.tu-muenchen.de/

Abstract. The OMIS project aims at defining a standard interface between tools for parallel systems and monitoring systems. Monitors act as mediators between tools and the parallel program running on some target architecture. Their task is to observe and manipulate the program according to the tool's commands. A standardized interface will allow different research groups to develop tools which can be used concurrently with the same program. OCM, an OMIS compliant monitoring system, is the first realization of such an environment. It is designed for PVM programs running on workstation clusters. The paper will give an outline of the goals of this project and describe important details of the monitoring system's design.

1   Introduction

The situation in the field of parallel and distributed computing is still characterized by the fact that productivity with implementing programs is much lower than with sequential programs. A reason can be seen in additional complexities like nondeterminism, race conditions, deadlocks etc. Furthermore, there are only few adequate tools available to handle these complexities. Even the most basic problem of parallel code debugging is not yet well supported.

If we take a closer look at available development tools, we find that many of them belong to the category of trace based off-line tools. They provide means to inspect a programs behavior after its completion. Thus, no manipulation of the running code is possible nor can the tool user interact in an adaptive way with his program. For e.g. debugging or program steering however, it is inevitable to have powerful on-line tools at hand. In more detail, tools should be capable of doing both to observe as well as to manipulate the running program.

This requires to have powerful monitoring systems which act as mediators between the tool and the software complex of application program, programming library, and operating system components. The monitoring system is responsible for accepting commands from the tool and providing it with corresponding results. In addition, it controls the program by instrumenting parts of the above mentioned software components.

Up to now, most of the approaches for monitoring systems are custom built, i.e. they are designed for dedicated purposes and tied to specific architectures or programming environments. Several works about monitoring methodologies were published ([1], [10], [11], [13], [2]), but no comprehensive approach summing up state-of-the-art techniques to a common on-line monitoring system has been proposed.

In order to avoid the complexity of on-line monitoring many tools are based on traces. This refers to debugging tools [5] as well as to performance analysis [6]. A main reason for the popularity of traces is the existence of several standard formats, like SDDF [12] or PICL [4].

In addition, the development of tools and monitor systems is almost never done by independent research groups. As a result, tools run only in environments which are especially adapted to their needs and no other tool can run with this environment.

The goal of the OMIS (on-line monitoring interface specification) project is to overcome these problems and separate tool and monitor development by providing a standard interface between them. It will also provide tool developers with on-line capabilities instead of only trace data to be processed. The project is a joint effort between groups at Technische Universität München and at Emory University (see [9]).

The remainder of the paper will give some insight into basic ideas of OMIS and present the design of OCM, an OMIS compliant monitoring system for PVM running on heterogeneous workstation clusters.

2 OMIS and OCM

The OMIS project started in summer 1995. Beginning of 1996 we published the OMIS document version 1.0 as a first definition of what we specified during that time. Version 1.0 serves as a basis for the design and implementation of an OMIS compliant monitoring system (OCM).

One of the main goals of the OMIS project is to decouple tool and monitor development. Groups being involved in one or the other research topic can now fully concentrate on their proper research. Furthermore, having a specified standard, allows to broaden the capabilities of parallel tools. Eventually we will come to

- **interoperable tools** where tools from different sources can be used concurrently and
- **unified tool environments** which can be used in identical form on different target architectures.

Figure 1 illustrates this fact and clarifies the decisive role of OMIS and OCM in this concept. OMIS defines an on-line mechanism of tool/program-interaction. We consider interactive tools like e.g. debuggers and performance analyzers and automatic tools like e.g. load balancing facilities. Off-line tools performing program trace analysis are supported by connecting an on-line trace generator tool to OMIS. After program completion trace data recorded can be fed into any appropriate off-line tool. This is identical to working principles of trace analysis tools. However, with OMIS manipulation facilities of an OCM will remain inactive.

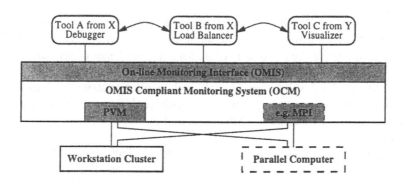

Fig. 1. Unified interoperable tools running on various target architectures via OMIS and OCM

The goal of the research group at LRR-TUM is to design and implement OCM for a specific target architecture and to integrate it with a sophisticated tool environment. This is covered with THE TOOL-SET approach (see [8]). OMIS and OCM will be freely available to the research community and may form the basis for advanced tool technology. Finally, the users will benefit from the ability to be able to select from various tools which can be applied concurrently to the same running program.

For the first step we will design and implement an OMIS compliant monitoring system for PVM running on workstation clusters. The following sections of the paper will outline general concepts for the realization of such a system. Concrete design decisions will be discussed as they form the basis for an implementation.

3 Design Concepts of OCM

Concepts for the design of an OMIS compliant monitoring system can be divided into two categories: First, many design concepts refer to aspects which can be found in the specification document itself. Methods of interaction with the tools belong to this category. Second, it is inevitable to design concepts which guarantee an efficiently working implementation of OCM on a certain target architecture.

The first category mainly influences the logical behavior and provides a frame for the software structure of the the implementation whereas the second category already fixes which software mechanisms provided by the system have to be used.

Let us first take a closer look at both classes before describing in more detail the cooperation between all components involved.

3.1 OMIS Related Design Concepts

The OMIS document tries to set up a monitoring interface which is as far as possible independent from the programming library used for applications. However, any instance of an OMIS compliant monitoring system must be implemented for a fixed programming library or parallel programming language and a dedicated target architecture. As the interface specification tries to be abstract, our goal is to identify and implement a generic monitoring system which can easily be ported among different combinations of software and hardware architecture. The design of the individual modules of the monitoring system shall always identify whether or not the modules belong to the core system. This also implies an extensive definition of interfaces between parts belonging to the core system and architecture specific parts.

Cooperation between tools and the monitoring system is specified to be composed of service requests and replies to them. A service request is sent to the monitoring system as a coded string which describes which activity has to be invoked on the occurrence of a specified event. As a matter of fact, service requests program the monitoring system to listen to event occurrences, perform actions, and transfer results back to the tool. The monitoring system inherently follows an event/action working model. Cooperation with tools is also covered by this model as the receipt of service requests represents an event to which an appropriate action belongs.

As the application program to be manipulated is distributed the monitoring system has to adopt this architectural principle. The monitoring system belonging to one application program consists of a set of communicating single-node monitors. They are functionally fully equipped and additionally perform the task of coordinating their activities in order to guarantee detection of global events and/or execution of global actions. Global here means non-local, i.e. any activity that involves more than one node of the application's node set.

3.2 Implementation Related Design Concepts

One of the main questions refers to the type and number of new components which have to be added to the already available software and hardware environment. Technically the question is of how to start up the monitoring system, how many components belong to it, how should they interoperate, how to manage dynamic changes, and finally, when will they be deleted from the system?

Currently, the monitoring system is designed to consist of one monitor process per node of the virtual machine of a user. Each monitor process runs under the same user identification as the application program. No special rights are necessary to install this software. In case of multiuser tools like e.g. load balancers we need an additional component on every node which can determine the node status. This so-called node status reporter is designed as a process which has to be integrated into the system by the super user. It is persistent in the sense that it runs also on nodes where PVM tasks are not yet or no longer existing. The node status reporter can be accessed by any single-node monitor in order to get up-to-date load values.

In addition to having monitor and node status reporter processes we also need to instrument libraries and daemons of all runtime environment software. In our systems these are the PVM library and daemon and PFSLib library and daemon. The latter is a parallel file system for workstation clusters developed at LRR-TUM [7]. Instrumentation is inevitable to efficiently observe the dynamic program behavior and to be able to manipulate it.

Appropriate communication mechanisms have to be chosen for component interoperation on a single node and between nodes.

3.3 The Software Module Structure

Let us now have a closer look at the software module structure of the complete system. On the top of Figure 2 we can see various tools running e.g. on a remote workstation. Via a communication library they send requests to the monitor process on any node of the system. The monitor handles request interpretation and starts appropriate activities. A more detailed description of its tasks will follow in the next section.

The monitor process is linked together with extensions which are necessary for controlling specialized software like the PFSLib daemon. A discussion of this functionality is beyond the scope of this paper. Please refer to the OMIS document for details [9].

Every monitor cooperates with tasks on its node, with daemons of the programming environment, with the operating system, and with monitors on other nodes. Task code is composed of the code performing the program's calculation, library code linked to it, and a monitoring library added to this complex. The monitoring library implements for instance monitor objects like timers and counters which are directly handled in the application context. Usually, manipulation of these objects is triggered by the execution of programming library calls. E.g. with performance analysis a send message call might add the call completion time to a clock counter. Potentially all calls of all programming libraries used can be instrumented by adding wrappers to the original function calls which can then perform the necessary activities. Technically, communication with the monitor process is designed (but not restricted) to be handled via shared memory segments.

Instrumentation of the PVM daemon is needed for special purposes only like e.g. getting a signal when a message is received from another node. All information which is not stored in the library's address space has to be requested from the daemon. By analyzing the structure of PVM we found that instrumentation of the daemon will be of minor importance as almost all necessary information are located outside of its address space. As with the task code interoperation is handled by using shared memory segments

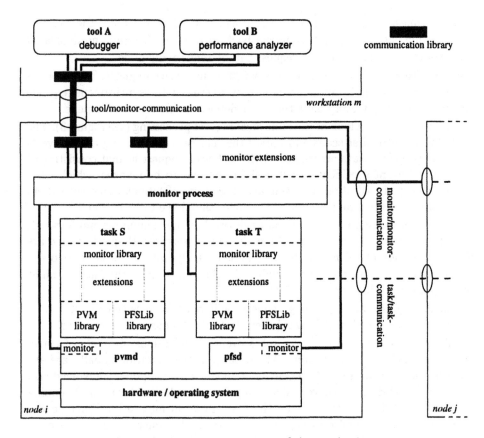

Fig. 2. The software module structure of the monitoring system

Finally, with the operating system there is no active instrumentation employed. It would require to have the source code of the operating system thus resulting in a non-portable approach for the monitoring system. Here all cooperation is handled via well defined system calls (e.g. the ptrace() interface) and devices (e.g. /dev/kmem).

For the PVM based implementation of OCM we decided to also use PVM communication for the cooperation of the individual node monitors. However, with another target architecture this module can easily be replaced by e.g. native Unix communication mechanisms.

The following section will discuss the processing of requests in the monitor process itself. We will not describe the module structure of the monitor but explain which stages a request passes until an answer gets back to the tool.

4   Request Processing in the Monitor Process

From the tool's point of view a request to the monitoring system is a function call which gets two input parameters: the service request and a callback function. The service request is represented as a string while an address denotes the callback function for passing result parameters back to the tool. There is no dedicated master monitor, thus a service request can be sent to any node.

We distinguish synchronous and asynchronous service requests. Synchronous requests are followed by an immediate reply whereas with asynchronous requests we will have a reply every time the event defined in the request happens.

Let us now have a look at the node local request processing (see Figure 3). The receiving monitor first scans and parses the service request string and converts it into appropriate data structures. OMIS defines a request format consisting of an optional event specification and a list of actions to be invoked, which must not be empty. In addition all event and action specifications carry with them node numbers which indicate on which nodes they have to be executed.

The request distribution compiler analyzes these node numbers and forwards local requests to the next stage, the node request receiver. Requests which are not only or not at all meant to be handled locally will be forwarded to the appropriate monitors on remote nodes. At the same time other nodes might send requests to this node.

Local requests and incoming requests will both be received by the node-request receiver which will forward them to the instrumentation manager.

The instrumentation manager splits event specifications into lists of corresponding basic instrumentations. These basic instrumentations finally are applied to the appropriate code segments of all software components involved (i.e. task codes and daemon codes). Measuring for example the duration of sending routines is handled by two events: the first starts an integrating timer on the event of entering a send routine; the second stops the timer if the process exits the routine. Activation of instrumentations is done by means of basic instrumentation services (in our example both instrumentation points are located in the communication library of PVM, i.e. in the task's address space).

The instrumentation manager not only activates instrumentation but also triggers the event/action management. The latter stores all defined events together with lists of actions to be invoked on their occurrence. Note that with tool manipulation getting more complex new actions will possibly be appended to already existing events. If actions are to be deleted the event/action manager checks whether or not the event has to be purged, too. In this case, all instrumentations belonging to this event will be deactivated.

Event/action processing might also be triggered by remote events (e.g. with requests like "stop task on node A if something happens on node B"). In this case the local monitor receives the event occurence information from a remote node and starts appropriate activities. The feature of involving several nodes in one service request is mainly used with debuggers and load balancers. It allows to specify breakpoint conditions which spawn several nodes.

In case of a synchronous service request sent by a tool the action list will be activated immediately by the instrumentation manager as there is no event we have to wait for.

With asynchronous service requests the actions are triggered as soon as the

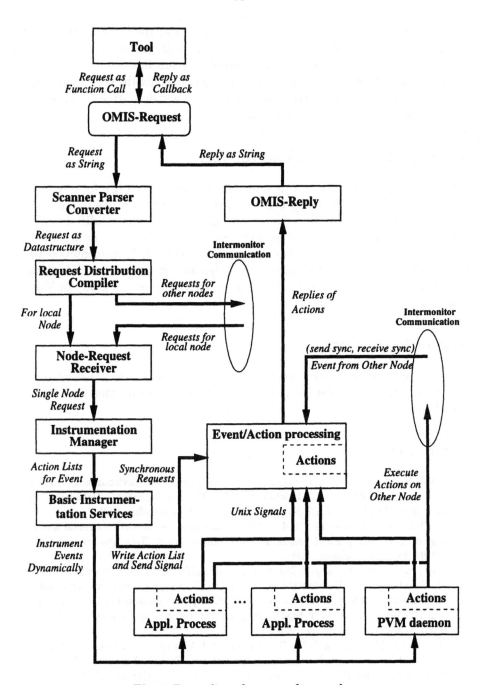

Fig. 3. Proceeding of requests from tools

event occurs. The event/action manager searches the appropriate action lists and invokes all specified actions. Their results are collected and passed back to the requesting tool by calling the specified callback function.

The design documents which are currently developed for the OCM implementation comprise a concise description of all modules and data structures involved in this cycle. Please refer to [3] for details.

5 Project Status and Future Work

Currently, the OMIS document is un-modified. We received helpful comments and criticism that will be integrated in a new version which is due to appear in late fall 96. With the OCM project – the realization of what OMIS describes – we started beginning of 96. The design phase will be finished in September and will result in publicly available design documents [3]. Implementation will start at that time. Currently, we already design and implement the node status reporter. This work is carried out at Emory University in the group of Vaidy Sunderam.

Acknowledgements

We would like to thank our team members Manfred Geischeder, Michael Uemminghaus, and Hans Zeller for their contribution within the framework of the OMIS and OCM project.

References

1. Bernd Bruegge. A portable platform for distributed event environments. In *Proceedings of the ACM/ONR Workshop on Parallel and Distributed Debugging*, volume 26 of *ACM SIGPLAN Notices*, pages 184–193, December 1991.
2. C. Clemencon, J. Fritscher, and R. Rühl. Visualization, execution control and replay of massively parallel programs within annai's tool. Technical Report CSCS TR-94-09, CSCS, Manno, 1994.
   http://www.cscs.ch/pub/CSCS/techreports/1995/CSCS-TR-94-09.ps.gz.
3. M. Geischeder, M. Uemminghaus, H. Zeller, T. Ludwig, Michael Oberhuber, R. Wismüller, and A. Bode. OCM: OMIS Compliant Monitoring System — Design Documents. Technical report, to appear at Technische Universität München, Munich, Germany, October 1996.
4. G.A. Geist, M.T. Heath, B.W. Peyton, and P.H. Worley. A User's Guide to PICL, a portable instrumented communication library. Technical Report ORNL/TM-11616, Oak Ridge National Laboratory, Oak Ridge, TN, October 1990.
5. S. Grabner, D. Kranzlmüller, and J. Volkert. Debugging parallel programs using atempt. In B. Hertzberger and G. Serazzi, editors, *High Performance Computing and Networking 1995*, number 919 in Lecture Notes in Computer Science, pages 235–240. Springer, May 1995.
6. M. T. Helth and J.E. Finger. Paragraph: A toll for visualizing performance of parallel programs. Technical report, Oak Ridge National Laboratory, 1993.

7. S. Lamberts, T. Ludwig, C. Röder, and A. Bode. PFSLib — a file system for parallel programming environments. Technical Report TUM-I9619, SFB-Bericht Nr. 342/10/96 A, Technische Universität München, Munich, Germany, May 1996. http:/~ludwig/papers/tl9602/WWW/report.html ftp:/~ludwig/papers/tl9602/PS/article.ps.gz.

8. T. Ludwig and R. Wismüller. THE TOOL-SET environment. In A. Bode, T. Ludwig, V. Sunderam, and R. Wismüller, editors, *Workshop on PVM, MPI, Tools, and Applications*, pages 28–32. Technische Universität München, November 1995. http:/~ludwig/papers/tl9505/WWW/article.html ftp:/~ludwig/papers/tl9505/PS/article.ps.gz.

9. T. Ludwig, R. Wismüller, V. Sunderam, and A. Bode. OMIS — On-line Monitoring Interface Specification. Technical Report TUM-I9609, SFB-Bericht Nr. 342/05/96 A, Technische Universität München, Munich, Germany, February 1996. http:/~ludwig/papers/tl9601/WWW/article.html ftp:/~ludwig/papers/tl9601/PS/article.ps.gz.

10. James E. Lumpp, Howard Jay Siegel, and Dan C. Marinescu. Specification and Identification of Events for Debugging and Performance Monitoring of Distributed Multiprocessor Systems. In *Proceedings of the 10th International Conference on Distributed Computing Systems*, pages 476–483, Paris, May 1990.

11. A. K. Petrenko. Methods for debugging and monitoring parallel programs: A survey. *Programming and Computer Software*, 20(3):113–129, may–june 1994.

12. Daniel A. Reed. Experimental analysis of parallel systems: Techniques and open problems. In Günter Haring and Gabriele Kotsis, editors, *Computer Performance Evaluation*, volume 794 of *LNCS*, pages 25–51. Springer, 1994.

13. Reinhard Schwarz and Friedemann Mattern. Detecting causal relationships on distributed computations: In search of the holy grail. Technical Report SFB 124-15, University of Kaiserslautern, December 1992.

State Based Visualization of PVM Applications*

Roland Wismüller

Lehrstuhl für Rechnertechnik und Rechnerorganisation (LRR-TUM)
Institut für Informatik, Technische Universität München, D-80290 München, Germany
Tel.: +49-89-289-28164, Fax: +49-89-289-28232
email: wismuell@informatik.tu-muenchen.de

Abstract. Understanding the dynamic behavior of parallel programs is a critical issue both for debugging and for optimization. A visualization tool displaying an animated sequence of the global states the program runs through offers valuable support for this process. The paper presents the features and the implementation of VISTOP, a state based visualizer for PVM applications. It supports program flow visualization based on various views and uses an event ordering algorithm to ensure consistent visualization without requiring a global clock.

1 Introduction

The complex interaction of a large number of components makes programming parallel and distributed systems a complicated and error-prone task. The dynamics supported by programming libraries like PVM can make this situation even worse, because the execution of programs where tasks (or other objects) are created and deleted during runtime can be hard to comprehend. Breakpoint-based debuggers, which are also available for parallel programs now (e.g. P2D2 [4]) offer programmers valuable help in understanding program behavior and in locating bugs. However, they basically allow only the inspection of the local states of single processes. But especially for finding bugs related to communication or synchronization, the global program state is of much more importance, e.g. which tasks are currently waiting for messages, which are sending, which groups do exist and what tasks do they contain? Although debuggers can somehow answer these questions, getting the answer is a rather complex job for the user. In addition, debuggers do not provide a coarse grain overview on a program's execution, e.g. its communication structure.

A solution to these problems is to complement debuggers with a visualization system that shows consistent global states of an application and animates state changes. Such a tool offers both a global overview and – by recording relevant information – also an animation of the program's coarse grain execution flow.

In the following, we present the concepts and some implementation issues of the visualization system VISTOP that animates communication and synchronization as well as dynamic creation, deletion, and group membership of tasks. The visualizer is an integral part of a larger environment called THE TOOL-SET for PVM [12], which among others will also include a parallel debugger.

---

* Partly funded by the German Science Foundation, Contract: SFB 342, TP A1

2 Visualizing the Execution of PVM Programs

The idea of providing tools visualizing the communication and synchronization behavior of parallel programs is not new. For PVM, a couple of such tools already do exist. Some examples are XPVM [5, 2], ParaGraph [7, 6], HeNCE [1], and Trapper [14]. However, these tools do not adequately address several problems, whose solution has been a primary design goal of VISTOP.

First, most of these systems only display an event stream, usually as a space-time diagram. From such a representation, it is, however, difficult to get information on a program's global state, e.g. the tasks blocked in receive requests, members of groups, contents of the tasks' receive queues, or information on barriers. VISTOP computes an internal model of a program's global state from the stream of events by including knowledge about the semantics of PVM. For instance, the members of a group are computed from "joingroup" and "leavegroup" events. Therefore, VISTOP can display global states instead of events. The model can also detect certain errors, e.g. if a message is unpacked incorrectly.

A second issue is consistency of visualization. Event visualization tools usually display events in the order in which they are received by the tool. However, due to local buffering and network delays, this strategy can lead to a visualization violating causality. VISTOP addresses this problem by sorting the received events in such a way that causal relationships are satisfied and consistency with the modeled state is guaranteed.

Finally, existing visualizers offer only limited scalability. Space-time diagrams use one line for each PVM task; some tools do not even allow these lines to be reordered. VISTOP allows to freely select the objects to be visualized and to rearrange them in the display area. It also supports a hierarchical visualization at different levels of abstraction by allowing tasks to be grouped together into a hierarchy of container objects according to the application's logical structure.

VISTOP visualizes a program's execution as a sequence of snapshots where each snapshot shows different views of the program's global state. A control panel allows forward or backward stepping through the sequence of snapshots, and a continuous play mode with selectable speed. The transitions between different snapshots are animated in order to accentuate the state change. Currently, the visualizer provides three different views of program state:

1. The **concurrency view** displays a dynamic, hierarchical task graph showing communication and synchronization between tasks. The nodes of this graph represent objects of the programming model (tasks, barriers) or container objects (subgraphs of the program). Arcs denote interaction between objects. The left side of Fig. 1 shows a screen dump of this view that is described in more detail below.

2. The **object creation view** displays the dynamic creation graph of tasks spawned and terminated during runtime. It consists of trees, whose root nodes are the initial tasks spawned by the PVM console. Other nodes represent tasks spawned during run-time; the parent entry defines the spawning task. An example of this view is shown in the upper right part of Fig. 1.

3. The **system view** shows the distribution of tasks over the different hosts. The display is organized as a matrix where each column contains all objects located on one host.

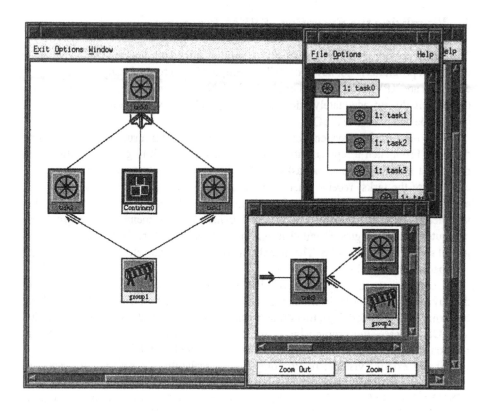

Fig. 1. Concurrency view and object creation view

We will now present the visualizer's concurrency view in more detail. The visualization patterns used in this view are depicted in Fig. 2. When visualizing a snapshot of a program's execution, each existing PVM task and each active barrier is represented by an icon, which may be slightly modified in color or shape to display some of the object's local state information (e.g. timeout or error conditions). Interactions are visualized as lines drawn between the objects involved. Additional arrow heads in different shapes and colors on either side of the line specify the kind of interaction: A green (filled) arrow head denotes an interaction that does not block the active partner, while a red (hollow) arrow head denotes a blocking one. The direction of the arrow head indicates the direction of the (requested) information transfer, while its position marks the active partner. In the middle left of Fig. 2, **task1** is sending to **task0**, since there is a filled arrow head attached to **task1** and directed towards **task0**. Broadcasts and multicasts are visualized in an intuitive way by showing that a task sends to multiple tasks at the same time (see **task0** in Fig. 1). The lower left of Fig. 2 shows the visualization pattern when **task4** waits for a message from **task3**. The same pattern is also used for barrier synchronization. In this case, the icon of **task3** is replaced by a barrier symbol, which is created when the first task in a group calls **pvm_barrier()** and is destroyed again when the last task leaves this call.

In addition to task and barrier objects, users can create container objects to realize a

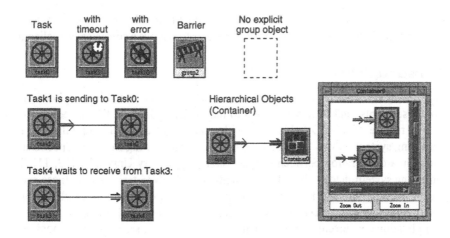

Fig. 2. Visualization patterns in the concurrency view

hierarchical visualization. A container consists of an icon and an associated concurrency window showing what is happening inside it. Objects can be moved into the container by simply dragging icons into the container window. The container's icon then represents the sum of all objects located in the container itself or a nested one. Interactions between a container and other objects are combined by overlaying the corresponding arrow heads, providing a comprehensive summary. Details about the individual interactions can be derived by inspecting the container's concurrency window. In the lower right of Fig. 2, **task0** broadcasts to **task1** and **task2**, both in the container, while **task1** receives a message from **task0**. **task2** didn't start receiving yet and is still waiting for a message from **task0**. By closing and opening concurrency windows associated with containers, the complexity of the visualization can be controlled. If a window is closed, events totally inside the container will be invisible and will therefore also be ignored when determining the next snapshot to be visualized.

The concurrency view does not include explicit objects representing groups. In principle, group objects could have been realized as containers, which combine all tasks in a group. However, containers in VISTOP form a strict hierarchy of tasks, while in PVM tasks can belong to multiple groups. Thus we decided to provide an additional view, called group manager. It displays a list of all groups existing at the current snapshot, where a single group, a union, or an intersection of groups can be selected from. All objects that do not belong to the specified group(s) will then loose parts of their coloring, thus allowing to quickly determine group memberships.

3 Implementation Issues

3.1 Structure of VISTOP

VISTOP is implemented as a hierarchy of three layers: A **data acquisition layer** responsible for gathering events, a **modeling layer** computing consistent global states from this event stream, and a **visualization layer**, which finally displays different views of the computed global state. In the following, we will provide some more information on the first two layers and the issue of sorting events to achieve a consistent visualization.

3.2 Data acquisition

The task of this layer is to monitor relevant events in an executing application, to store these events, and to forward them to the modeling layer in the correct order. In the final release of VISTOP, this layer will make use of an OMIS compliant on-line monitoring system. OMIS is an emerging On-line Monitoring Interface Specification defined in a cooperation between our laboratory and Emory University [13, 11]. The monitoring system detects the relevant events in each PVM task, buffers these events in local trace buffers and forwards the buffers to the tool either periodically or upon request. It also provides mechanisms for other tools, e.g. a parallel debugger, to attach to the application concurrently to the visualizer. In such a case, services of the monitoring system can inform the visualizer on relevant modifications done by other tools. Thus, the visualizer can instruct the monitoring system to flush its trace buffers when the application is stopped by a debugger, and can correctly display its state at the breakpoint. Furthermore, the on-line monitoring system offers services to obtain state information of an application. This allows the visualizer to be started at some breakpoint, which requires to initialize its state model with the application's current state.

However, as the monitoring system is not fully implemented yet, the data acquisition layer currently works with the trace facility integrated into the PVM library since PVM 3.3. It allows to obtain a trace of PVM library calls usually containing one event entry for the call and its input parameters and one for the return and the result parameters. For portability reasons, we do not receive the traced events directly from the PVM system, but read the files written by XPVM, which use the standardized SDDF trace format. However, the traces do not fully support the needs of state oriented visualization tools.

There are currently two major limitations we had to overcome. First, the event indicating the execution of a broadcast only specifies the group where the broadcast is sent to. But for a correct visualization it is important to know to which tasks the message has actually been sent. We therefore have slightly modified the PVM library, so that this information is included in the event trace.

Another limitation is that the PVM tracing mechanism does not record an event when a message is inserted into a task's receive queue. For the modeling of global states, this has three consequences:

1. The VISTOP animation is designed to show whether a task blocks in a receive call due to an empty message queue. Without the event, this is not possible, since we do not know the point in time, when a message is put into the queue. As a current solution, we assume that messages are inserted as soon as the sender finishes the send operation.
2. The receive queue of a task cannot be computed accurately. All the model can provide is a *set* of receivable messages. It cannot determine the order in which messages are stored in the queue[2].

---

[2] Modeling the queue of receivable messages of a task is also complicated by PVM's direct message routing scheme. However, discussing this topic is beyond the scope of this paper.

3. It is impossible to visualize receive operations in the desired way, if no sender is specified in **pvm_recv()**, i.e. the wild-card is used. Since the receive queue cannot be modeled properly, it is then unknown, which task is the sender of the message being received. We currently overcome this problem by delaying the visualization of receive operations until we have seen the event generated by the end of the receive call. This event always contains the sender task.

In addition to event gathering, the data acquisition layer must also ensure that the event stream is forwarded to the modeling layer in an order consistent with causality and the modeled state. Since this topic touches both layers, we delay its discussion until we have presented the modeling layer in more detail.

3.3 Modeling Layer

The modeling layer computes information on the global state of the monitored application from the sequence of events received from the data acquisition layer. It is implemented as an extended state machine, which computes a new state s' from the current state s when given an event e and a visualization direction d: $s' = f(s, e, d)$. The state transition function f obeys the rules $f(f(s, e, \text{forward}), e, \text{backward}) = s$ and $f(f(s, e, \text{backward}), e, \text{forward}) = s$, which ensure that forward and backward animation will be consistent to each other. Being able to incrementally compute states in both directions frees the modeling layer from storing the whole history of snapshots and greatly reduces memory consumption.

The application's global state is modeled by a collection of C++ objects, representing hosts, tasks, groups, and barriers. Host objects only contain the host's name; task objects store a task's id, its name, a reference to the host, task state (e.g. sending, computing, ...), its parent's id, a list of groups where the task is member of, and lists containing information on currently receivable messages and the messages that have been sent and received since the program has been started. Message information includes the sender task, the tag, length, and the sequence of data types packed into the message. The messages' contents are, however, not stored. Group objects provide information on the group's name and the list of member tasks. Barrier objects finally contain a reference to the proper group, the number of tasks that must synchronize, and the list of tasks currently waiting at the barrier. All information contained in these objects is available to the user in a detailed view which is popped up by clicking at the object's graphical representation in a visualization window. Interactions between objects are internally represented by a fifth kind of objects containing the interaction type (e.g. send, broadcast, barrier wait) and references to the interacting partners.

In addition to computing state information and passing it to the visualization layer, the modeling layer also performs extensive error checking. For instance, it can check the correct usage of barriers, examines whether messages are unpacked correctly, and can provide detailed information on why a PVM library call has failed.

3.4 Consistent Ordering of Events

The modeling layer computes state information from the sequence of events passed from the data acquisition layer. For a correct visualization it is therefore important to pass the events in a correct order. However, the data acquisition layer may receive events generated on different hosts in an incorrect order due to event buffering done by the hosts or due to network delays. It is impossible to order these events correctly by using their time stamps, since their resolution is usually not good enough, and clocks on different hosts cannot be synchronized with the required accuracy. However, the order of events must obey the causality (or "happened before") relation [10], i.e. an event e_1 must be prior to an event e_2 if there is a causal dependency from e_1 to e_2 (written as $e_1 \rightarrow e_2$). As an example, consider e_1 to be the sending of a message and e_2 to be the receipt of this message on a different host. Even if the data acquisition layer receives e_2 before e_1 due to buffer or network delays, it must forward e_1 to the modeling layer first, because a message cannot be received before it has been sent. In the current implementation, we store incoming events in different buffers for each host used by the application, assuming that the monitoring system ensures correct event ordering within a single host. Whenever the next event is requested by the modeling layer, we have to determine the buffer from which to read that event, thus incrementally merging the local buffers into a consistent global sequence.

The problem of merging local event traces to a global one according to causal relationships is well known; there are also several solutions yet (e.g. [8, 9]). However, with a state based visualization approach it turns out that just considering causal dependencies between events is not sufficient for achieving a consistent visualization. As an example, consider the situation depicted in the space-time diagram in Fig. 3. Note that there are no causal dependencies (in the sense of "happened before") between events in different tasks. However, it is obviously incorrect to order these events as e_1, e_2, e_3, e_4, because the broadcast actually sent the message to task t_1, but with this sequence t_1 would not be in the group when the broadcast is executed. In contrast, the orders e_2, e_1, e_3, e_4 and e_1, e_3, e_2, e_4 are both correct.

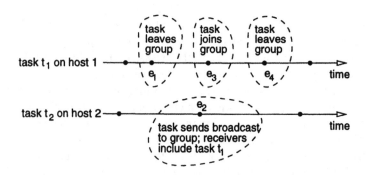

Fig. 3. Example illustrating consistency constraints between events and modeled state

This implies that in addition to dependencies between events, we have constraints imposed by the current state in the model when selecting the next event to process. Therefore, we use an extended sorting algorithm that works as follows: The input are n event buffers, one for each host. We read and remove the next event e_i to be processed from buffer i, only if the following conditions are met:

1. there is no event e_j in an other buffer j with $e_j \rightarrow e_i$ (causality)
2. e_i can happen in the currently modeled state of the application (consistency)

By construction, this strategy will never result in an inconsistent visualization, however, it can lead to deadlock or starvation, i.e. cases where no further events can be selected from one or more event buffers. In our example, according to this strategy we could select three events in the sequence e_1, e_3, e_4. If t_1 never joins the group again, we can not select event e_2 anymore and therefore also no succeeding event in the buffer of t_2. We therefore use a heuristics guiding the selection based on the following considerations:

1. If there is an event e_i that cannot be processed due to condition 2, the modeled state differs from the state where e_i actually occurred in the program execution.
2. In such a situation, we favor events that make this difference smaller.
3. Events that increase the difference are processed only if there are no other selectable events. When considering such an event (say e_k), we perform a look-ahead search in the event-buffers to make sure that the "negative effect" of e_k on the modeled state will be undone by a later event in the buffer. We process e_k only if this search is successful.
4. If there are two events e_i and e_j that both could be processed, but e_j would change the state in a way that inhibits subsequent processing of e_i due to condition 2, e_i is processed first.

The heuristics ensures that the sorting algorithm will never select events that would result in starvation or deadlock. In our example, it leads to the consistent event sequence e_2, e_1, e_3, e_4. In cases where the heuristics leaves multiple candidates to choose from, which must be concurrent events for which no uniquely defined ordering exists, we use the events' time stamps for selecting the next one being processed. If the events' timely separation is greater than the clock skew between different hosts, this ensures that the events are displayed in the order in which they have been monitored.

4 Conclusion and Future Work

VISTOP is a state based visualization system that offers a consistent global view on the execution of PVM applications. Experience with a previous version of this tool [3] for an other message passing library has proved that state based visualization offers valuable help in understanding the dynamic behavior of parallel programs, debugging message passing and synchronization errors, and even for locating performance bottlenecks.

VISTOP is a part of THE TOOL-SET for PVM [12], which also will include other tools, such as a parallel file system, checkpointing and load balancing facilities, a parallel debugger, a performance analyzer and a deterministic execution controller. Future work on the visualizer includes extensions for the support of THE TOOL-SET's automatic components, i.e. an animation of I/O operations using the parallel file system, and task migrations performed by the load balancer. In addition, we will support a tightly integrated usage of VISTOP with the debugger, the deterministic execution controller and the checkpointer. Using these tools together will allow to start the debugger on any state snapshot selected in the visualizer. This will be achieved by automatically restarting the application from the last available checkpoint prior to the selected snapshot and by using the deterministic execution controller to ensure that the selected state is reached even in the presence of nondeterminism due to message races. In this way debugging large programs with high execution times will be efficiently supported.

References

1. A. Beguelin, J. Dongarra, A. Geist, R. Manchek, and K. Moore. HeNCE: A Heterogeneous Network Computing Environment. Technical report UT-CS-93-205, Univ. of Tennessee, Computer Science Department, Aug. 1993.
2. A. Beguelin, J. Dongarra, A. Geist, R. Manchek, and V. Sunderam. Recent Enhancements to PVM. *Intl. J. of Supercomputing Applications and High Performance Computing*, 1995.
3. A. Bode and P. Braun. Monitoring and Visualization in TOPSYS. In *Proc. Workshop on Monitoring and Visualization of Parallel Processing Systems*, pages 97 – 118, Moravany nad Váhom, CSFR, Oct. 1992.
4. D. Cheng and R. Hood. A Portable Debugger for Parallel and Distributed Programs. In *Proc. Supercomputing'94*, pages 723–732, Washington D.C., Nov. 1994. IEEE.
5. A. Geist, J. Kohl, and P. Papadopoulos. Visualization, Debugging, and Performance in PVM. In *Proc. Visualization and Debugging Workshop*, Oct. 1994.
6. M. T. Heath. Recent Developments and Case Studies in Performance Visualization using ParaGraph. In *Proc. Workshop on Monitoring and Visualization of Parallel Processing Systems*, pages 175 – 200, Moravany nad Váhom, CSFR, Oct. 1992.
7. M. T. Heath and J. A. Etheridge. Visualizing the Performance of Parallel Programs. *IEEE Software*, 8(5), Sept. 1991.
8. D. P. Helmbold, C. E. McDowell, and J. Z. Wang. Analyzing Traces with Anonymous Synchronization. In *Proc. Intl. Conf. on Parallel Processing*, pages 70–77, Aug. 1990.
9. D. Kimelman and D. Zernik. On-the-Fly Topological Sort – A Basis for Interactive Debugging and Live Visualization of Parallel Programs. In *Proc. ACM/ONR Workshop on Parallel and Distributed Debugging*, pages 12–20, San Diego, CA, May 1993.
10. L. Lamport. Time, Clocks, and the Ordering of Events in a Distributed System. *Commun. ACM*, 21(7):558–565, July 1978.
11. T. Ludwig, M. Oberhuber, and R. Wismüller. An Open Monitoring System for Parallel and Distributed Programs. To appear in Proceedings of EuroPar'96, Lyon, France, Aug. 1996.
12. T. Ludwig, R. Wismüller, R. Borgeest, S. Lamberts, C. Röder, G. Stellner, and A. Bode. THE TOOL-SET – An Integrated Tool Environment for PVM. In *Proc. EuroPVM'95 Short Papers*, Lyon, France, Sept. 1995. ENS Lyon. Technical Report 95-02.
    http://wwwbode.informatik.tu-muenchen.de/~wismuell/publicatio ns/europvm95.ps.gz.
13. T. Ludwig, R. Wismüller, V. Sunderam, and A. Bode. OMIS – On-line Monitoring Interface Specification. Technical Report TUM-I9609, SFB-Bericht Nr. 342/05/96 A, Technische Universität München, Feb. 1996.
    http://wwwbode.informatik.tu-muenchen.de/~omis/HTML/OMIS/Vers ion-1.0/version-1.0.ps.gz.
14. C. Scheidler and L. Schäfers. TRAPPER: A Graphical Programming Environment for Industrial High-Performance Applications. In *Proc. PARLE'93, Parallel Architectures and Languages Europe*, volume 694 of *LNCS*, pages 403–413, München, June 1993.

Workstation Cluster as a Parallel Hardware Environment for High Performance Computing

Martin Kahlert[1], Utz Wever[2] and Qinghua Zheng[2]

[1] Technical University of Munich, D-80333 Munich, Germany
[2] Siemens AG, Corporate Research and Development, D-81730 Munich, Germany

Abstract. In this paper we discuss concepts for remote procedure calls specially suited for heterogeneous workstation clusters. Many computer programs for industrial applications, for example in nuclear security, VLSI design and medical diagnosis can be mapped into this programming model. Based on concepts such as dynamical load balancing, fault tolerance and a classification of input- and output parameters the code generator TAPAS[3] is developed. The generation of parallel code offers the possibility to run parallel programs efficiently on workstation clusters. The performance of the generated codes are demonstrated by three large scale applications.

1 Introduction

Because of the fast development of several RISC architectures, workstation clusters are becoming a high performance and low cost computing environment. However, this environment is characterized by having some significant disadvantages: The low communication rate, the heterogeneous environment, the different load situation and the high possibility of failure. In order to overcome these disadvantages, we present concepts such as dynamical load balancing, fault tolerance and a classification of input- and output parameters in order to minimize the amount of communication. The portability is ensured by using the communication software PVM [4]. Based on these concepts the code generator TAPAS is developed. The generation of parallel code offers the possibility to run parallel programs efficiently on workstation clusters.

The paper is organized as following: First of all, three large scale application fields are introduced, which could be parallelized by using a parallel remote procedure call (section 2):

- Three-dimensional dynamic analyses of pile foundations (pile-soil-pile interaction) for nuclear security (section 2.1);
- Harmonic Balance for nonlinear frequency analyses by VLSI design (section 2.2);
- and the maximum intensity projection algorithm for imaging processing, which often used by medical diagnosis (section 2.3).

---

[3] Tool for Automatic Parallization of Arbitrary Subroutines

In section 3 concepts such as parameter characterization, dynamical load balancing and fault tolerance are presented. The target of these concepts is to overcome the preceding disadvantages, or, in other words to fulfill the requirements of workstation cluster. Based on these concepts the code generator TAPAS is developed (see section 3.4). It generates automatically portable high performance parallel code for workstation cluster. The portability of the generated code is guaranteed by the parallel software interface PVM. Finally, speedup results obtained for the three large scale applications are presented.

2 Three industrial applications and their parallel structure

In this section we present three industrial applications, which could be parallelized by using the parallel *master-slave* programming model. Each of the time consuming parts of the algorithms could be described as a repetitive call of a procedure.

2.1 Earthquake Analysis of Pile Foundations

The software package PINTER was developed for conducting three-dimensional dynamic analyses of pile foundations, including full pile-soil-pile interaction (see [9, 10, 11]). The analysis is carried out in the frequency domain.

The equation of dynamic equilibrium for a viscoelastic body is of the form $Kr + C\dot{r} + M\ddot{r} = p$ where K, C, M are stiffness, damping and mass matrix respectively, r, $\dot{r}$, $\ddot{r}$ are displacement, velocity and acceleration vectors and p is a applied force vector.

Applying the Fourier transformation a set of linear equations is achieved: $(K + iC\omega - M\omega^2)R(\omega) = P(\omega)$. The solution of the linear simultaneous equations may simply be written as $R(\omega) = S^{-1}(\omega)P(\omega)$. The inverse Fourier transform gives then $r(t)$.

Application 1 *Program PINTER*

1. *After discretization:* $Kr(t) + C\dot{r}(t) + M\ddot{r}(t) = p(t)$
2. *Fourier transform:* $p(t) \Rightarrow P(\omega)$
3. *Calculate:* $S(\omega_k) = K + iC\omega_k - M\omega_k^2$, $\quad k = 0, 1, 2, ...$
4. *Solve:* $S(\omega_k)R(\omega_k) = P(\omega_k)$, $\quad k = 0, 1, 2, ...$
5. *Inverse discrete Fourier transform:* $R(\omega_k), k = 0, 1, 2, ... \Rightarrow r(t)$

For large examples more than 98% of the total CPU time is spent for solving large linear systems (step 4 in Application 1). The remaining 2% include all other operations such as model evaluation and inverse Fast Fourier Transform.

The solution of the linear systems depends only on the frequency ω. It can be performed simultaneously by several processes in a natural way.

2.2 Harmonic Balance for nonlinear frequency analyses

The circuit simulation program TITAN is developed by Corporate Research and Development at Siemens AG for VLSI design (see [3]). In TITAN there are two main types of analyses: Time domain analysis for the verification of large memory circuits (e.g. DRAMs) and frequency domain analysis for analyzing analog circuits. Here we discuss the parallelization of frequency analysis.

The network equations generated by Kirchhoff's laws may be written as a system of nonlinear differential algebraic equations (DAE) $f(x(t), \dot{x}(t), t) = 0$, where $f : \mathbb{R}^{2n+1} \to \mathbb{R}^n$ describes the circuit and $x(t) \in \mathbb{R}^n$ the state vector. For periodic solutions we make the foundation that $x(t)$ can be expressed in terms of a Fourier series. Because the Fourier series is truncated we cannot fulfill the DAE exactly. Instead, the approximation of Galerkin is used. This method is called *Harmonic Balance*. The solution of these nonlinear equations results in an iterative solving of a very large sparse linear block system. These linear block equations can be solved by the following Block-Gauss-scheme:

Application 2 *Block Gauss Algorithm*

$$For \quad i = 1, ..., n \quad do:$$
$$\left\{ \begin{array}{l} (A) \ QR\text{-}Decompose \ M_{i,i} = Q_i R_i \ where \ Q_i^T Q_i = I \\ For \quad j \in \{i+1, ..., n\} \quad do: \\ \quad \{ \ (B) \ Determine \ B_i : Y R_i = M_{j,i}, \quad B_j = Y Q_i^T \ \} \\ For \quad j, k \in \{i+1, ..., n\} \quad do: \\ \quad \{ \ (C) \ M_{j,k} = M_{j,k} - B_j M_{i,k} \} \end{array} \right\}$$

For larger applications, more than 99 % of the whole CPU time is spent in the Block-Gauss-scheme. This is due to the fact that the number of operations increases with $O(k^3)$ where k is the number of frequencies. The QR-decomposition of the diagonal block (operation (A) in Algorithm 2) has to run sequentially while the two loops (B and C) are able to run in parallel (for each Gaussian step). Since the QR-decomposition needs less than 5% of the total CPU time, most of the algorithm runs in parallel.

2.3 Imaging Processing for Medical Diagnosis

An other application for parallelization is the digital image processing. For given intensity imagings $(Imag_k(x, y), k = 1, ..., n)$ of a 3D object, the maximun intensity of projection for a given direction is calculated by the MIPS algorithm.

Each input image $Imag_k$, $k = 1, ..., n$ is a 2D raster imaging. It describes the intensity of a 3D object (a head for example) on the 2D plane $z = const_k, k \in \{1, 2, ..., n\}$. In order to calculate the maximum intensity of projection $V(x^*, y^*)$, $x^*, y^* = 1, ..., m$ (also a 2D raster imaging), one has to draw a line on each point (x^*, y^*) parallel to the projection and calculate the intersections between this line and the planes of the $Imag_k$. Using an interpolation algorithm the intensity of the intersection points are calculated. Their maximum is the value of $V(x^*, y^*)$. Because

$$\max_{x \in U_1 \cup U_2} \{f(x)\} = \max \left\{ \max_{x \in U_1}\{f(x)\}, \max_{x \in U_2}\{f(x)\} \right\}$$

for every index sets U_1, U_2 of function $f : U_1 \cup U_2 \to \mathbb{R}$, this algorithm can be parallelized by using domain decomposition:

Application 3 *MIPS Algorithm*

1. *Decomposition the 3D domain:* $D = D_1 \cup D_2 \cup \cdots \cup D_p$
2. *Evaluation the domain MIPS:* $V_i = MIPS(D_i)$, $i = 1, ..., p$
3. *Calculated the domain offset:* $(x_i^0, y_i^0) = Offset(D_i)$, $i = 1, ..., p$
4. *Assemble the images:* $V(x^*, y^*) = \max_{i=1,...,p}\{V_i(x^* + x_i^0, y^* + y_i^0)\}$

Here the input images $D = \{Imag_k\}_{k=1,...,n}$ are divided into p parts D_i, $i = 1, ..., p$. This is also simple to parallize, but each job has to share one output V.

3 Concepts for workstation clusters

The three applications discussed before have a similar structure: The CPU-time consuming part of the algorithms can be handled by a procedure called in a loop. Using communication software packages such as PVM (see [4]), this structure can be parallelized by a *remote procedure call*. The slaves are able to run in parallel on different CPU's. All arguments of the procedure have to be communicated among the master and slave processes (for more detail see sections 3.2 and 3.3).

In order to get high performance, communication overhead has to be minimized. In this section we discuss concepts which are able to overcome the disadvantages when using a cluster of workstations as a parallel computing environment.

On workstation cluster, the generated code must be portable. Therefore we decided to make use of the PVM parallel software platform for communicating among the workstations. PVM has established to be a quasi standard for the communication among UNIX-workstations and is supported by the most parallel computers vendors (e.g. Cray, Convex, SGI, etc), see [4].

3.1 Classification of Parameters

In order to minimize the amount of communication and the overhead for fault tolerance, see also sections 3.2 and 3.3, the parameters of those procedures which will be parallelized, have to be classified very precisely. A rough classification subdivides the parameters in input and output parameters. FORTRAN which is one of the most important languages for high performance computing, requires another type of parameter (working-space) because there is no possibility of allocating memory dynamically.

The input parameters are further divided into parameters with constant content during the loop and parameters with variable contents (can not be overwritten during the loop and can be overwritten). For the constant parameters just *one* communication is needed for each slave. For fault tolerance techniques only a copy of content is needed for parameters with variable content, in other cases just a remark of address is necessary (the details are described in section 3.3). Also the output parameters are subdivided into the three classes: Parameters with separate domain and parameters with shared domain (standard operations (min, max, sum, sub, ...) and user defined operations). The parameters types are summarized in the following Table.

Type	Subtype	Remark	abb.
	constant contents	only one communication	Ia
Input-parameter	variable contents not overwritten	backup not necessary	Ib
	variable contents can be overwritten		Ic
Output-parameter	separate domain		Oa
	shared domain standard post processing		Ob
	shared domain user def. post processing		Oc
working-field	fixed size		Wa
	variable size	allocated in run time	Wb

3.2 Dynamical Load Balancing

To order to exploit the different performance of workstation architectures the tasks have to be divided such that the size of the task corresponds with the resource provided by the machine. In many applications, there are much more tasks than machines. In this case the actual load situation of a workstation during the run of parallel software can also be taken into account. Many statistical models for load balancing have already been introduced. In the general case of different sized tasks, these models offer no satisfying answer. Therefore we decided to work with the well-known procedure given in the following chart (n: number of jobs, p: number of processors, $n > p$).

---

dynamical load balancing
do i=1,p: { *send* Job(i) *to* Processor(i) }
do i=p+1,n:
    {   *receive* Results, j=Job_no, k=proc_no
        send Job(i) *to* Processor(k) }
do i=p+1,n: { *receive* Results, j=Job_no, k=proc_no }

This principle can be characterized by the statement: *The master keeps all slaves busy.* With this approach, not only the actual load of the workstation is exploited but also the general performance of the machine and the load of the network. The approach is very similar to the "job's pool" used by LINDA [5], but here the jobs are actively distributed by the master.

3.3   Fault tolerance

A user of (parallel) software wants to be sure that his hardware environment guarantees a reliable run of his program. On workstation clusters this requirement is no longer valid because the individual workstation is beyond the control of the programmer. Many concepts have already been introduced to keep a parallel program fault tolerant toward workstation failure. E.g. the ISIS software package [13] for communication among workstations sends copies of the task to other hosts. The main target of our approach was to keep the overhead for fault tolerance technique as low as possible. The master keeps records of all tasks which are sent. If one task is missing, another host is spawned and the task is sent again. Much work has been invested to achieve efficient book-keeping by the master. If possible, only the addresses of the arguments of one task are stored (e.g. using the parameter type Ib, see section 3.1). Of course this concept implies that the master process is not allowed to fail. It must be spawned at one's own workstation or at a server. The concept is summarized in the following scheme:

---

fault tolerance
>   for a given time raster
>>   *for each* slave s *do*
>>>   *if* (status(s)$\neq$o.k.|run_time(s)>T_max)
>>>   kill process s;
>>>   reconstruct all input parameter of
>>>   Job, which was running on s;
>>>   spawn new process with name s;
>>>   repeat the Job on s;
>>   *end if*

In this concept a very slow task is handled like a failing task. In other words: *A failing task is a "infinite" slow task.*

3.4   The code generator TAPAS

Many concepts have already been introduced for generating remote procedure call applications: E.g. the NIDLE Compiler from Hewlett Packard [12] or extensions of operating systems and computer languages like Mach [1] and High Performance Fortran [8]. The Heterogeneous Network Computing Environment (HeNCE) [2] from the PVM team and ENTERPRISE [6] produce portable code for UNIX workstation. However, the concepts of section 3 let our approach be producing much better speedup results on a workstation cluster for the presented examples (more than 2 times better). Therefore we decided to develop the code generator TAPAS, which based on these concepts. Input dates for the generator TAPAS are the sequential source code and an input language specifying the classification of the arguments of those procedures to be run in parallel. Outputs are a master and a slave program communicating by PVM. The number of slaves can be determined interactively. Besides the advantages discussed in the cuurent section, there are also some additional advantages of code generation in point of view of software engineering aspects:

- Parallel software based on remote procedure call generation can be standardized.
- Only the sequential software must be serviced. The parallel executable can be generated automatically.
- The user does not need any know-how about PVM. The communication interface is generated by TAPAS.

4   Results and conclusion

In this section we present speedup results obtained by the parallel code of the remote procedure call generator TAPAS for the three applications considered in section 2.

Two results are presented for the Pinter-Algorithm. The large one (Exp.2) is the analysis of a pile group consisting of 43 piles. Here a set of 34 driving frequencies up to 10 Hz is used.

The over-linear increase of speedup for Harmonic Balance is a well known phenomena for parallel implementations. Linear speedup just starts when the size of the subproblem equals the size of the main memory (For the serial run we have used the largest machine in the cluster with 128MB. However, the executable needs more than 250MB for this problem) (see also [7]).

By using the output parameter type Ob (output parameters with shared domain and standard operation max, see section 3), the MIPS-Algorithm can be parallelized automatically by TAPAS. In this way the images V are assembled directly. The following Table shows the results of parallelization achieved by an example with 256 images sized by 128 × 128. 9 directions of projection are evaluated. Again a cluster of SUN4 Workstations is used. The achieved speedup results of the automatically generated code are presented:

	Number of Processors	1	2	4	5	6	10	15	20	30
Pinter	Elapsed time (min)	86.33	48		19.2		9.7		6.5	
Exp.1	Speedup	1	1.8		4.5		8.9		13.3	
Pinter	Elapsed time (h)	40.5					5.87			
Exp.2	Speedup	1					6.9			
Harm. Bal.	Elapsed time (h)	51.2	13.1	6.4		3.5	3.0	2.8		1.85
	Speedup	1	3.9	8.0		14.6	17.1	18.6		27.7
MIPS alg.	Elapsed time (sec)	871	482	265		180				
	Speedup	1	1.8	3.3		4.8				

Elapsed time and speedup by parallelism for the automatically generated code discussed in section 2. Pinter: Earthquake Analysis, Harm. Bal.: Harmonic Balance and MIPS: Imaging Processing. The over-linearly speedup for Harmonic Balance occurs due to paging effects.

In conclusion, we have presented a tool for high performance remote procedure call generation specially for heterogeneous workstation clusters. The main ideas to achieve high performance for this hardware environment can be summarized as following: Minimization of communication by using parameter classification, dynamical load balancing, restart for loosing processes and portability by using PVM. The performance of the parallel codes generated by TAPAS was demonstrated by three large scale applications. For all cases satisfactory speedup results are achieved. Thanks to fault tolerance techniques, the parallel program runs as reliably as a serial program.

References

1. Accetta et. al.: *MACH: A New Kernel Foundation for UNIX Development.* Usenix Conference Atlanta, (1986)
2. Beguelin, A., Dongarra, J., Geist, A., Manchek, R. and Sunderam, V.: *HeNCE: A Users' Guide Version 2.0.* (1994)
3. Feldmann, U., Wever, U., Zheng, Q., Schultz, R. and Wriedt, H.: *Algorithms for Modern Circuit Simulation.* Archiv für Elektronik und Übertragungstechnik. Vol. 46, No 4, pp 274-285 (1992)
4. Geist, A., Beguelin, A., Dongarra, J., Jiang, W., Manchek, R. and Sunderam, V.: *PVM 3.0 User's Guide and Reference Manuel.* Oak Ridge National Laboratory, Tennessee (1993)
5. Gelernter, D.: *Information managemeni in LINDA* Parallel processing and artificial intelligence (eds. Reeve, M. and Zenith S.E.) J.Wiley (1989)
6. Schaeffer, J., Szafron, D., Lobe, G. and Parsons, I.: *The Enterprise Model for Developing Distributed Applications* IEEE Parallel & Distributed Technology Systems & Applications, Vol. 1, No. 3, pp. 85–96, (1993)
7. Schneider,M., Wever,U. and Zheng, Q.: *Parallel Harmonic Balance.* VLSI'93, IFIP Transactions A-42, pp. 251-260, North-Holland, Amsterdam (1994)
8. Steele, G.: *High Performance Fortran.* Workshop on Distributed Computing, Sigplan Notices, (1993)
9. Trbojevic, V.M.,: *PINTER - Computer program for Dynamic Analysis of Pile Foundations, Theoretical Manual, Version B.1,* Principia Mechanica Ltd., Report No. 161/82, London, October 1982.
10. Trbojevic, V.M., Marti, J., Danisch, R. and Delinic, K.: *Pile-Soil-Pile Interaction Analysis for Pile Groups.* 6th Int. Conf on Structural Mechanics in Reactor Technology, Paris, Paper K5/4.
11. Wolf, J.P.: *Dynamic Stiffness and Seismic Input Motion of a Group of Battered Piles,* (ASCE National Convention), Conference on Structural Design of Nuclear Power Plants, Boston, April 1979.
12. *Network Computing System (NCS) Reference.* Apollo Computer Inc., (1987)
13. *The Isis Distributed Toolkit, Version 3.0, User Reference Manual* Isis Distributed Systems, Inc. (1992)

Using PVM to Implement PPARDB/PVM, a Portable Parallel Database Management System

N. Papakostas, G. Papakonstantinou, P. Tsanakas

National Technical University of Athens
Digital Systems and Computers Laboratory
Department of Electrical and Computer Engineering
Zographou Campus, Zographou, GR-15780, Greece
Tel.: +30 1 772 1529, Fax: +30 1 772 1533
E-Mail: nass@dsclab.ece.ntua.gr

Abstract
Although parallel databases have been an active research topic for several years, portable parallel databases represent a relatively new concept. They are required in order to make parallel systems available and appealing to an application-demanding user community. PPARDB / PVM, a portable parallel database management system that employs a "shared nothing" architecture and uses operator parallelism has been built on a heterogeneous workstation network. PPARDB / PVM uses PVM for portable communication primitives. Each workstation in the network is treated as a separate database node.

Keywords
Parallel Databases, PVM, Portability, Heterogeneous Network, Shared Nothing

1 Introduction and Motivation

Parallel computers, especially distributed memory multiprocessors, currently seem to be the only viable solution to a number of processing power intensive problems, especially when cost limitations restrict use of very expensive systems such as supercomputers.

Nevertheless, a parallel system still is a specialised computer that, not only is quite expensive but requires special programming skills and methodologies as well. Issues such as function and data decomposition and networking set-up arise when one tries to use a parallel computer to solve any problem efficiently. These issues, along with the lack of architectural, communications or programming standards limit both the scope and acceptance of parallel systems and the number of applications available for them.

Databases have always been very important applications for any computing environment. Parallel databases share the advantages and disadvantages of parallel systems. The various parallel databases that have been developed, e.g. Bubba [BORA90] and Gamma [DEWI90], are not compatible with each other, require different programming approaches, sometimes run on specialised hardware and in general are not yet capable of boosting widespread adoption and use of parallel systems the same way serial databases undeniably did for Von Neuman computers.

Making parallel databases portable is a serious step towards the creation of parallel databases that will play a significant role in the general acceptance of parallel systems. To the best of our knowledge, portability issues of parallel databases have not yet been extensively investigated ([HU95], [FRIE90]). In our work we do not focus in the traditional parallel databases research issues such as performance and efficiency, data partitioning and availability, optimisation etc. Instead, we are looking for methodologies to make parallel databases portable and, therefore, available to a wider range of parallel systems, consequently making them available and appealing to a wider range of users and applications.

In this paper we present the porting of a parallel database management system to a different architecture using a standard parallel programming environment as the portability layer. The resulting portable parallel database management system is named PPARDB / PVM, for Portable Parallel Database Management System / PVM based.

2 Parallel Database Management Systems

During the previous years, parallel database management systems have been a topic of active research. Several such systems have been proposed, with various parallelism, network topology, sharing and partitioning characteristics. One common characteristic of those systems is that they adopt the relational model, since relational queries are ideally suited to parallel execution [DEWI92]. Bubba [BORA90] employs operator parallelism in a shared-nothing (SN) architecture, using a hypercube network topology and range and hash horizontal partitioning. Gamma [DEWI90] runs on a SN hypercube and incorporates a combination of horizontal partitioning schemes. XPRS [STON88] uses a shared memory (SM) architecture and both horizontal and vertical partitioning to achieve query, operator and tuple parallelism. PARDB [THEO92b] is a SN system (although the existence of one disk per node is emulated) that employs operator parallelism and horizontal partitioning. The Sybase Navigational Server [HU95] employs a SN architecture with special availability and portability features and is built on top of the Sybase SQL server.

The selection of the relational parallel database management system to initially port to another architecture is instrumental to the success of the task in hand. Meanwhile, the guidelines used for the selection apply to the "portable parallel databases" in general. Therefore, the selected parallel database management system should not rely heavily on special characteristics of the underlying hardware or software. It should run on parallel systems with a general purpose processor per node. Specialised node processors (e.g. bit-slice) perhaps prove faster for specific tasks but are not acceptable for portability. Also, the operating system kernel in each node should provide process, memory, disk I/O and communications allocation and management. Extra features of operating systems (e.g. barriers) may be useful but also make porting more difficult. Another issue is the sharing architecture. Shared nothing seems to be more adaptable to various architectures. Also, indexing must be used to ensure good database performance per processing node. Indexing is best exploited using horizontal partitioning, since partial indexes on each node's data set can be created independently of other nodes and aggregate keys are properly handled. Finally, the parallel database should be relatively simple, at least for the initial porting, so that any extra complexity does not interfere with portability.

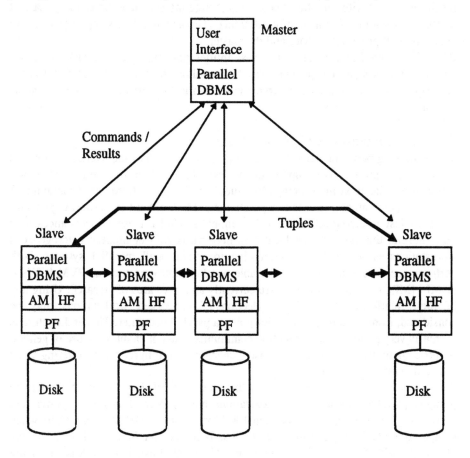

Figure 1: PARDB architecture

According to the previous guidelines, PARDB [THEO92] is a very good choice. It runs on a Transputer network on top of the Helios [PERI89] operating system. It has a simple layered architecture which is depicted in figure 1. A master node interfaces with the user, accepts queries and sends them to the slave - processing nodes using a broadcast mechanism. Also, the master holds the system catalogues / database schema, including the distribution of relations and tuples to the slaves. The slaves have a layered architecture: a paged file (PF) layer handles files organised as pages and is the layer that interacts directly with the system disks. All other layers finally use the PF layer to access the data. Above the PF layer, a heap file (HF) layer handles data organised as unindexed fixed-sized records, whereas an access method (AM) layer handles data organised as indexed (hashed) pages. These three layers manage the data stored at any specific node. A fourth layer, the Parallel DBMS layer, uses their services to implement the relational (e.g. project, select, join, union, difference etc.) and auxiliary (e.g. create, delete, balance, remove-duplicates) operators in parallel. That layer communicates with the master to receive commands and return result tuples and with the other nodes to exchange intermediate tuples in the case of multiscan operators such as join or difference. The actual network topology is a ring connecting all processing nodes and the master.

PARDB's layered architecture greatly simplifies porting, since network, hardware and operating system dependencies can be consolidated in a single layer that is the one to be adapted for each new architecture, while the rest of the system remains unchanged.

3 Communications and Networking

The processing power of state-of-the-art processors, as well as the overall capabilities of the computers based upon them, especially scientific and engineering workstations, keep on increasing, while they remain quite affordable. Programming such a system is much easier than a parallel one. Modern built-in networking enables seamless fast interconnection of such computers. Under these circumstances it is quite normal that an Ethernet local area network (LAN) of modern heterogeneous workstations is a very good alternative to "traditional" parallel systems, both performance-wise and cost-wise. Portable parallel platforms such as PVM [GEI94] or MPI [MPI94] implementations are currently an active area of research. Using any of these platforms, the whole collection of networked workstations may be considered, from an application's viewpoint, a parallel computer. This architecture is currently very popular to the scientific community, since it enables the use of general purpose computers to emulate expensive specialised multiprocessors, offering great cost and versatility benefits.

A set of networked workstations under the control of PVM is a good choice for the portable parallel database. The workstations have plenty of processor power and networking bandwidth, e.g. 10 or even 100 Mbps Ethernet or FDDI, especially when compared to the Transputer network that has been used for PARDB. They also have fast hard disks that naturally fit into the SN model. Last, but not least, they are generally available.

The parallel programming platform to enable the use of a set of networked workstations as a parallel computer is PVM, a mature and widely used programming environment. Although the majority of PVM applications so far are scientific / computational ones, PVM offers the primitives required for the operation of the database, i.e. for sending, receiving, broadcasting and multicasting a message, creating and handling processes and signals, process synchronisation etc. Also, the crowd computational model supported by PVM nicely fits the one master / multiple slaves relationship of the PARDB processing elements.

4 Portable Parallel Database Management System Architecture

The architecture of the parallel database management system used, PARDB, has been modified to allow for easier porting. The new architecture is shown in figure 2.

PARDB uses the SN model of data sharing. A Transputer network is a classical example of a distributed memory architecture but, by the time of the development of PARDB only one disk was available. To implement SN, PARDB uses an elaborate scheme of simulation of multiple disk operation over this single disk. PPARDB doesn't, of course, use this scheme, since multiple disks are available on the various networked workstations, so that true SN operation can be achieved.

The most significant adaptation of PARDB to portability requirements was, as would be expected, the removal of all machine (Transputer network) or operating system (Helios) dependencies. Standard POSIX calls and a proper Helios portability library containing the original calls are now used wherever Helios specific calls had previously been used. POSIX calls are portable across a great variety of operating systems and hardware combinations.

Multithreading has been an issue addressed during the development of the system. PARDB slaves use the multithreading facilities within a node of Helios [PERI89] for process control. Unfortunately, such facilities are not yet available to most operating systems, especially UNIX-derived ones, introducing a serious portability problem. To avoid that, initial PARDB code has been carefully checked and, when required, rearranged or rewritten to remove any non-portable programming practices, like liberal use of shared variables, and make the system work in the context of a single thread, thus making it portable to both the Helios and the UNIX process models. Communication between layers is accomplished by simple function calls with the appropriate parameters. All communication to the other processes, the master or any other slaves, takes place via message passing from a single point in a predefined way. This model fits very well in the PVM environment. PVM considers the whole slave as a single process with which usual message passing is performed.

It has been a design choice not to use threaded or "lightweight" process versions of PVM, e.g. TPVM [FERR94] or LPVM [ZHOU95], not only because of the lack of a portable threads package across the different platforms, but also because of the relative immaturity of those packages for the task in hand. LPVM for instance still runs only on single Sun and IBM SMP systems and is considered experimental. We have thus chosen to sacrifice the extra performance and low memory requirements of multithreading operation to achieve better portability and robustness.

Network topology handling is a part of the system that has undergone major changes. PARDB connects its slaves in a ring, so that two communication channels (pipes) per slave node, one to the "next" and one to the "previous" neighbour, plus one channel to the master are required. This architecture simplifies parallel operator algorithm design but is not suitable for portability, since parallel systems use various topologies, for instance hybercubes. Special care must be taken for the current Ethernet workstation version of the system, since Ethernet is a multi-access network capable of direct broadcasting and multicasting. Retransmission of the same data over the network is possible, e.g. when sending commands from the master to the slaves or when sending intermediate multiscan operator data to other nodes, resulting in slower performance. To avoid such problems and make the system independent of network topology a new layer, the Network Topology (NT) layer is introduced. To transmit a message to a specific other node (point-to-point), to a set of other nodes (multicast) or to all nodes (broadcast), the Parallel DBMS layer on each node requests so from the NT layer. The latter is aware of both the network topology and the communications subsystem, in this case PVM, and arranges transmission of the message accordingly. To port the system one must simply inform the NT layer of the new topology and of the communications layer. This is done by means of an NT configuration file.

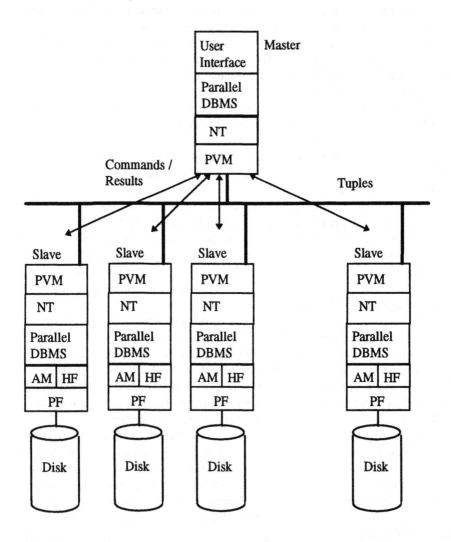

Figure 2: PPARDB / PVM architecture

Of the various parallel database system layers, only the NT layer needs to be ported, for the whole system to be ported to a new parallel system architecture. The other layers use standard POSIX calls and are thus portable as long as the architecture implements these calls. For the NT layer to be ported, it must be adapted to accommodate new network topologies and communications layers.

The communications layer, currently PVM, can be whatever native message passing capabilities the underlying architecture offers.

5 PVM Issues

Unlike PARDB, PVM has not undergone any design changes. In fact, there are not any required PVM changes and the system can run on top of standard PVM, for obvious compatibility and portability reasons. All required parallel database portability modifications have been done on the database management system itself and not on the communications layer. Moreover, PVM is used only as a communications subsystem for message passing between the parallel database nodes. The configuration of the nodes is static and determined at system start up. Therefore, the PVM process control facilities are not really needed.

An important function of PVM for operation over heterogeneous networks must be noted, though, because currently there is no such subsystem in PPARDB and the built-in capabilities of PVM are used. It is the XDR support in PVM for byte ordering and data representation issues.

An enhancement to PVM, better multicasting support, is under way. PVM normally implements multicast by sending the message to all recipients, each of them in turn. Such an operation obviously duplicates messages over the network, especially a multi-access one. Many UNIX operating systems currently support multicasting, so it's natural for PVM to include such support. Multicasting support for a PVM host is included as a hostfile option, so that PVM knows which hosts support it. When a multicast message is sent, all recipient hosts supporting multicasting are included in a multicast group address (Internet class D). The message is sent to each non-multicast host, as well as to this group address, decreasing the number of messages sent over the network by the number of members in the multicast group address (minus 1).

6 Conclusions and Future Development

The system is in its final debugging stage. Performance figures will be available soon. It is implemented on top of a 10 Mbps Ethernet network consisting of 12 Sun SPARCstation and Silicon Graphics Indy workstations running Solaris, SunOS and IRIX.

Our work on PPARDB / PVM has shown that parallel databases can run on inexpensive and standard configurations, like a set of networked workstations. They can also be portable to various diverse architectures, making them more appealing, from an application availability point of view, to a wider user community.

To accomplish this goal, the database must be carefully designed to eliminate any non-portable constructs and platform dependencies. Source code function calls standards conformance is required. It must also be layered to limit changes to the layer closer to machine dependencies. A proper message passing subsystem must also be selected to ensure proper communication.

PPARDB / PVM is only the first step towards creating a really architecture independent and portable parallel database. The system is already scheduled to be ported to a number of other architectures, among them an Intel Paragon and a Parsytec GC. A significant upgrade to the NT layer is also planned. Its goal is to allow NT to dynamically adjust the logical network topology according to the relational operator processed. For instance, operators such as select or project are best suited by a farm model, where all slave nodes send partial results to the master, whereas a join may require a ring or a hypercube. NT can be altered to support such changes.

7 References

[BORA90]   Boral H. et al., "Prototyping Bubba, A highly parallel database system", IEEE Transactions on Knowledge and Data Engineering, Vol. 2 No. 1, March 1990.

[DEWI90]   DeWitt D. et al., "The Gamma database machine project", IEEE Transactions on Knowledge and Data Engineering, Vol. 2 No. 1, March 1990.

[DEWI92]   DeWitt D. and Gray J., "Parallel database systems: the future of high performance database systems", Communications of the ACM, Vol. 35 No. 6, June 1992.

[FERR94]   Ferrari A. and Sunderam V.S., "TPVM: Distributed concurrent computing with lightweight processes", Dept. of Mathematics and Computer Science, Emory University, 1994.

[FRIE90]   Frieder O., "On the design, implementation and evaluation of a portable parallel database system", Proceedings of the PARBASE-90 International Conference on Databases, Parallel Architectures and their Applications, Los Alamitos, USA, 1990.

[GEI94]   Geist A. et al., "PVM: Parallel Virtual Machine", MIT Press, Cambridge, USA, 1994.

[HU95]   Ron-Chung Hu and Stellwagen, R., "Navigation Server: A highly parallel DBMS on open systems", Proceedings of the 11th International Conference on Data Engineering, Taipei, Taiwan, March 1995.

[MPI94]   Message Passing Interface Forum, "MPI: A Message-Passing Interface standard", International Journal of Supercomputer Applications, Vol. 8 No. 3 & 4, 1994.

[PERI89]   Perihelion Software Ltd., "The Helios operating system", Prentice Hall International, 1989.

[STON88]   Stonebraker M. et al., "The design of XPRS", Proceedings of the 14th VLDB Conference, Los Angeles, USA, 1988.

[THEO92]   Theoharis T., Papakonstantinou G. and Tsanakas P., "The design of PARDB: a parallel relational database management system", Proceedings of 7th ISCIS Conference, Antalya, Turkey, November 1992.

[ZHOU95]   Zhou H. and Geist A., "LPVM: A step towards multithread PVM", Oak Ridge National Laboratory, Oak Ridge, USA, 1995.

SPTHEO - A PVM-Based Parallel Theorem Prover

Christian B. Suttner*

Institut für Informatik
TU München, Germany
Email: suttner@informatik.tu-muenchen.de

Abstract. SPTHEO is a parallelization of the sequential theorem proving system SETHEO, based on the SPS model for parallel search. The SPS model has been designed to allow efficient parallel search even in comparatively low bandwidth and high latency environments, such as workstation networks. In order to obtain a portable and efficient implementation, the PVM message passing system has been used for implementing the communication part of the system. This report describes the basic outline of the system, and presents evaluation results for the communication aspects as well as the performance as a proof system.

1 Introduction

The goal in automated theorem proving (ATP) is, given a set of axioms and a sentence to be proven, to show that the sentence is a logical consequence of the axioms. This general paradigm can be used in many applications, e.g. software verification [Sch95]. ATP is based on search: the solution to a problem is found by a systematical trial-and-error of possible combinations between inference rules, axioms, and deduced formulas. A common way to describe a search space is the OR-search tree. Figure 1 gives an example for an OR-search tree. The top node denotes the original problem (start situation), from which alternative search paths extend. In theorem proving, the OR-branches result from different clauses that can be used to solve a particular subgoal.

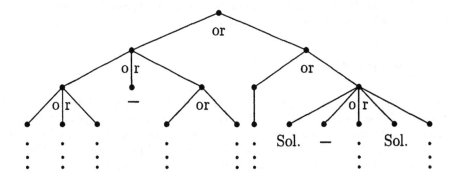

Fig. 1. OR-search-tree.

* This research has been funded by the DFG SFB 342 subproject A5 (Parallelization in Inference Systems).

The predominant problem in theorem proving (and search in general) is the performance limitation that arises due to a combinatorial explosion of the search space: if a proof consists of many individual steps, the number of possible paths becomes intractable for exhaustive search. Even though advanced systems use various search pruning techniques, many interesting problems require enormous computing time or are well beyond tractability.

In order to improve the usefulness of ATP systems in practical applications, a significant performance increase, esp. with respect to solving harder problems, is required. An approach to achieve such an improvement is the use of parallel instead of sequential search. Parallelism offers two advantages in that respect. First, the use of hundreds of processors significantly increases the hardware resources, both in computational power and in main memory capacity. This allows up to a linear reduction in the runtime of problems and makes problems tractable which previously were out of reach due to memory requirements. Second, and even more importantly, parallel search allows the avoidance of early bad search decisions. The exploration of several paths in parallel ensures that (viewed from the top of the search tree) at least one process is guaranteed to follow the best search path initially. This can lead to an exponential reduction in the amount of search required, compared to sequential depth-first exploration of the search tree. In sequential search, a similar reduction can only be achieved with breadth-first search, a technique that proved undesirable in practice due to memory requirements that are usually too large for a single processor.

A thorough examination of previous work in parallel automated theorem proving [SS94] has lead to the establishment of guidelines which seem necessary to ensure the construction of a parallel search system which lasts for more than one hardware generation and can adopt improvements in sequential search technology adequately:

- coarse-grain parallelization $\rightarrow$ minimize the synchronization overhead
- start with good sequential systems $\rightarrow$ use their sophistication and functionality
- use powerful processors $\rightarrow$ be competitive with best sequential system (SW + HW)
- avoid specialized hardware $\rightarrow$ increases lifetime and portability
- simple model for parallelization (esp. wrt. communication) $\rightarrow$ good for implementation/maintenance and integration of new sequential improvements

The SPS (Static Partitioning with Slackness) model [Sut95, Sut96] provides a parallelization model which follows these guidelines.

2   The SPTHEO System

The Underlying Sequential System. SPTHEO (Static Partitioning THEOrem prover) [Sut95] is based on the SETHEO (SEquential THEOrem prover) system [LSBB92, LMG94, GLMS94]. SETHEO is a sound and complete theorem prover for first order predicate logic. It is based on the model elimination calculus (similar to PROLOG) and implemented as an extended Warren Abstract Machine [Sch91].

The Computational Model of SPTHEO. The execution of SPTHEO is based on the SPS-model, and can be separated into the following three phases:

$$\boxed{\text{Task Generation}} \rightarrow \boxed{\text{Task Distribution}} \rightarrow \boxed{\text{Parallel Execution}}$$

Briefly, tasks are generated in an initial, sequential search phase. These tasks are then distributed and executed in parallel. The motivation for this is to minimize the communication overhead, with the idea to avoid load imbalance by proper task generation and by supplying more than one task to each processor (slackness). Below the individual phases are described in more detail.

The Task Generation Phase. In the first phase (task generation), an initial, finite segment of the search space is explored. During this phase, tasks for parallel execution are generated. Task generation occurs only during this phase and independently of the processing of the tasks later (static partitioning). The number of generated tasks (denoted by the letter m) is equal to or larger than the number of processors (denoted by the letter n). The term *slackness* is used to express the relation between the number of tasks and the number of processors.

The task generation phase is defined by the rules used for partitioning the search space into tasks, a generation strategy, and the desired number of tasks. SPTHEO allows OR- and independent-AND partitioning (iAND-partitioning for short). With OR-partitioning, a separate task is created for each alternative clause that can be used to solve the current subgoal. With iAND-partitioning, a separate task is created for each independent group of subgoals (i.e., no variables shared between the groups) that remains to be solved. The generation strategy used is iterative-deepening search over the depth of the proof tree. This means that for each particular deepening level, the search is pruned as soon as the resulting proofs would exceed a certain depth. The task generation thus operates as follows. The levels defined by iterative-deepening are successively explored, counting the number of search cut-offs due to reaching the depth limit. Each such point where the search has been artificially stopped can be used as a task for parallel search. As soon as a level is reached which allows to create enough tasks (m is specified by the user), task generation is enabled and the desired number of tasks is approximated by appropriately switching to a previous deepening level (where the number of tasks that remain to be produced is already known).

The Task Distribution Phase. In the second phase (task distribution), the tasks generated in phase one are distributed among the available processors. SPTHEO makes beneficial use of the availability of all tasks prior to their distribution: in the iAND-processing option, redundant tasks may occur, which detected and eliminated. The task distribution affects the potential for load imbalance: if tasks with very long run time accumulate at one processor, while small tasks accumulate on another, imbalance results. Experiments showed that the cumulation of similar sized tasks is heuristically minimized, if the distance in the OR-search tree between the tasks mapped to the same processor is as large as possible. Further experiments [Hub93] showed that a very good approximation of the optimal solution is obtained by employing a simple modulo mapping: $\text{Task}_i \rightarrow \text{Prozessor}_{i \bmod n}$, where Task_i is the i-th task generated.

In order to keep the communication overhead low, an efficient task encoding is used. Since each task represents a location in the OR-search tree, simply the choices made during the search that lead to the task-defining position in the search tree are stored. This is simply a list of numbers, for example "3 1 ...$^{'}$ for saying: take the third choice among the first set of alternatives, take the first choice among the second set, etc. From these numbers, each processor then recomputes deterministically the state that is represented by the task, and proceeds searching for a solution from there. For iAND-partitioning, the tasks do not comprise full OR-search tree nodes, but are further restricted to solving a particular set of open subgoals within such an OR-search tree node. Therefore, the task encoding for an iAND-task consists of the previously described list of numbers plus a list of numbers that denotes which subgoals (e.g., the second and the fifth) have to be processed. Also, each processor needs to read in the problem clauses, which are stored in a file (typically 1-3 kB) accessed by the processors via NFS.

The Parallel Task Execution Phase. Finally, in the third phase (task execution), the tasks are executed independently on their assigned processors. Note that no interaction with other tasks is necessary in order to carry out a task. Since possibly more than one task has to be processed per processor, a service strategy is required. For this, quasi-parallel processing is preferable over serial processing. The reason for this is that a task may not terminate within a given runtime limit. With serial processing, this would lead to an infinite delay of all tasks remaining on that processor, which causes search incompleteness. Timesharing of tasks at a processor is realized by starting for each task a modified version of the SETHEO system (extended by the ability to receive and process a task) as a PVM process.

The Process Structure of SPTHEO. SPTHEO employs a simple master/slave process structure. A modified SETHEO process (master) performs the initial sequential task generation, including redundancy elimination for iAND-partitioning. The generated tasks are then one after the other spawned on the available processors. For this, a load-dependent performance index for all processors is established at the time the virtual machine is built. Tasks are then distributed in a round-robin manner, with processors ordered according to their performance index.

Each terminating task returns a result status to the master. For OR-partitioning, each task simply returns a status flag denoting success (a proof has been found) or failure (the search space has been completely exhausted with no proof found). As soon as the first success message is received, the master terminates all still active processes. Furthermore, each process terminates itself after its runtime limit is exhausted.

For iAND-partitioning, the same basic structure is used. However, for each success message received, the master checks if the solution of the respective task already provides a solution to the full problem, or if further open subgoals remain. In case further work remains, a message is printed summarizing the current proof advance. Regarding control, iAND-partitioning allows two options. Either all iAND-tasks are distributed in the beginning, or only a selected subset. Based on the membership of particular iAND-subgoals in different OR-nodes (the same iAND-subgoal can be part of several OR-nodes), partial orderings can be constructed which enforce as much parallelism as necessary, but as little as possible. This has been shown to be able to improve the efficiency of parallel search [Sut95]. Therefore, initially only a small set of tasks are started, and further tasks are spawned by the master whenever a success message is received and work remains to be done, depending on the partial task ordering.

3 Evaluation of SPTHEO

The SPTHEO system has been extensively evaluated using the TPTP library [SSY94, SS95]. Altogether, 2571 TPTP v1.1.3 problems were evaluated, thereby covering a broad application spectrum of 25 scientific domains (such as algebra, set theory, and program verification), with problem difficulties ranging from very easy to very difficult (including open) problems.

In order to determine the performance improvement compared to the SETHEO system and a previous parallelization, called RCTHEO (for Random Competition parallelization of SETHEO [Ert93]), both these systems have been tested as well. Since all three systems share most of their code (all are versions of the same program), the comparison is highly accurate in the sense that differences are mainly due to differences in the computational models, and not due to implementation differences.

All experiments have been performed on a network of 121 HP workstations connected by Ethernet. The network consists of 110 HP-720 workstations (58 Drystone MIPS) with 32 MByte main memory each (200 MByte disk for swap space), and 11 HP-715/50 workstations (62 Drystone MIPS) with 80 MByte main memory each (2GByte storage disk and two 500MByte disks for system and NFS data). Each HP-715/50 workstation operates as file server for 10 HP-720 clients. The network is partitioned with bridges into 5 segments with two servers each, and one segment with one server. In order to ensure that equivalent hardware is used in all experiments, no processes were run on server machines.

Runtime Assessment in a Distributed Environment. A major difficulty for the evaluation of a network-based parallel system is the influence of other users on the available processing capacity, both in terms of processor load and available communication bandwidth. As a result, the wall-clock runtime of a parallel job can vary significantly. Since an important purpose of the SPTHEO evaluation is an analysis of the performance of the SPS-model, such variations are undesirable. Ideally, exclusive usage of the network would be possible, allowing to obtain the true runtimes on such a network. However, measurements show that the employed network is highly utilized, even during the night and on weekends. Therefore, extensive experiments based on exclusive usage are not possible. Fortunately, both the SPS-model and the random competition model (RCTHEO) allow an analysis which reliably approximates exclusive usage results. For SPTHEO, this is achieved by the following.

Instead of the standard SPTHEO operation where termination occurs as soon as a proof has been found, each task is independently processed until it fails, succeeds, or is terminated due to the runtime limit. The performance statistics (number of search steps and runtime) for all tasks are then collected in a file. Also, at the beginning of this file the performance statistics for the task generation phase are included.

Given these data, the runtime on some hardware platform can be estimated. Assume a particular task σ is the first task that leads to a proof in a standard SPTHEO run under exclusive hardware usage. The runtime of SPTHEO using OR-partitioning then consists of the runtime T_{gen} for task generation, the time $T_{dist}(\sigma)$ until σ is distributed, the runtime T_σ for processing task σ, the time T_{spp_delay} for the processing delay of σ due to time sharing, and the time $T_{termination}$ until the success message from σ is received and processed by the master[2] (runtime $= T_{gen} + T_{dist}(\sigma) + T_\sigma + T_{spp\_delay} + T_{termination}$). T_{gen} and T_σ are already contained explicitly in the logging file. T_{dist} can be upper bounded by the worst-case assumption that σ is always the last task distributed. In practice, for many problems a proof is reported to the master even before all tasks have been distributed. For example for $m_{desired} = 256$, in arithmetic (geometric) average the best task is encountered after distributing 50% (26%) of the tasks. T_{spp_delay} depends on the number of tasks, the times of processing start, and runtimes of the tasks that are processed on the same processor as σ. Given a particular distribution scheme for tasks, the number and runtimes of these tasks can be extracted from the logging file. For an

---

[2] One may also add the time until the master terminated all tasks still being processed.

upper bound on T_{spp_delay}, it can be assumed that all these tasks start at the same time as σ. In case the task switching overhead due to quasi-parallelism is not negligible, measurements need to be performed. It is then straightforward to compute T_{spp_delay} from the logging data and measurements. Finally, $T_{termination}$ can be obtained by appropriate communication time measurements. In case no proof is found within the runtime limit, the runtime can be computed as above, based on the particular task σ which terminates last.

Figure 2 shows the wall-clock time required for distributing a single task (including process startup) using the PVM message passing library on the HP network under non-exclusive usage. The measurements have been obtained for SPTHEO, but similar values can be assumed for RCTHEO. It shows arithmetical average values for the average and maximal distribution time over 20 repetitions, for 20 different problems with different numbers of tasks to distribute. The average time for up to 20 tasks is approximately 0.1 seconds, with maximal values from 0.2 to 0.7 seconds. The figure shows a slight increase of the average value as the number of tasks increases, and a significant increase of the maximal value that is observed (here up to 2.1 seconds). The reason for this is that as the number of tasks increases, the probability increases that a processor with a high load is used. Such a processor responds slowly, leading to significantly increased maximal values for some tasks, which in turn leads to an increase in the average distribution time. The figure suggests that for exclusive network usage an average distribution time of approximately 0.1 wall-clock seconds can be expected, for at least up to 60 tasks.

Below the output regarding task distribution of an actual SPTHEO run is shown. The run has been performed during daytime with a high load on the network.

```
Entering PVM! Start Parallel Processing...
Spawned tasks:  1  2  3  4  5  6  7  8  9 10 11 12 13 14 15 16 17 18 19 20 21 22
Spawned tasks: 23 24 25 26 27 28 29 30 31 32 33 34 35 36 37 38 39 40 41 42 43 44
Spawned tasks: 45 46 47 48 49 50 51 52 53 54 55
Task 37 (iand-task) successful ... solves part of OR-node  31 (1 tasks left)
Spawned tasks: 56 57 58 59 60
Task 39 (iand-task) successful ... solves part of OR-node  32 (1 tasks left)
Spawned tasks: 61 62 63 64 65 66 67 68 69 70 71 72 73 74 75 76 77 78 79 80 81

Task distribution time (wall-clock): 61.40 seconds
Task spawning times   (wall-clock): min= 0.04 max= 5.27 average= 0.79 seconds
```

The example shows that two iAND-tasks (37 and 39) are solved before the task distribution is finished. The time required to distribute all tasks has been quite high: 61 seconds. The reason for this is that some processors respond slowly (the maximal time required for a single spawn has been over 5 seconds) due to other load. The maximal time for a spawn observed in a run has been 191 seconds. High spawning times can significantly decrease the performance: while the particular slow-spawning task may not even be necessary for a proof, the processing of all tasks remaining to be spawned is blocked. This can be counteracted either by a modified (but more complicated) system design using a hierarchical task distribution, or preferably by a different spawning primitive which is non-blocking.

Proof-finding Performance of SPTHEO. The most relevant criteria in automated theorem proving is the number of problems that can be solved with given resources. Most problems solvable by SETHEO are solved within a few seconds (using 1000 seconds instead of 100 seconds increases the number of solved problems by 6%). For the comparison, a runtime limit of 1000 seconds ($\approx$ 17 min) for each problem has

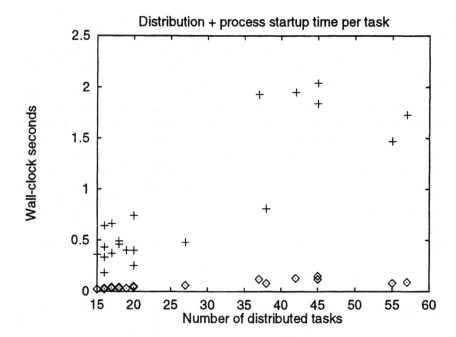

Fig. 2. Distribution and process startup time per task, averaged over 20 repetitions for 20 different problems (one vertical pair of ◇, + for each problem).
"◇" denotes the arithmetical average, "+" denotes the maximal value observed.

been used for SETHEO, in order to capture most problems which can be solved by SETHEO in a reasonable time. For SPTHEO and RCTHEO, a runtime limit of 20 seconds per task has been used. A comparatively small value has been necessary for practicability (the runtime for one parallel system with 256 processors is roughly $1800 \times 256 \times 20 seconds = 2560 hours$).

It is found that SPTHEO achieves a substantial performance improvement compared to both systems. As an indication of the achieved advance in ATP, with 256 processors (simulated) SPTHEO solves 167 more problems than SETHEO, and 150 more problems than RCTHEO from the problems in the TPTP library. Neglecting all trivial problems in the TPTP (i.e., problems solved by the sequential system in less than one second), this represents an increase of 83% of solved problems. Even compared to the parallel RCTHEO systems, 69% more non-trivial problems are solved.

Speedup Performance of SPTHEO. Figure 3 shows the relative speedup for SPTHEO based on the number of search steps, for $m_{desired} = 16$ and $m_{desired} = 256$. It shows nearly linear speedup up to 16 processors, and an increasing deviation from linear speedup for larger n. It should be noted that the plots display the geometric averages; the arithmetic averages are much closer to being linear for large numbers of processors. The are two main reasons why the relative speedup does not remain linear for large n. Firstly, usually slightly more tasks are generated than specified by $m_{desired}$. As a consequence, there are cases where the number of tasks at the processor which first terminates successfully for some n is not decreased by using twice as many processors, due to an unfortunate task distribution. Inspection of some individual problems with low relative speedup indeed showed that adjusting the number of processors such that $spp = \frac{m}{n}$ are whole numbers (or as close as possible), better speedup is obtained. Secondly, the influence of the serial fraction given by the task generation overhead increases with decreasing spp. For large n, the number of search steps required to solve a task is frequently smaller than the number of search steps required for task generation, which limits the achievable speedup considerably. As discussed before, the relative speedup metric does not provide a useful judgment in these cases, because it ignores the absolute values. Reducing the task generation overhead would improve the relative speedup for large n significantly. However, since the task generation overhead is typically quite small in absolute terms, the actual performance of the system in terms of theorems provable under typical constraints or in terms of runtime (in the order of seconds) would not improve noticeably. This shows the danger of using a relative speedup for the evaluation of parallel systems: an improvement of the relative speedup for large n would be possible, but ineffective for the absolute system performance.

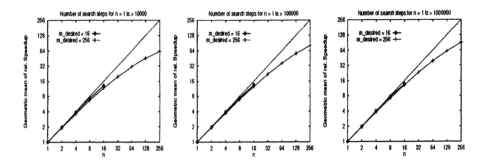

Fig. 3. Geometric average of relative speedup of SPTHEO based on the number of search steps performed for different numbers of processors, for three subsets of the TPTP problems based on the number of search steps required by one processor.

A more interesting evaluation, in particular for potential users of the system, is given in Figure 4. It shows lower bounds for the geometric average of the absolute speedup of SPTHEO compared to SETHEO. The runtime measurements for this include the input overhead. Furthermore, the task distribution overhead is included based on the assumption that the task which terminates the computation is the $m/2$-th task distributed (where m is the number of generated tasks), and assuming a distribution time of 0.1 seconds per task. According to the presented data about the occurrence of the best task within the task set (see p. 5), the use of $m/2$ in this calculation provides a pessimistic average case assumption.

Figure 4 gives lower bounds on the average speedup because there is a large number of problems for which SETHEO did not find a solution (28%). The true speedup for these problems is therefore unknown, and the runtime limit of 1000 seconds is used as a lower bound of the SETHEO runtime. Therefore, the decreasing speedup slope for large n is not necessarily an indication of a performance saturation, but is significantly influenced by the SETHEO runtime limit. Again, it may be noted that the arithmetic average

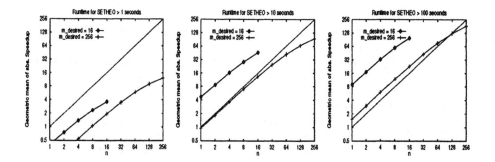

Fig. 4. Lower bounds on the geometric average of the runtime speedup of SPTHEO compared to SETHEO including input and task distribution times, for three TPTP problem subsets based on the runtime required by SETHEO.

speedup values are significantly larger in all cases. Moving from the left to the middle to the right plot shows a strong dependency of the speedup on the problem difficulty for SETHEO. Including many of the simple problems, low absolute speedup is achieved. This is little surprising, since the overhead of generating and distributing tasks cannot pay off for problems which require only a few seconds sequentially (the estimate of the distribution overhead alone leads to a runtime of $128 \times 0.1 = 12.8$ seconds for 256 processors). For increasingly difficult problems (middle and right plots), the absolute speedup becomes superlinear in average (up to some n for $m_{desired} = 256$). The plots also reveal a significantly better performance for $m_{desired} = 16$ than for $m_{desired} = 256$ for an equivalent number of processors. This is mainly due to the task distribution overhead. For the applied estimate, the geometric averages are $T_{dist} = 1.1$ seconds for $m_{desired} = 16$ and $T_{dist} = 13.3$ seconds for $m_{desired} = 256$, regardless of n. These times are part of the serial fraction of the computation, and therefore reduce the speedup noticeably. Assuming faster communication hardware would therefore lead to a significant shift upwards of the curves for $m_{desired} = 256$, and a small shift upwards for $m_{desired} = 16$. This shows how the net improvement for some number of processors depends on the communication performance of the hardware. For good absolute speedup on systems with low communication performance, it is sufficient to use a comparatively small number of processors.

4 Conclusion

In this paper, the SPTHEO system has been presented. SPTHEO is the first ATP system utilizing OR– and independent-AND search space partitioning. It is implemented in C and PVM and thereby provides a portable system running on most parallel hardware as well as networks of workstations. An evaluation has been presented which is based on extensive experiments performed on a network of 121 HP-Workstations. At the time of its construction, this represented the largest PVM application (in terms of workstation nodes) in operation.

The availability of a message-passing library significantly simplified the implementation of the parallel system. With respect to the further development of such libraries, three issues were noticed that would improve their use in SPTHEO-like systems:

- best processor spawn

Spawn the next process on the most powerful processor available.
- non-blocking spawn

Do not wait for a process spawn to finish, but continue working immediately. For SPTHEO, the blocking of individual spawns provides a significant performance limitation, as it delays the start of all following tasks.
- environment evaluation suite

A program which produces performance statistics for the current virtual machine, such as the minimal, maximal, and average time required to spawn a process and to send messages of various sizes. This would support developers when deciding about tradeoffs depending on the communication performance and would be generally useful for obtaining information about the currently available performance.

References

[Ert93]   W. Ertel. *Parallele Suche mit randomisiertem Wettbewerb in Inferenzsystemen*, volume 25 of *DISKI*. Infix-Verlag, 1993.

[GLMS94] C. Goller, R. Letz, K. Mayr, and J. Schumann. SETHEO V3.2: Recent Developments (System Abstract). In *Proceedings of CADE-12*, pages 778–782. Springer LNAI 814, 1994.

[Hub93]   M. Huber. Parallele Simulation des Theorembeweiser SETHEO unter Verwendung des Static Partitioning Konzepts. Diplomarbeit, Institut für Informatik, TU München, 1993.

[LMG94]   R. Letz, K. Mayr, and C. Goller. Controlled Integrations of the Cut Rule into Connection Tableau Calculi. *J. of Autom. Reas.*, 13(3):297–337, 1994.

[LSBB92] R. Letz, J. Schumann, S. Bayerl, and W. Bibel. SETHEO: A High-Performance Theorem Prover. *J. of Autom. Reasoning*, 8(2):183–212, 1992.

[Sch91]   J. Schumann. Efficient Theorem Provers based on an Abstract Machine. Dissertation, Institut für Informatik, TU München, Germany, 1991.

[Sch95]   J. Schumann. Using SETHEO for Verifying the Development of a Communication Protocol in FOCUS - A Case Study. In *Proc. of Workshop Analytic Tableaux and Related Methods*, pages 338–352. Springer LNAI 918, 1995.

[SS94]   C.B. Suttner and J. Schumann. Parallel Automated Theorem Proving. In *Parallel Processing for Artificial Intelligence 1*, Machine Intelligence and Pattern Recognition 14, pages 209–257. Elsevier, 1994.

[SS95]   C.B. Suttner and G. Sutcliffe. The TPTP Problem Library (TPTP v1.2.0 - TR Date 19.5.95), 1995. Technical Report AR-95-03, Institut für Informatik, TU München, Germany. Accessible via http://wwwjessen.informatik.tu-muenchen.de/~suttner/TPTP.html

[SSY94]   G. Sutcliffe, C.B. Suttner, and T. Yemenis. The TPTP Problem Library. In *Proceedings of CADE-12*, pages 252–266. Springer LNAI 814, 1994.

[Sut95]   C.B. Suttner. *Parallelization of Search-based Systems by Static Partitioning with Slackness*, 1995. Dissertation, Institut für Informatik, TU München. Published by Infix-Verlag, volume DISKI 101, Germany.

[Sut96]   C.B. Suttner. Static Partitioning with Slackness. In *Parallel Processing for Artificial Intelligence 3*. Elsevier, 1996. to appear.

Parallel Construction of Finite Solvable Groups*

Anton Betten

Department of Mathematics
University of Bayreuth
Germany

Abstract. An algorithm for the construction of finite solvable groups of small order is given. A parallelized version under PVM is presented. Different models for parallelization are discussed.

1 Group Extensions

A finite group G is called *solvable*, if there exists a chain of normal subgroups $1 = G_0 < G_1 < \ldots < G_s = G$ such that each group G_i is normal in its succesor G_{i+1} and that the index $[G_{i+1} : G_i]$ is prime for $i = 0, \ldots, s-1$.

Consider the simplest situation of a *group extension* G with a normal subgroup N of prime index $p = [G : N]$ (compare HUPPERT I, 14.8 [3]). Take an arbitrary $g \in G \backslash N$. Then $G = \langle N, g \rangle$ and the factor group G/N consists of the cosets $N, Ng, Ng^2, \ldots, Ng^{p-1}$. Each coset Ng^i with $i \not\equiv 0 \bmod p$ generates the factor group. Because of $(Ng)^p = Ng^p = N$ one gets $g^p = h \in N$. As N is normal in G, $g^{-1}ng = n^g$ is an element of N for any $n \in N$. Conjugation with the fixed element $g \in G$ defines an automorphism of N since $(n_1 n_2)^g = g^{-1} n_2 g g^{-1} n_2 g = n_2^g n_2^g$. This *inner automorphism* of G considered as an automorphism of N is not generally an inner automorphism of N. Define $\alpha_g : N \to N, n \mapsto n^g$, the associated automorphism. What is known about α_g?

i) $g^p = h \in N \cap \langle g \rangle$, a cyclic and therefore abelian subgroup, so $h^g = h^{\alpha_g} = h$.
ii) $g^p = h$ implies for any $n \in N$: $n^{g^p} = n^{\alpha_g^p} = n^h$, so that $\alpha_g^p = \mathrm{inn}_h$, where inn_h is the inner automorphism of N induced by conjugation with h.

On the other hand, one easily verifies that for any group N and any pair of elements $h \in N$ and $\alpha \in \mathrm{Aut}(N)$ there exists a group G of order $p \cdot |N|$, if

i) $h^\alpha = h$ and
ii) $\alpha^p = \mathrm{inn}_h$.

One obtains the group extension by introducing an element g with $g^p := h$ and $g^{-1}ng := n^\alpha$. $\langle N, g \rangle$ defines a group G of order $p \cdot |N|$ and, by definition, N is normal in G.

Elements $h \in N$ and $\alpha \in \mathrm{Aut}(N)$ where N is a fixed group are called *admissible*, if the conditions i) and ii) of above are fullfilled. Let now h and α

---

* This research was supported by PROCOPE

be an arbitrary admissible pair for prime p and group N. Define the group extension $\text{Ext}(N, p, h, \alpha)$ to be the group generated by the elements of N and an element g with $g^p := h$ and $n^g := n^\alpha$ for each $n \in N$. Define the set of all possible group extensions $\text{Ext}(N, p)$ to be $\{\text{Ext}(N, p, h, \alpha) | h \in N, \alpha \in \text{Aut}(N), h \text{ and } \alpha \text{ admissible for } N \text{ and } p\}$. If $\mathcal{G}$ is a set of groups one defines $\text{Ext}(\mathcal{G}, p) := \cup_{G \in \mathcal{G}} \text{Ext}(G, p)$.

Clearly, the construction of such group extensions needs a good knowledge of the automorphism group.

Because subgroups of solvable groups are solvable too, it is possible to construct solvable groups by iteration of the procedure just indicated: Assume one wants to find all solvable groups of a given order n. Because the order of subgroups always divides the order n, the lattice of divisors of n comes into play. By induction, one can assume that all strict subgroups have already been determined. Then, one has to construct all group extensions for all possible subgroups of prime index (in the group of order n). Therefore, consider the prime factorization of n. Any rearrangement of the primes might possibly occur as a sequence of orders of factor groups in the normal chain of a solvable group of order n. So, one starts at the bottom (the trivial group) constructing all groups of prime order. Then one goes up and determines all groups with order a product of two primes, three primes and so on. Note that the lattice of subgroups of a group of order n can be divided into layers according to the number of primes of the corresponding group orders (counting multiplicities).

Denoting the set of solvable groups of order n by $\mathcal{AG}_n$ (German: "auflösbare Gruppe") one has

$$\mathcal{AG}_n = \bigcup_{p|n, p \text{ prime}} \text{Ext}(\mathcal{AG}_{n/p}, p). \tag{1}$$

In the prime power case this reduces to the formula $\mathcal{AG}_{p^k} = \text{Ext}(\mathcal{AG}_{p^{k-1}}, p)$. Because groups of prime power order are solvable, any group of prime power order will be obtained by this way.

In the following we will specialize our notation to the case of solvable groups $\mathcal{AG}_n$. Let p and q be primes dividing n. Then it might happen that extensions $G_1 \in \text{Ext}(\mathcal{AG}_{n/p}, p)$ and $G_2 \in \text{Ext}(\mathcal{AG}_{n/q}, q)$ define isomorphic groups. So, the task is to *find* all the extensions for a given order and afterwards to *reduce* the set of groups up to isomorphism.

As example, consider the two groups of order 4 and their extensions to groups of order 8:

$$4\#1 \simeq Z_2 \times Z_2:$$

$A^2 = id$
$B^2 = id,\ A^B = B^{-1}AB = A$

id	A	B	BA
A	id	BA	B
B	BA	id	A
BA	B	A	id

$$4\#2 \simeq Z_4:$$

$A^2 = id$
$B^2 = A,\ A^B = B^{-1}AB = A$

id	A	B	BA
A	id	BA	B
B	BA	A	id
BA	B	id	A

Next, we show the groups together with their subgroup lattice (where only selected cover-relations are drawn). In the text below one can find generators and relations, the column to the right gives a label for the group, the isoclinism class, the order of the first central factor group, the order of the derived subgroup, the order of the automorphism group, their factorization, the Sylow type and the number of conjugacy classes of subgroups with respect to conjugation by the full automorphism group (first line), by the group of only inner automorphisms (second line) and by the trivial group (third line). So, this line also gives the number of groups in any layer of the lattice. To the right of the closing braces the sums of the entries are given. Here is not enough space to explain this in details, the interested reader may have a look at:

http://btm2xd.mat.uni-bayreuth.de/home/research.html

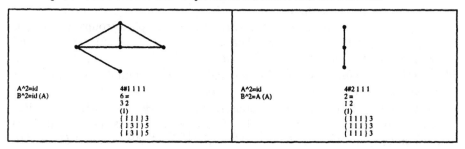

Fig. 1. The two groups of order 4

To define group extensions, it is necessary to study the automorphism groups of 4#1 and 4#2. The group $Z_2 \times Z_2$ admits any permutation of its non-trivial elements as an automorphism. So, $\mathrm{Aut}(4\#1) \simeq S_3$. In the other case, there is only one non-trivial automorphism, namely the map $B \mapsto BA = B^{-1}$ (mapping A onto itself). In this case $\mathrm{Aut}(Z_4) \simeq Z_2$.

In the following, we often substitute the elements of the groups by their lexicographic numbers, always counting from 0 on for convenience. Thus we have $0 = id$, $1 = A$, $2 = B$, $3 = BA$ (because $B^2 = id$). As permutation groups, $\mathrm{Aut}(4\#1) = \langle (1\,2), (2\,3) \rangle$ and $\mathrm{Aut}(4\#2) = \langle (2\,3) \rangle$.

In order to compute admissible pairs $\alpha \in \mathrm{Aut}(N)$ and $h \in N$ (where $N \in \{4\#1, 4\#2\}$) we define the *extension matrix* of a group N:

$$E(N,p) = (e_{\alpha,h})_{\alpha \in \mathrm{Aut}(N), h \in N} = \begin{cases} 1 \Leftrightarrow h^\alpha = h \wedge \alpha^p = \mathrm{inn}_h \\ 0 \quad \text{otherwise} \end{cases}, \qquad (2)$$

i.e. $e_{\alpha,h}$ is 1 iff (α, h) is an admissible pair (for N and p). The elements of $\mathrm{Aut}(4\#1)$ (left column) and $\mathrm{Aut}(4\#2)$ (automorphisms listed by their images on the generators and as permutations of the elements) are:

no. $\alpha \in \mathrm{Aut}(4\#1)$ ord(α)
0 [1, 2] id 1
1 [1, 3] $(2\,3)$ 2   no. $\alpha \in \mathrm{Aut}(4\#2)$ ord(α)
2 [2, 1] $(1\,2)$ 2   0 [1,2] id 1   (3)
3 [2, 3] $(1\,2\,3)$ 3   1 [1,3] $(2\,3)$ 2
4 [3, 1] $(1\,3\,2)$ 3
5 [3, 2] $(1\,3)$ 2

The extension matrices are:

$$E(4\#1,2) = \begin{pmatrix} X X X X \\ X X \, . \, . \\ X \, . \, . \, X \\ . \, . \, . \, . \\ . \, . \, . \, . \\ X \, . \, X \, . \end{pmatrix} \quad E(4\#2,2) = \begin{pmatrix} X X X X \\ X X \, . \, . \end{pmatrix} \tag{4}$$

So, there are 10 possible extensions of 4#1 and 6 extensions of 4#2. As noted before, one cannot expect 16 different groups of order 8 since some of the candidates may be isomorphic.

By computing Ext(4#1) one gets three non-isomorphic groups: the first is $Z_2 \times Z_2 \times Z_2$ and will be called 8#1. The second is $Z_4 \times Z_2$ (8#2), the third 8#3 is non-abelian: $A^2 = id$, $B^2 = id$, $C^2 = id$ with relations $A^B = A$, $A^C = A$ and $B^C = AB$. This defines a dihedral group and is also an example that computers may give other presentations of a group than one would expect.

In computing Ext(4#2) one obtains the groups 8#2 and 8#3 again. But there are two new groups: 8#4 is cyclic of order 8 and 8#5 is the (non-abelian) quaternionic group: $A^2 = id$, $B^2 = A$, $C^2 = A$, $A^B = A$, $A^C = A$ and $B^C = AB$. It is clear that – for example – one cannot get 8#1 as an extension of $4\#2 \simeq Z_4$: the elementary abelian group has no subgroup isomorphic to Z_4.

Generally, it is desirable to compute a list of groups which is both *complete* and *irredundant* – that is, a representative of each isomorphism type is present and no two groups are of the same isomorphism type. Completeness is guaranteed by the introductory remarks of this section; irredundancy involves solving the *isomorphism problem* for groups. This is yet another story which definitively cannot be solved in this paper. One would have to look at invariants of groups and talk about canonical forms to solve the isomorphism problem.

The amount of computational effort needed for constructing all groups of given order mainly depends on the number of candidates of groups to be tested. There are methods which reduce the number of candidates drastically. For instance when considering conjugacy classes in $\mathrm{Aut}(G)$ it is easily seen that it suffices to compute extensions using just a system of representatives of the classes. Another reduction can be made by letting the centralizer of each fixed automorphism act on the entries 1 in the extension matrix. We do not notice further details. Let us move straightforward to parallelization.

2 Parallelization

We are going to parallelize the computation of $\mathrm{Ext}(\mathcal{N}, p)$. Just before starting, one has to look at the program in order to discover the possible approaches for parallelization. In our case two attempts were made to parallelize at different positions: the crucial point is to know where in the program most of the work is located. One has to recognize subproblems in the algorithm that can be computed widely independent, i.e. without the need of too much communication between the parts. An important concept is the notion of *task granularity* which is the ratio of computation to network activity of a task. This parameter can decide on success or failure of a parallelization in many cases. So, keeping this value in mind is very important.

The serial version of the program acts in the following way (compare figure 2): For any group $N \in \mathcal{N}$ all possible p-extensions are computed by the generator (using reduced extension matrices). These are the candidates which have to be filtered up to isomorphism. An important tool for parallelism is a so called

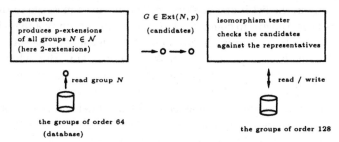

Fig. 2. The serial program

"pool of tasks". Here one collects all the subproblems which shall be processed in parallel. A one-to-many relationship between the involved programs is also included in the model: one certain program acts as a controller and administrates the pool (it is therefore called "master"). He also controls the subordinate part, namely those programs doing the parallel work (therefore called "slaves") (compare figure 3). The "pool of tasks" model can be compared with the situ-

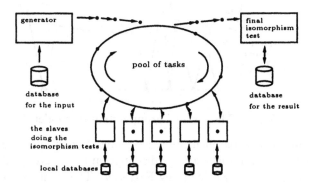

Fig. 3. The pool of tasks model

ation at the airport: The distribution of the luggage to the travellers is solved by circulation; the suitcases (which can be compared with the tasks here) run upon a cyclic band until they are fetched by some traveller. In our language of distributed computing, the task is assigned to a particular slave for processing. But what is a task and what shall the slaves do with it?

There are two approaches. Figure 3 just shows the first one by the labels (the model is independent and will also be used in the second algorithm).

The first way is to parallelize the isomorphism part of the problem, i.e. the right hand side of figure 2. For any newly generated group it has to be tested whether the group is an isomorphic copy of another group already computed or not. This work is done by the slaves, each of them holding a local database of all groups which already have been computed. The newly generated groups are deposited at the pool of tasks where they reside until some slave is idle (and willing to do the job, just to say it in human language). At this moment such a group is transferred to the slave. If it proves to be isomorphic to one of the groups already in the list it can be skipped (this means a message to the master). In the other case, the group comes back to the master (or only a tag which group is meant because the group is already held at the master). The slave is marked to be idle again. It will get another task if there is any. But what about the group sent back to the master with the information: "non-isomorphic to groups $1,\ldots,l$" where l is the number of groups at the slave's list? Now we have a difficulty: In the meantime, the master might have defined new groups and so further tests are necessary. But maybe there will not be too many new groups, so one can do this instantly. One has to be careful at this point to avoid possible bottlenecks in the computation. Experience shows that the ratio of definition of new groups is relatively low (compared to the number of candidates which have to be tested). So, estimatedly, there will not be too many late isomorphism tests needed at the master. Finally, if the group passes these tests too it is in fact new: The master adds it to his own list of groups and distributes it to all isomorphism slaves so that they can complete their lists.

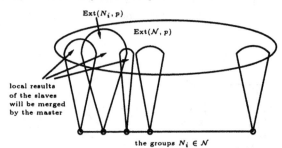

Fig. 4. The parallel group extension model

A second approach is parallelization according to equation (1) (see figure 4). Again, one uses a pool of tasks for the distribution of the groups $N \in \mathcal{N}$. Each slave gets his own group for calculating $\text{Ext}(N,p)$. The results of the slaves will be sent back to the master. There they will get merged together. One might suppose a parallel merging in form of a binary tree but one has to admit that the results of the slaves will come back in no predictable order: the computation of $\text{Ext}(N)$ may sometimes be very difficult, sometimes very easy (the amount of work depends for instance on the size – or better the number of entries 1 – of the extension matrix). Thus a linear merge was preferred in the actual implementation.

Some remarks should be added concerning the aspect of holding the data in the algorithm. As the number of groups can be quite large for higher n, the amount of storage needed by the slaves should not be underestimated. At the moment (in the current implementation) each slave keeps his own database for ease of access (no collisions during write). One has to admit that the access to common disk space via NFS can cause another bottleneck. Local disks will help here. But since the list of groups is the same for all slaves, an approach of a commonly shared database seems to be a good idea. The *PIOUS* [5] system for parallel file IO could support this aspect of the problem. PIOUS is able to use network wide distributed disk space for its files. The total amount of storage would be widely reduced. Maybe the network traffic increases a little. This question is not yet tested but PIOUS seems to be an interesting approach.

3 Results: Serial vs. Parallel Version

At first, let us try to compare the speed of different architectures (see table 1). This seems to be necessary because results of general benchmarks cannot easily be transferred to the specific program which is run here. For the test, we just run a very small example. Note that this program uses only integers (no floating points) and is very special in its kind because it merely does not "compute" in the narrow sense. It is more a collection of lots of (deeply nested) loops. Moreover, the program does not take advantage of DECs 64 bit processors. This might help to explain why hardware based on Intel Pentium is so well suited for the program (considering also prices !). On each platform, a high grade of optimization was tried (cxx supports optimization only up to -02).

machine type	cmplr.	hh:mm:ss	P90 speed
PentiumPro 200 MHz	g++ -03	10:52	299 %
DEC AlphaStation 600, 333 MHz	cxx -02	11:06	293 %
SGI PowerChallenge chip: R10000, 190 MHz	g++ -03	12:21	263 %
Intel Pentium 90 MHz	g++ -03	32:30	100 %
DEC Alpha 3000 / 600, chip 21064, 175 MHz	cxx -02	34:55	93 %
Silicon Graphics Indy	g++ -03	38:29	84 %
DEC Alpha 3000 / 400, chip 21064, 130 MHz	cxx -02	49:49	65 %
Intel 486 DX2/50 MHz VL	gcc -03	1:36:08	34 %

Table 1. The serial version

Testing the PVM version of the program involves some difficulties. It only makes sense to measure the real time of the run, i.e. the life-time of the master. There is no way of measuring "user-time" as it was done in the serial version. Thus, the load of the computer imposed by other users has a negative impact on the evaluation. The values presented here were obtained during a week-end's night with no other processes running on the machines. It is also desirable to test the PVM program on a pool of homogeneous machines so that one can study the effect of succesively increasing the number of processors. The test runs presented here were made on a pool of (equal) SGI workstations at the computing center of Bayreuth (see figure 5).

For optimal behaviour of the parallelized version, the test was limited to the computation of the sets of extensions $Ext(G, p)$ (which was the primary aim

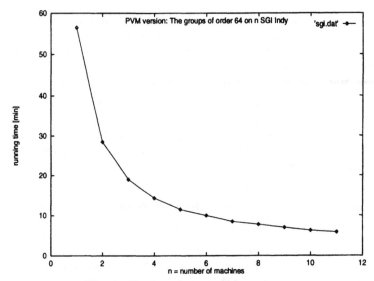

Fig. 5. Testing the parallel version

of the parallelized version of the program); the second step (merging these lists together) was left out for the test. Another possible source of friction comes from task granularity. It has already been discussed that the amount of computation should not be too low compared with the time spent for sending messages and for administrating the distributed system. Here we have the other side of the medal: if one task is computing for a very long time, in the meantime many (or all) other tasks might have terminated. So in this case the computation is unnecessarily lengthened at the end with only few working tasks. The problem chosen for this particular test was the computation of all 267 groups of order 64 realized as extensions of the 51 groups of order 32.

References

1. BETTINA EICK: Charakterisierung und Konstruktion von Frattinigruppen mit Anwendung in der Konstruktion endlicher Gruppen. Thesis, RWTH Aachen, 1996.
2. MARSHALL HALL, JAMES K. SENIOR: The groups of order 2^n ($n \leq 6$). MacMillan Company, New York, London 1964.
3. BERTRAM HUPPERT: Endliche Gruppen I. Springer Verlag, Berlin, Heidelberg, New York, 1967.
4. REINHARD LAUE: Zur Konstruktion und Klassifikation endlicher auflösbarer Gruppen. *Bayreuther Math. Schr.* **9** (1982).
5. STEVEN A. MOYER, V. S. SUNDERAM: PIOUS for PVM, Version 1.2, User's Guide and Reference Manual. http://www.mathcs.emory.edu/Research/Pious.html
6. JOACHIM NEUBÜSER: Die Untergruppenverbände der Gruppen der Ordnungen ≤ 100 mit Ausnahme der Ordnungen 64 und 96. Thesis, Kiel 1967.
7. EDWARD ANTHONY O'BRIEN: The groups of order 256. *J. Algebra*, **143** (1991), 219-235.

Proving Properties of PVM Applications — A Case Study with CoCheck

Jürgen Menden and Georg Stellner

Institut für Informatik der Technischen Universität München
D-80290 München

email: {menden,stellner}@informatik.tu-muenchen.de

Abstract. The results of a case study where we applied a formal method to prove properties of CoCheck, an extention of PVM for the creation of checkpoints of parallel applications on workstation clusters. Although the functionality of CoCheck had been demonstrated in experiments, there was no proof of the desired properties. Consequently, a formal method had to be applied which allows to prove those properties.

1 Introduction

Networks of workstations (NOW) are widely used to run parallel applications [1]. Currently, many parallel programming environments for workstations, such as p4 [2], NXLib [14], the de facto standard PVM [6] and the proposed standard MPI [7] exist. But none of them allows to create checkpoints or migrate processes. CoCheck is an extension to PVM which allows both.

The design and implementation of CoCheck was pragmatic, so no formal specification or verification methods were applied. After the implementation had been finished and the functionality of the prototypes had been shown in experiments, we nevertheless were concerned about the correctness of the approach. During the work, it was particularly important to find a representation of the CoCheck protocol that was easy to understand for the designers of CoCheck and both formal and precise enough to allow the proof of the desired properties. In this paper, we present a suitable model and give an example of a proof.

In the next section we present the CoCheck environment and explain how checkpoints are taken. Then we briefly introduce the model we have used to formally describe the protocol. An example of a proof follows. Finally, we conclude with a brief summary and an outlook on future projects.

2 Overview of CoCheck

The problem of checkpointing a parallel application or migrating a process of a parallel application is based on the problem of determining global states in a distributed system, which is elaborately discussed in [12] and [4]. The basic idea of CoCheck is to achieve a situation where no more messages are in transit. In programming environments that guarantee message delivery in the sending

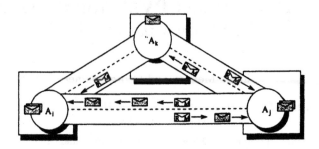

Fig. 1. Clearing communication channels.

sequence so called "ready messages" (RM) can be used to flush the messages off the network (cf. figure 1).

After each process has received a notification that a checkpoint has to be taken or processes must migrate, each process sends a RM to all the other processes. Besides, each process receives any incoming message. If it is a RM from a process, then there should be no further messages from there, i.e. this channel has been cleared. If the message is not a RM, it is stored in a buffer area of the process so that it is available to the application later on.

Finally, after a process has collected the RMs from all other processes, there are no further messages in transit which have its address as destination. Hence, the process can be seen as an isolated single process for which it is easy to create a checkpoint or migrate it to a new host with an existing single process checkpointer [8]. As no messages are in transit at checkpoint time, there is no need to forward messages after restart which is necessary in other systems [3, 5]. A more elaborated description of the protocol can be found in [13, 15]

3 Description of the Model

We modeled only a single communication channel, consisting of a *Sender* and a *Receiver* process as the channels are independent from each other and the protocol for the remaining channels behaves similarly. The *Coordinator*, which initiates the checkpoint and the restart is a third process needed.

To model the processes we need a language which is easy to understand by a programmer and on the other hand provides a clear semantics which we need for the reasoning about properties of the system. We solved these tradeoffs by using SPL (Simple Programming Language) proposed in [11] (see also [9]).

3.1 The *Coordinator*

The *Coordinator* initiates the checkpointing by sending a stop signal to the *Sender* and the *Receiver*, after that it is waiting for a done message from the *Receiver*. When it arrives the *Coordinator* knows that the message queue is empty and it might dump the processes. The restart of the processes is initiated by the *Coordinator* sending a restart command to both, the *Sender* and the *Receiver*. The *Coordinator* is then ready for creating the next checkpoint.

```
        program Coordinator ::
        var ctrl : {stop, done, restart}
        begin
m0 :        while forever do
m1 :            (c2s, c2r) ≪ stop
                do
m3 :                r2c ≫ ctrl
m4 :            while ctrl ≠ done
m5 :                (c2s, c2r) ≪ restart
                end
        end
```

Table 1.1. The *Coordinator* process.

```
        program Sender ::
        var t : T
        var ctrl : {stop, done, restart}
        var cont : {false, true}
        begin
s0 :        while forever do
s1 :            while cont do
                switch
s2 :                in ≫ t
s3 :                intern ≪ t
                or
s2 :                c2s ≫ ctrl
s4 :                if ctrl = stop then
s5 :                    cont = false
                    end
                end
            end
s6 :        intern ≪ RM
            do
s7 :            c2s ≫ ctrl
s8 :        while ctrl ≠ restart
            end
        end
```

Table 1.2. The *Sender* process.

Table 1.1 shows the *Coordinator* in SPL. $c2s$ and $c2r$ are the communication channels from the *Coordinator* to the *Sender* and the *Receiver*. $r2c$ is the channel from the *Receiver* to the *Coordinator*. stop and restart are broadcasted by the same command to both, the *Sender* and the *Receiver*.

3.2 The *Sender*

The data which has to be sent by the algorithm is represented by the queue *in*, the data which is already received is in the queue *out*. Normally, the *Sender* receives some message from *in* and puts it on the internal line to the *Receiver* ($s_{2,3}$). But if the *Sender* receives a stop signal from the *Coordinator* it has to stop forwarding messages and send the RM to the *Receiver*. Having done this, the *Sender* is ready for being dumped. On restart the *Sender* is waiting for a restart message from the *Coordinator* and continues processing of *in*.

Table 1.2 shows the *Sender* in SPL. The **switch**... **or**... **end** statement is a nondeterministic choice. The label s_2 is therefore used twice, once after the **switch** and again after the **or** to indicate that from position s_2 it could either take a message from *in* and forward it, or receive a message from the *Coordinator*.

```
        program Receiver ::
        var read : boolean
        var got_ready : boolean
        var ctrl : {stop, done, restart}
        var t : T
        var buffer : list of T
        begin
r₀ :        while forever do
r₁ :            read = true
r₂ :            got_ready = false
r₃ :            while read do
                    switch
r₄ :                    c2r ≫ ctrl
r₅ :                    if ctrl = stop then
r₆ :                        read = false
                        end
                    or
r₄ :                    intern ≫ t provided buffer = []
r₇ :                    if t = RM then
r₈ :                        got_ready = true
                        else
r₉ :                        out ≪ t
                        end
                    or
r₄ :                    when buffer ≠ []
r₁₀ :                       t = first(buffer)
r₁₁ :                       buffer = rest(buffer)
r₁₂ :                       out ≪ t
                        end
                    end
                end
r₁₃ :           while ¬got_ready do
r₁₄ :               intern ≫ t
r₁₅ :               if t = RM then
r₁₆ :                   got_ready = true
                    else
r₁₇ :                   buffer = buffer
                    end
                end
r₁₈ :           r2c ≪ done
                do
r₁₉ :               c2r ≫ ctrl
r₂₀ :           while ctrl ≠ restart
            end
        end
```

Table 2. The receiver process

3.3 The Receiver

The model of the receiver process is larger,[1] but not more complicated. Essentially the *Receiver* puts any messages it receives from the *Sender* on the *out* channel, thereby remembering but not forwarding any RM message. As long as the local *buffer* is empty, this is done by receiving a message from *intern* ($r_{4,7,8,9}$). If there exist some older, buffered messages, then these should be put on *out* first ($r_{4,10,11,12}$). This behavior continues until a stop message from the *Coordinator* is received ($r_{4,5,6}$).

After receiving a stop message from the *Coordinator*, the *Receiver* has to read further messages from the *intern* queue and buffer them, until it receives a RM (if not already received prior to the stop). After that the *Coordinator* is informed (via a done message) that the line is empty and the process is ready to be dumped. At restart there is really nothing to do for the *Receiver*, it just waits for the restart message from the *Coordinator* and continues.

4 Temporal Logic

The semantics of SPL is given in terms of a transition system ([11]). Any state in this transition system is given by the set of values the variables have at that time.

Definition 1 (State) *Let* $V = \{v_1, \ldots, v_n\}$ *be a finite set of typed variables. Each state* $s \in \Sigma$ *is an interpretation of* V, *assigning to each variable of* V *a value over its respective domain.*

Here V includes all program variables, if a variable is local to a process A, it is written $v_{(A)}$, but if it is clear from the context, to which process the variable belongs, this index is omitted. Also each process A has a (local) process counter $\pi_{(A)}$ denoting the label of the next statement to process.

Definition 2 (Transition) *Let* T *be a finite set of transitions. Each transition* $\tau \in T$ *is associated with an assertion* ρ_τ, *called the transition relation, which refers to both, unprimed and primed versions of the variables. Thus the transition relation expresses the relation between a state* s *and it's successor* s'.

Now we introduce some abbreviations: at_a_i stands for $\pi = a_i$, at'_a_i for $\pi' = a_i$, and $at\_a_{i,\ldots,j}$ means that a process A is in $a_i, \ldots a_j$. The derivation of the transitions from the program syntax is fairly simple. For example an assignment statement

$$a_i : \quad v = F(v_1, \ldots, v_n)$$
$$a_{i+1} :$$

leads to the assertion

$$at\_a_i \wedge at'\_a_{i+1} \wedge v' = F(v_1, \ldots, v_n).$$

---

[1] This is because it might receive a RM message from *intern* even before the stop message from the *Coordinator* arrives.

The condition statement

$$
\begin{aligned}
&\text{a}_i: \quad \textbf{if } Con \textbf{ then} \\
&\text{a}_{i+1}: \quad \ldots \\
&\qquad \textbf{else} \\
&\text{a}_j: \quad \ldots \\
&\qquad \textbf{end} \\
&\text{a}_k:
\end{aligned}
$$

leads to the following transitions:

$$\rho_t : at\_a_i \wedge Con \wedge at'\_a_{i+1}$$

$$\rho_f : at\_a_i \wedge \neg Con \wedge at'\_a_j$$

The **while–** and **do … while**–loops are equally straightforward. The non-deterministic **switch** statement is even easier, as it does not have a separate transition associated with it. But because some labels are duplicated, there is, at the start of the **switch** statement, always more than one transition possible.

Definition 3 (Transition system) *A transition system is a 4–tuple* (V, Σ, T, Θ), *where* V *is a finite set of variables,* Σ *is a set of states over* V, T *is a set of transitions and* Θ *is an assertion (state formula) which characterizes all initial states.*

We asume that one of the transitions $\tau_i \in T$ is the idling transition, with the transition relation $\rho_i : V' = V$.

Definition 4 (Computation) *Given a transition system* (V, Σ, T, Θ) *a computation is an infinite sequence of states* $\sigma : s_0, s_1, \ldots$ *where*

- s_0 *is initial, i.e.,* $s_0 \models \Theta$
- *for each* $i \geq 0$ *there exists a* $\tau \in T$ *with* $(s_i, s_{i+1}) \models \rho_\tau$

To express properties of programs, or of the corresponding transition system we build on state formulae (assertions). These state formulae are first order formulae which describe program states which can arise in a computation. The only temporal operator we use in this paper is the always–operator, where

$$P \models \Box\phi$$

means, that in any state s_i of any computation of program P the assertion ϕ is true. To prove this property we use the basic invariance proof rule ([9, 10])

1. $\Theta \rightarrow \phi$
2. $\{\phi\} T \{\phi\}$

---

3. $\Box\phi$.

Here $\{\phi\} T \{\phi\}$ means, that for every transition $\tau \in T$ we have to prove that

$$(\phi \wedge \rho_\tau) \rightarrow \phi'.$$

where ϕ' is obtained from ϕ by replacing all variables v by v'.

5 Proving Correctness

The property we want to show is, that all messages are correctly delivered from *in* to *out*. This includes, that no message is lost or duplicated, and that the order of the messages is preserved. The idea is, that if we concatenate *in*, *intern* and *out*, then this list will always remain unchanged over time.

Unfortunately this is not enough. The first problem is, that additional RM messages are put on *intern*, which must be removed. To solve this, we define $\overline{intern}$, which maps *intern* to the corresponding list with all RM messages removed. Another problem is, that the messages, which are buffered in *buffer* have to be considered. Third we observe, that in some states (actually at_s_3, at_r_7 and at_r_{15}) a message is already received from *in* or *intern*, but not yet forwarded to *intern*, *out* or *buffer*. The correct property which has to be shown is therefore

$$\Box u = in \cdot (at\_s_3 \ ? \ [t_{(S)}]) \cdot \overline{intern} \cdot (at\_r_{15} \ ? \ \overline{[t_{(R)}]}) \cdot buffer \cdot (at\_r_7 \ ? \ \overline{[t_{(R)}]}) \cdot out, \quad (1)$$

where u is a rigid variable. To prove this property we can ignore most transitions. No transition of the *Coordinator* for example changes any of the variables of (1) and must therefore be considered.

The transitions of the *Sender* which we have to consider are $s_2 \to s_3$, $s_3 \to s_1$ and $s_6 \to s_7$. We will only show $s_2 \to s_3$ here, the remaining are similar and are not presented to due space limitations. Let $p = \overline{intern} \cdot (at\_r_{15} \ ? \ \overline{[t_{(R)}]}) \cdot buffer \cdot (at\_r_7 \ ? \ \overline{[t_{(R)}]}) \cdot out$.

1. $u = in \cdot (at\_s_3 \ ? \ [t_{(S)}]) \cdot p$
2. $at\_s_2 \wedge in = [q_1, \ldots, q_n, q]$
3. $at'\_s_3 \wedge in' = [q_1, \ldots, q_n] \wedge t'_{(S)} = q$

---

4. $u = [q_1, \ldots, q_n, q] \cdot p$
5. $u = in' \cdot [t'_{(S)}] \cdot p'$
6. $u = in' \cdot (at'\_s_3 \ ? \ [t'_{(S)}]) \cdot p'$

The transitions of the *Receiver*, which have to be considered, are $r_4 \to r_7$, $r_7 \to r_8$ (because at_r_7 changes value), $r_7 \to r_9$, $r_9 \to r_3$, $r_{10} \to r_{11}$, $r_{11} \to r_{12}$, $r_{12} \to r_4$, $r_{14} \to r_{15}$, $r_{15} \to r_{16}$, $r_{15} \to r_{17}$, and $r_{17} \to r_{13}$. All of them are fairly straightforward and are not presented here due to space restrictions.

6 Conclusion and Further Work

The main goal of this work was to prove the correctness of CoCheck. We have modeled the protocol in an easy way and have proved the most important safety property. The proof of this global property was fairly easy. Local safety properties (which are only valid at some points in the computation) are much more extensive and tool support would be desirable. Liveness requirements of CoCheck are not covered in this paper. It remains to be proven that, if some message arrives at *in*, then it will always be delivered to *out* at some time in the future.

References

1. T. E. Anderson, D. E. Culler, and D. A. Patterson. A case for NOW (networks of workstations). *IEEE Micro*, 15(1):54–64, February 1995.
2. R. M. Butler and E. L. Lusk. Monitors, messages and clusters: The p4 parallel programming system. *Parallel Computing*, 20(4):547–564, April 1994.
3. J. Casas, D. Clark, R. Konuru, S. Otto, R. Prouty, and J. Walpole. MPVM: A Migration Transparent Version of PVM. Technical Report, Dept. of Computer Science and Engineering, Oregon State Institute of Science and Technology, 20000 NW Walker Road, P.O.Box 91000, Portland, OR 97291-1000, 1994.
4. K. M. Chandy and L. Lamport. Distributed snapshots: Determining global states of distributed systems. *ACM Transactions on Computer Systems*, 3(1):63–75, February 1985.
5. L. Dikken. DynamicPVM: Task Migration in PVM. Technical Report, Shell Research, Amsterdam, November 1993.
6. A. Geist, A. Beguelin, J. Dongarra, W. Jiang, R. Manchek, and V. Sunderam. *PVM: Parallel Virtual Machine — A Users' Guide and Tutorial for Networked Parallel Computing*. Scientific and Engineering Computation. The MIT Press, Cambridge, MA, 1994.
7. W. Gropp, E. Lusk, and A. Skjellum. *Using MPI — Portable Parallel Programming with the Message-Passing Interface*. Scientific and Engineering Computation Series. The MIT Press, Cambridge, MA, 1994.
8. M. Litzkow, M. Livny, and M. Mutka. Condor — A Hunter of Idle Workstations. In *Proceedings of the 8th International Conference on Distributed Systems*, pages 104–111, Los Alamitos, CA, 1988. IEEE Computer Society Press.
9. Z. Manna and A. Pnueli. Completing the temporal picture. *Theoretical Computer Science*, 83:97–130, 1991.
10. Z. Manna and A. Pnueli. *The Temporal Logic of Reactive and Concurrent Systems: Specification*. Springer Verlag, 1991.
11. Z. Manna and A. Pnueli. *Temporal Verification of Reactive Systems: Safety*. Springer Verlag, 1995.
12. R. Schwarz and F. Mattern. Detecting causal relationships in distributed computations: In search of the holy grail. SFB-Bericht, Institut für Informatik, Universität Kaiserslautern, Postfach 3049, D-6750 Kaiserslautern, December 1992.
13. G. Stellner. CoCheck: Checkpointing and Process Migration for MPI. In *Proceedings of the IPPS'96*, pages 526–531, Honolulu, HI, April 1996. IEEE Computer Society Press, Los Alamitos.
14. G. Stellner, A. Bode, S. Lamberts, and T. Ludwig. NXLib — A parallel programming environment for workstation clusters. In C. Halatsis, D. Maritsas, G. Philokyprou, and S. Theodoridis, editors, *PARLE'94 Parallel Architectures and Languages Europe*, volume 817 of *Lecture Notes in Computer Science*, pages 745–748, Berlin, July 1994. Springer Verlag.
15. G. Stellner and J. Pruyne. Resource Management and Checkpointing for PVM. In *Proceedings of the 2nd European PVM Users' Group Meeting*, pages 131–136, Lyon, September 1995. Editions Hermes.

Molecular Dynamics Simulations on Cray Clusters Using the SCIDDLE-PVM Environment

Peter Arbenz[1], Martin Billeter[2], Peter Güntert[2],
Peter Luginbühl[2], Michela Taufer[1,3], and Urs von Matt[1]

[1] Institute of Scientific Computing, Swiss Federal Institute of Technology (ETH), 8092 Zürich, Switzerland, email: [arbenz,vonmatt]@inf.ethz.ch
[2] Institute of Molecular Biology and Biophysics, Swiss Federal Institute of Technology (ETH), 8093 Zürich, Switzerland, email: [billeter,guentert,lugi]@mol.biol.ethz.ch
[3] Università degli Studi di Padova, Dipartimento di Elettronica ed Informatica, Via Gradenigo 6/a, Padova, Italy, email: taufer@dei.unipd.it

1 Introduction

The computer simulation of the dynamical behavior of large molecules is an extremely time consuming task. Only with today's high-performance computers has it become possible to integrate over a realistic time frame of about a nanosecond for relatively small systems with hundreds or thousands of atoms. The simulation of the motion of very large atoms from biochemistry is still out of reach. Nevertheless, with every improvement of hardware or software computational chemists and biochemists are able to attack larger molecules.

In this paper we report on an effort to parallelize OPAL [6], a software package developed at the Institute of Molecular Biology and Biophysics at ETH Zürich to perform energy minimizations and molecular dynamics simulations of proteins and nucleic acids in vacuo and in water. OPAL uses classical mechanics, i.e., the Newtonian equations of motion, to compute the trajectories $\mathbf{r}_i(t)$ of n atoms as a function of time t. Newton's second law expresses the acceleration as

$$m_i \frac{d^2}{dt^2}\mathbf{r}_i(t) = \mathbf{F}_i(t), \tag{1}$$

where m_i denotes the mass of atom i. The force $\mathbf{F}_i(t)$ can be written as the negative gradient of the atomic interaction function V:

$$\mathbf{F}_i(t) = -\frac{\partial}{\partial \mathbf{r}_i(t)} V\big(\mathbf{r}_1(t), \ldots, \mathbf{r}_n(t)\big).$$

A typical function V has the form [10]

$$V(\mathbf{r}_1, \ldots, \mathbf{r}_n) = \sum_{\text{all bonds}} \frac{1}{2}K_b(b - b_0)^2 + \sum_{\text{all bond angles}} \frac{1}{2}K_\Theta(\theta - \theta_0)^2 +$$

$$\sum_{\text{improper dihedrals}} \frac{1}{2}K_\xi(\xi - \xi_0)^2 + \sum_{\text{dihedrals}} K_\varphi\big(1 + \cos(n\varphi - \delta)\big) +$$

$$\sum_{\text{all pairs } (i,j)} \Big(\frac{C_{12}(i,j)}{r_{ij}^{12}} - \frac{C_6(i,j)}{r_{ij}^6} + \frac{q_i q_j}{4\pi\epsilon_0\epsilon_r r_{ij}}\Big).$$

The first term models the covalent bond-stretching interaction along bond b. The value of b_0 denotes the minimum-energy bond length, and the force constant K_b depends on the particular type of bond. The second term represents the bond-angle bending (three-body) interaction. The (four-body) dihedral-angle interactions consist of two terms: a harmonic term for dihedral angles ξ that are not allowed to make transitions, e.g., dihedral angles within aromatic rings or dihedral angles to maintain chirality, and a sinusoidal term for the other dihedral angles φ, which may make 360° turns. The last term captures the non-bonded interactions over all pairs of atoms. It is composed of the van der Waals and the Coulomb interactions between atoms i and j with charges q_i and q_j at a distance r_{ij}.

In a numerical simulation the equations (1) are integrated in small time steps Δt, typically 1–10 fs for molecular systems. Therefore a realistic simulation of 1000 ps requires up to 10^6 integration time steps.

The last term of V, i.e., the sum over all pairs of atoms, consumes most of the computing time during a simulation. Van Gunsteren and Mark report [10] that in 1992 the evaluation of the interactions between the pairs of atoms of a 1000-atom system required about 1 s. Therefore at least 300 h were needed to execute 10^6 time steps. These order-of-magnitude estimates are still applicable today.

Fortunately, these calculations also offer a high degree of parallelism. In OPAL we evaluate the non-bonded interactions in parallel on p processors. Each processor receives the coordinates $\mathbf{r}_i(t)$ of all the atoms, and it computes a subset of the interactions. Thus we can execute $O(n^2)$ arithmetic operations on each server, whereas only $O(n)$ data items need to be communicated. Consequently we may expect a significant speedup for a large number n of atoms.

2 SCIDDLE

The SCIDDLE environment [2, 3, 9] supports the parallelization of an application according to the client-server paradigm. It provides asynchronous remote procedure calls (RPCs) as its only communication primitive. In SCIDDLE, an application is decomposed into a client process and an arbitrary number of server processes. Servers are special processes that are waiting to execute RPC requests from their client. Servers can also start other servers themselves. Thus the topology of a SCIDDLE application can be described by a tree structure.

The interface between client and server processes is described by a SCIDDLE interface definition (cf. Sect. 3). A compiler translates this interface definition into communication stubs for the client and server processes. Error checking is performed both at compile-time and at runtime.

SCIDDLE-PVM uses the PVM system to implement asynchronous RPCs. An application only needs to use PVM to start the server processes. No explicit message passing is necessary any more since all the communication is performed through SCIDDLE. Thus SCIDDLE applications benefit from the safety and ease of use of RPCs. They are also exceedingly portable as PVM becomes available on more and more platforms. In [1] it is shown that the overhead introduced by SCIDDLE is minimal and can be neglected for applications with large messages.

In recent years, PVM has become a de-facto standard for distributed applications. Its wide acceptance has lead numerous computer vendors to provide high-performance implementations. For the Intel Paragon and the IBM SP/2, say, PVM implementations exist that sit directly on top of the native message passing interface [5]. Cray Research offers a version of PVM tuned for the Cray J90 SuperCluster [8]. Applications based on SCIDDLE-PVM will be able to exploit these optimized PVM implementations.

3 Parallelization

SCIDDLE uses a simple declarative language to specify the RPCs exported by a server. In Fig. 1 we present a simplified version for the OPAL server that computes the non-bonded interactions (nbint).

```
INTERFACE OPAL_Server;

  CONST Max = 10000;

  TYPE  Matrix = ARRAY [3, Max] OF LONGREAL;

  PROCEDURE nbint (IN nval                   : INTEGER;
                   IN atcor [1:1:3, 1:1:nval] : Matrix;
                   OUT atfor [1:1:3, 1:1:nval]: Matrix): ASYNC;

  END
```

Fig. 1. Remote Interface Definition

Constants may be defined and used as symbolic array dimensions. User-defined types may be constructed from arrays and records. Procedures come in two flavors, synchronous and asynchronous. Each procedure parameter is tagged with a direction attribute. Parameters are always copied in the respective directions.

As SCIDDLE is designed in particular for parallel distributed numerical applications, it provides special array handling support for easy distribution of subarrays to multiple servers. A subarray can be selected by attributing array parameters with views. A view is, like an array section in MATLAB [7], a triple of the form [begin-index:stride:end-index] (cf. Fig. 1). The view components are either constants or the names of other integer parameters passed in the same call.

The SCIDDLE stub compiler translates a remote interface definition into the appropriate PVM communication primitives. In the example of Fig. 1 the procedure **nbint** is declared asynchronous. Thus the client stub provides the two subroutines

```
int invoke_nbint (int nval, Matrix atcor, int sid);
int claim_nbint  (Matrix atfor, int cid);
```

A call of **invoke_nbint** initiates an asynchronous RPC on the server **sid**. The input parameters are sent to the server, and a call identifier **cid** is returned as the function result. **invoke_nbint** returns as soon as the parameters are safely on their way to the server.

As soon as the client is ready to receive the results from an asynchronous RPC it calls the subroutine **claim_nbint**. The call identifier obtained from **invoke_nbint** is consumed, and the output parameters are retrieved.

If the procedure **nbint** had been declared synchronous the server stub would only provide the subroutine

```
int call_nbint (int nval, Matrix atcor, Matrix atfor, int sid);
```

A call of **call_nbint** blocks the client until the results have been returned from the server. Such a blocking RPC may be useful for quick calculations, but no parallelism can be obtained in this way.

The server process must implement the procedure

```
void nbint (int nval, Matrix atcor, Matrix atfor);
```

The server stub receives the input parameters from the client, calls **nbint**, and sends the results back to the client.

The SCIDDLE runtime system also offers additional subroutines for starting and terminating server processes. Multiple ongoing asynchronous RPCs can be managed conveniently by means of call groups [9].

4 Results

In this section we present the results obtained from a first parallel implementation of OPAL. Our test case consisted of an abbreviated simulation of the complex between the Antennapedia homeodomain from Drosophila and DNA immersed in 2714 water molecules [4]. We executed 20 time steps of 2 fs, resulting in a total simulation time of 40 fs. Pictures of these molecules can be found in Fig. 2.

We conducted our experiments on one and two nodes of the Cray J90 SuperCluster at ETH Zürich. Each node features a shared memory of 2 Gigabytes and 8 vector processors. One Cray CPU delivers a peak performance of 200 Megaflops. The nodes are connected by a fast HIPPI network with a bandwith of 100 Megabytes/sec.

Table 1 gives an overview of the execution times and speedups obtained from the parallel version of OPAL using various numbers of servers. The experiments with 6 and 8 servers were executed on two Cray J90 nodes, whereas all the

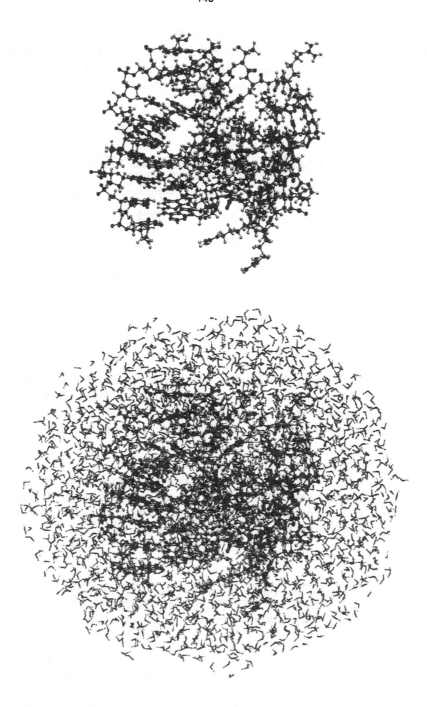

Fig. 2. Homeodomain and DNA without Water (top) and with Water (bottom)

other results were obtained on a single node. The column labelled "Execution Time" reports the wall clock time needed for a simulation run, including the time required for reading input files, pre-, and postprocessing. The speedups were computed as the ratio of the parallel execution time and the sequential execution time.

OPAL Version	Number of Servers	Execution Time	Speedup
sequential		432.6 sec	
parallel	1	547.4 sec	0.79
	2	302.9 sec	1.43
	3	241.3 sec	1.79
	4	174.6 sec	2.48
	6	146.8 sec	2.95
	8	144.0 sec	3.00

Table 1. Results on Cray J90 SuperCluster

The experiments on the Cray J90 were performed during the normal operation of the system. Therefore the execution times include overhead due to the operating system and the timesharing environment. If the experiments were run on a dedicated system we would obtain faster and better reproducible execution times.

We also ran the same benchmarks on a set of Silicon Graphics Indy workstations at ETH in Zürich. These workstations are connected by an Ethernet, and they were also used by other jobs during our benchmarks. However we performed our tests during times when the machines were only lightly loaded. Table 2 summarizes the results for this environment.

OPAL Version	Number of Servers	Execution Time	Speedup
sequential		4669 sec	
parallel	1	5171 sec	0.90
	2	2751 sec	1.70
	3	1842 sec	2.53
	4	1413 sec	3.30
	5	1240 sec	3.77
	6	972 sec	4.80
	7	882 sec	5.29

Table 2. Results on Network of Silicon Graphics Workstations

The sequential version of OPAL performs extremely well on a vector computer like the Cray. The most time-consuming calculations during the evaluation of the non-bonded interactions can be entirely vectorized. This explains the performance advantage of the Cray over the Silicon Graphics workstations.

The slowdown of the parallel version with one server compared to the sequential version can be explained by the overhead of the parallelization. This includes the time necessary to start the server processes as well as the communication time between UNIX processes. Some of this overhead only occurs during the initialization of the simulation, and thus is less significant for longer runs.

On the Cray the efficiency of the parallel version drops significantly if more than four servers are used. To a large extent this can be attributed to the timesharing operating system which does not offer gang scheduling for a set of PVM processes. But the Cray also loses some efficiency due to shorter vector lengths.

On the other hand the parallel implementation of OPAL scales very well on the network of workstations. This confirms that the computation is the main bottleneck of an OPAL simulation and not the communication between the client and its servers. We could expect an even higher performance if the server program were better tuned to the pipelined arithmetic and to the hierarchical memory structure of modern workstations.

The current version of OPAL uses a static load balancing scheme where each server receives tasks of equal size. This works well if all the processors in the network provide the same computing power. A more dynamic approach will be necessary if the machines are loaded very unevenly.

References

1. P. Arbenz, W. Gander, H. P. Lüthi, and U. von Matt: Sciddle 4.0, or, Remote Procedure Calls in PVM. In High-Performance Computing and Networking, Proceedings of the International Conference and Exhibition, ed. H. Liddell, A. Colbrook, B. Hertzberger and P. Sloot, Lecture Notes in Computer Science, Vol. 1067, Springer, Berlin, 1996, pp. 820–825.
2. P. Arbenz, H. P. Lüthi, J. E. Mertz, and W. Scott: Applied Distributed Supercomputing in Homogeneous Networks. International Journal of High Speed Computing, 4 (1992), pp. 87–108.
3. P. Arbenz, H. P. Lüthi, Ch. Sprenger, and S. Vogel: Sciddle: A Tool for Large Scale Distributed Computing. Concurrency: Practice and Experience, 7 (1995), pp. 121–146.
4. M. Billeter, P. Güntert, P. Luginbühl, and K. Wüthrich: Hydration and DNA recognition by homeodomains, Cell, 85 (1996), pp. 1057–1065.
5. H. Casanova, J. Dongarra, and W. Jiang: The Performance of PVM on Massively Parallel Processing Systems. Tech. Report CS-95-301, University of Tennessee, Computer Science Department, Knoxville, TN, August 1995, http://www.cs.utk.edu/~library/TechReports/1995/ut-cs-95-301.ps.Z.
6. P. Luginbühl, P. Güntert, M. Billeter, and K. Wüthrich: The new program OPAL for molecular dynamics simulations and energy refinements of biological macromolecules, J. Biomol. NMR, (1996), in press.

7. The MathWorks Inc., MATLAB, High-Performance Numeric Computation and Visualization Software, Natick, Massachusetts, 1992.

8. H. Poxon and L. Costello: Network PVM Performance. Cray Research Inc., Software Division, Eagan, MN, unpublished manuscript, June 1995.

9. Ch. Sprenger: User's Guide to Sciddle Version 3.0, Tech. Report 208, ETH Zürich, Computer Science Department, December 1993,
   `ftp://ftp.inf.ethz.ch/pub/publications/tech-reports/2xx/208.ps`.

10. W. F. van Gunsteren and A. E. Mark: On the interpretation of biochemical data by molecular dynamics computer simulation. Eur. J. Biochem., 204 (1992), pp. 947–961.

Parallelization of the SPAI Preconditioner in a Master-Slave Configuration

José M. Cela[1] and José M. Alfonso[1]

Centro Europeo de Paralelismo de Barcelona (CEPBA),
c/ Gran Capitán s/n, 08034 Barcelona, SPAIN,
(cela@ac.upc.es).

Abstract. In this paper we describe the parallelization of the SPAI preconditioner using a master-slave configuration in PVM. We propose three approaches to perform the communications in this algorithm, a centralized approach using the master process, a distributed approach based in a pooling mechanism, and a distributed approach based in a signaling mechanism.

1 Introduction

At the present the solution of large sparse linear systems is one of the most important numerical kernels in engineering simulations. Iterative methods are extensively used as solvers for such linear systems. However, iterative methods use to need some preconditioner to accelerate the convergence. The usual general purpose preconditioners were incomplete factorizations like ILUt [3]. Incomplete factorizations have a main drawback, they are sequential bottlenecks when the linear solver is parallelized. This bottleneck is not only in the factorization process which build the preconditioner, also the application of the preconditioner at each iteration is a sequential bottleneck. This is due to the fact that the application of these preconditioners requires the solution of triangular sparse systems which are a well known sequential bottleneck.

SPAI preconditioner was defined recently in [1] [2]. This preconditioner is based on the computation of a sparse matrix which is a roughly approximation of the inverse matrix in the linear system. SPAI is a general purpose preconditioner. It has a main advantage in front of incomplete factorizations, it produces directly an approximation of the inverse matrix, then only the matrix times vector multiplication is required when the preconditioner is applied at each iteration. Matrix times vector multiplication is a high parallel operation, so we expect a dramatic increase in the speed-up of the iterative method. Of course, this is not enough, for a complete application the main objective is to reduce the total time spent, a trade off between the number of iterations and the time per iteration should be considered for each specific problem in order to select the proper preconditioner. In this paper we are focused only in the parallelization of the algorithm to build the SPAI preconditioner.

The remainder of the paper is organized as follows, in section §2 we briefly review the sequential algorithm to compute the SPAI preconditioner. In section §3 we describe the three parallel approaches to parallelize this algorithm. Finally, in section §4 we present some experimental results about the performance of the different approaches.

2 SPAI algorithm

The SPAI preconditioner, M matrix in the equations, is computed solving the following minimization problem:

$$\|AM - I\|_F^2 < \epsilon_1 \qquad (1)$$

where ϵ_1 is a tolerance parameter, A is a $N \times N$ non singular matrix, and $\|\cdot\|_F$ is the Frobenius norm. Rewriting (1) using the definition of the Frobenius norm, we obtain

$$\|AM - I\|_F^2 = \sum_{k=1}^{N} \|Am_k - e_k\|_2^2 \qquad (2)$$

where e_k is the k-th vector of the canonical basis and m_k is the k-th column of M. This leads to inherent parallelism, because each column m_k of M can be compute independently solving the following problem,

$$\|Am_k - e_k\|_2^2 < \epsilon \qquad k = 1, \ldots, N \qquad (3)$$

where ϵ is a parameter given by the user. In order to control the memory spent by the algorithm, the user provides a second parameter, the maximum fill in allowed in each column m_k, we call this parameter `maxFill`. Then, the computation of m_k stops when ϵ is reached or the `maxFill` in is reached.

The difficulty lies in determining a good sparsity structure in m_k which captures the significative entries of the k-th column in A^{-1}. At the starting phase of the algorithm a simple structure for m_k is supposed, only $m_k(k) \neq 0$. In the next steps there is an automatic criteria to select new indices to from m_k.

Let be $\mathcal{J} = \{j \mid m_k(j) \neq 0, j = 1, \ldots, N\}$, and let be $\mathcal{I} = \{i \mid A(i,j) \neq 0, \forall j \in \mathcal{J}, i = 1, \ldots, N\}$. We define the matrix $\hat{A} = A(\mathcal{I}, \mathcal{J})$ and the vector $\hat{e}_k = e_k(\mathcal{I})$. Then, the following least squares problem must be solved to compute m_k,

$$min \left\| \hat{A}\hat{m}_k - \hat{e}_k \right\|_2^2 \qquad (4)$$

Thus, the QR decomposition of $\hat{A}$ must be computed to obtain $\hat{m}_k$. If the stop criteria is not reached with the computed $\hat{m}_k$, the set $\mathcal{J}$ must be extended to compute a new $\hat{m}_k$ vector. We must select indices that will lead to the most profitable reduction in $\|r\|_2 = \|A(\cdot, \mathcal{J})\hat{m}_k - e_k\|_2$. This is done selecting the set $\mathcal{L} = \{l \ / \ r(l) \neq 0, \ l = 1, \ldots, N\}$. For each index l there is a set of indices $\mathcal{N}_l = \{n \ / \ A(l,n) \neq 0, \ n \notin \mathcal{J}, \ n = 1, \ldots, N\}$. The potential new candidates which might be added to $\mathcal{J}$ are

$$\mathcal{G} = \bigcup_{l \in \mathcal{L}} \mathcal{N}_l$$

Then, for each $g \in \mathcal{G}$ the following coefficient is computed

$$\rho_g^2 = \|r\|_2^2 - \frac{(r^T A e_g)^2}{\|A e_g\|_2^2} \tag{5}$$

and the indices g which smallest ρ_g are selected, for more details see [1] [2]. Finally, the SPAI algorithm can be described as follows:

SPAI ALGORITHM:
```
    Do k = 1, ..., N              /* Loop in columns */
        Do While ||r||₂ > ε and maxFill in not reached
            1. Select the indices to be added to 𝒥
            2. Update the QR factorization of Â
            3. Compute m̂ₖ from (4)
            4. Compute ||r||₂
            5. Define the set 𝒢
            6. Compute ρ²_g from (5)
        EndWhile
    EndDo
```

3   Parallel Implementations

Several approaches are possible to parallelize the SPAI algorithm. We have selected a master-slave configuration, because our final purpose is to include an iterative method (GMRES(m) [4]) using the SPAI preconditioner in a sequential industrial application, which is used to analyze contaminant ground diffusion problems. This is a quite complex simulation package which solves several non linear PDEs. The 70% of the time in the sequential version is consumed by the linear system solver which is a LU decomposition. In a first phase only the linear solvers are going to be parallelized, in order to minimize the cost of the migration to the parallel version. A master-slave configuration is a suitable model for this kind of parallel code.

The master process generates the A matrix and the right hand side vector. Then, the master process distributes the A matrix and the rhs vector between the slaves processes. Each slave process computes a set of columns of the SPAI preconditioner and then the iterative method is started between the slaves. The parallelization of the iterative method is quite trivial, taking into account that the A matrix is distributed by columns between the slaves in the same way that the M matrix. The main difficulty to develop the parallel code is the parallelization of the SPAI algorithm.

To compute a column of M some communications between slaves are needed in order to read new columns/rows of the matrix. This is the case in the points 1. and 5. of the SPAI algorithm. The problem is that the pattern of these communications can not be defined a priori, it is only known in run time, and it is quite random.

In our parallelization approach there will be three different kinds of messages, each one of this messages will be distinguish by a flag:

- Data Request Message (DRM): A message with this flag contains the indices of the columns/rows needed by some slave process.
- Data Answer Message (DAM): A message with this flag contains the values of columns/rows previously requested by some slave process.
- End of Work Message (EWM): A message with this flag simply notifies the finalization of a specific work.

Multiple send/receive buffers are managed. We define two different buffers for each process. The first buffer is used to send/receive DRMs and EWMs. The second buffer is used to send/receive DAMs.

Two communication schemes can be used to parallelize SPAI algorithm:

1. Centralized: In this scheme the master process keeps the complete A matrix. A slave sends a DRM each time it needs a new set of columns/rows of A. The master answers this request sending a proper DAM.

2. Distributed: In this scheme the A matrix is completely distributed between the slaves processes. Then, all DRMs and DAMs are between slaves processes. We have three options to perform a distributed communication scheme.
   (a) Pooling: Each slave answers pending DRMs only when itself sends a DRM.
   (b) Signaling: Each slave answers pending DRMs when a user defined signal arrives.
   (c) Timing: Each slave answers pending DRMs when an alarm signal arrives.

We have considered only the options 1, 2.a and 2.b, in the following subsections we describe in detail how to perform all this options. We have omitted some details, as packing/unpacking routines, to clarify the algorithms.

3.1 Centralized Scheme

In this scheme the master answer all the DRM sent by the slaves. The main drawback of this scheme is that the master becomes a bottleneck. Then, when the number of slaves grows the speed-up is limited. The algorithm executed by the master and the slaves are showed in the following:

MASTER ALGORITHM:
```
    Tell to each slave the columns of M to be computed by it
    Do While (TRUE)
        Do p= 1, nb_slaves
            pvm_nrec(p, flag)          /* Non blocking receive */
            If (flag = DRM) pvm_send(p, DAM)
            If (flag = EWM) Mark this work as done
        EndDo
        If (all the work is completed) Break
    EndWhile
    pvm_barrier
```

SLAVE ALGORITHM:
```
    Do for all the columns given by the master
        Do While ||r||₂ > ϵ and maxFill not reached
            1. Select the indices to be added to J
            1.1. pvm_send(master, DRM)
            1.2. pvm_recv(master, DAM)     /* Blocking receive */
            2. Update the QR factorization of Â
            3. Compute m̂ₖ from (4)
            4. Compute ||r||₂
            5. Define the set G
            5.1. pvm_send(master, DRM)
            5.2. pvm_recv(master, DAM)     /* Blocking receive */
            6. Compute ρ²_g from (5)
        EndWhile
    EndDo
    pvm_send(master, EWM)
    pvm_barrier
    Start the iterative method
```

3.2 Distributed Scheme

In this scheme the master only distributes the matrix A between the slaves. After this phase all the communications are between slaves. Both the pooling and the signaling schemes have the same algorithm executed by the master and the slaves. The algorithm executed by the slaves is the following:

SLAVE ALGORITHM:
    Do for all the columns to be computed
        Do While $\|r\|_2 > \epsilon$ and maxFill not reached
            1. Select the indices to be added to $\mathcal{J}$
            Do for all the data that I need
                REQUEST to the process "proc" which stores this data
            EndDo
            2. Update the QR factorization of $\hat{A}$
            3. Compute $\hat{m}_k$ from (4)
            4. Compute $\|r\|_2$
            5. Define the set $\mathcal{G}$
            Do for all the data that I need
                REQUEST to the process "proc" which stores this data
            EndDo
            6. Compute ρ_g^2 from (5)
        EndWhile
    EndDo
    Do p= 1, nb_slaves and (p $\neq$ mypid)
        pvm_send(p, EWM)
    EndDo
    END_OF_WORK routine
    Do While (TRUE)
        Do p= 1, nb_slaves and (p $\neq$ mypid)
            pvm_nrec(p, flag)   /* Non blocking receive */
            If (flag = DRM) pvm_send(proc, DAM)
            If (flag = EWM) Mark this work as done.
        EndDo
        If (all the work is done) Break
    EndWhile
    Start the iterative method

The differences are in the routines REQUEST and END_OF_WORK. If a pooling mechanism is used the routine END_OF_WORK makes nothing. The routine REQUEST is the following in the pooling scheme:

REQUEST ROUTINE TO PROCESS "PROC":
    pvm_send(proc, DRM)
    Do While (TRUE)
        Do p= 1, nb_slaves and (p $\neq$ proc)
            pvm_nrec(p, flag)   /* Non blocking receive */
            If (flag = DRM) pvm_send(p, DAM)
            If (flag = EWM) Mark this work as done
        EndDo
        pvm_nrec(proc, flag)   /* Non blocking receive */
        If (flag = DRM) pvm_send(proc, DAM)
         If (flag = DAM) Take out data from buffer and Break

```
      If (flag = EWM) Mark this work as done
EndDo
```

However, if a signaling mechanism is used the routines REQUEST and END_OF_WORK are the following:

REQUEST ROUTINE:
```
    pvm_send(proc, DRM)
    pvm_sendsig(proc, SIGUSR1)
    pvm_recv(proc, DAM)        /* Blocking receive */
```
END_OF_WORK ROUTINE:
```
    system call signal(SIGUSR1, Ignore)
```

Moreover, in the signaling scheme an additional routine must be defined. This routine is executed when SIGUSR1 signal arrives.

SIGNAL ATTENTION ROUTINE:
```
    system call signal(SIGUSR1, Ignore)
    Do p= 1, nb_slaves and (p ≠ mypid)
        pvm_nrec( p, flag ) /* Non blocking receive */
        If (flag = DRM) pvm_send(p, DAM);
        If (flag = EWM) Mark this work as done
    EndDo
    system call signal(SIGUSR1, Signal Attention routine)
```

The main drawback of the pooling mechanism is the lack of flexibility in the answers to the DRMs. This increases the idle times until receive a DAM. The signaling mechanism gives the maximum flexibility, however it must to pay an additional cost of two system calls each time that the signal attention routine is activated. Note that the Signal Attention routine ignores the SIGUSR1 when it is working, then if a process sends a DRM and the attention routines does not receive it, the request will be answer the next time that the Signal Attention routine is started. The implementation of the signaling mechanism is more sensible to bugs in the implementation of PVM, because we suppose that the implementation of PVM guarantees a safe execution of send/receive if a signal arrives when they are been running, i.e., the EINTR error is considered inside these routines.

4   Experimental results

A priori is difficult to know what will be the optimum scheme in terms of time. Clearly, a distributed signaling scheme is more flexible, so we expect to obtain the maximum speed-up with it. It is expected that the second scheme in terms of speed-up will be the Distributed pooling scheme. Finally we expect that the centralized communication scheme works only for a few number of slave processes.

In figure 1 we show the speed-up as a function of the number of slaves processes. This figure confirm us that the distributed signaling scheme is clearly the best one, and that the centralized scheme has a bottleneck when the number of slaves grows upon 16. We have observed that the quantitative difference between the signaling and the pooling schemes can variate significantly depending on the problem. Although, we have never observe a problem where the signaling scheme is worse than the pooling scheme. Moreover, we have observed that the centralized scheme can be superior to the pooling scheme for small number of processes.

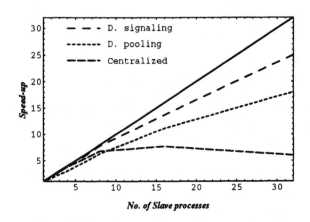

Fig. 1. Speed-up of the three schemes

References

1. Grote, M. and Simon, H.: Effective Parallel Preconditioning with Sparse Approximate Inverses. Proceedings of the 7th SIAM Conference on Parallel Processing for Scientific Computing, (1995), pp. 466-471.
2. Grote, M. and Simon, H.: A new approach to parallel preconditioning with Sparse Approximate Inverses. Tech. report SCCM-94-03, Department of Sci. Comput. and Comput. Math., Stanford University (1994).
3. Saad, Y: ILUT: a dual Threshold Incomplete LU factorization. Numerical Linear Algebra with applications, Vol. 1, No. 4 (1994), pp. 387-402.
4. Saad, Y., Sultz, M. H.: GMRES: A generalized Minimal Residual Algorithm for Solving Non symmetric Linear Systems. SIAM J. Sci. Stat. Comput., Vol 7, No. 3 (1986).

Parallel Lanczos Algorithm on a CRAY-T3D Combining PVM and SHMEM Routines

Josef Schüle*

Institute for Scientific Computing, P.O. Box 3329, 38023 Braunschweig, Germany

1 Introduction

The metal insulator transition (MIT) of cubic tungsten bronzes Na_xWO_3 has been studied in detail as a function of composition x and temperature [6]. One theoretical approach to address the question of the origin of the MIT in the Na_xWO_3 bronzes incorporating electron interaction and correlated disorder is a two-dimensional Anderson-Mott-Hubbard (AMH) model [7]. The resulting Hamiltonian matrix H has to be solved in an unrestricted Hartree-Fock (UHF) approach, t.m. independently for electrons in the two spin states α and β. At each step of an UHF iteration the matrix H has to be diagonalized. Only conduction band (CB) states are taken into account, t.m. approximately 3% of all eigenstates have to be calculated. The eigenvectors of CB states determine the electron distribution and introduce a new matrix H until self consistency is reached. Algorithmic details are given in section 2.

To exclude resp. investigate local order and disorder effects several stochastic distributions of sodium atoms have to be investigated for each composition x. This implies a master - slave type programming paradigm. The master generates random numbers to represent the distribution of sodium atoms. He collects and writes final results.

The UHF program part implies a master - slave, resp. a group leader - group member, structure as well (s. figure 1), where the leader executes non parallelizable code, prepares results and sends them to the master. It should be mentioned, that UHF calculations to different sodium distributions widely may differ in the number of iterations required to achieve self consistency. Therefore the overall program is designed to allow dynamic load balancing in the sense, that as well the master PE as group members of different groups may join and change group membership to assist the slowest group. This requires dynamic group management like PVM [8] does provide. But, as will be pointed out in section 3, exclusive usage of PVM does not lead to a speedup, but a speeddown.

SHMEM routines, very fast point-to-point communication routines supplied by CRAY [1], are necessary to achieve speedup but they provide not the flexibility in group management as needed. Velocity and flexibility is achievable combining SHMEM routines and PVM, as is described in section 4.

A global numbering of tungsten and oxygen atoms simplifies the calculation of the electronic density, allows an equal data distribution onto PEs, but

* j.schuele@tu-bs.de, phone: +49-531-391-5542

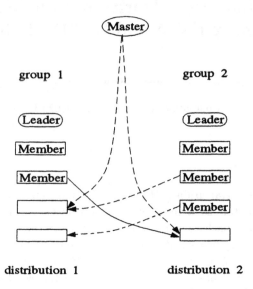

Fig. 1. Guided by a master each group, consistent of a leader and several members, calculates a stochastic sodium distribution. Master and group members may change group membership to assist an other group in an UHF cycle as indicated.

shows to be very communication intensive. Replacing it with a lattice orientated numbering reduces communication costs and runtime as discussed in section 5.

Finally, a conclusion is given in section 6.

2 Algorithm

The diagonalisation at each step of an UHF iteration is performed via the Lanczos algorithm introduced by Cullum and Willoughby [3, 4]. It avoids any reorthogonalisation of the Krylov sequence. Appearing spurious eigenvalues of the tridiagonal matrix T are identified and withdrawn. This procedure allows the calculation of parts of the spectrum characterized by an upper (end of the CB) and a lower (end of the valence band) boundary. Very close lying eigenstates, typical for this kind of applications, do not cause any problems.

Due to rounding errors in limited precision arithmetic, the initial dimension of the tridiagonalmatrix T has to be in the order of the original matrix H. Based on Sturm sequencing an appropriate size for the Lanczos matrix in the Ritz eigenvector computation is guessed. This size proves to be appropriate as upper boundary for the size of T in subsequent UHF iterations. Using these sizes reduces the computing time dramatically.

An outline of the used Lanczos algorithm is given in figure 2 together with a rough operation count estimation. Thereby is N the dimension of the Hamiltonian H which is in production code in the order of 10^4. M is the dimension of the tridiagonal matrix T and varies from $3N$ to cN where c lies in the range $[0.1, 0.5]$ depending on the varying size of T mentioned above. L is the number of eigenvectors required. Depending on the composition x it is in the order of $N/3 \times 0.1$. To avoid usage of secondary storage the Krylov sequence is calculated twice - in the tridiagonalisation as well as in the solution step.

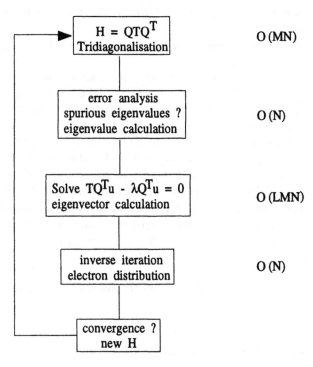

Fig. 2. Outline of an UHF cycle using the Lanczos variant proposed by Cullum and Willoughby. N is the order of H, M may vary each iteration. L is the number of eigenvectors required.

The tridiagonalisation consists of a recursive construction of a Krylov sequence as outlined in by the following scheme:

for i=1,...
$$v = u$$
$$u = r/\beta$$
$$r = Hu$$
$$\alpha = <u, r>$$
$$r = r - \alpha u - \beta v$$
$$\beta = \|r\|$$
end

This scheme may easily be modified to reduce the number of synchronisation points[5] at the cost of N additional flops.

As it is, this scheme requires $\mathcal{O}(2log_2 PE)$ startups for the reductions. Depending on the pattern of H, the matrix vector multiplication Hu causes considerable communication costs. This is discussed in more detail in section 5.

Error analysis and withdrawal of spurious eigenvalues is not parallelized. They consume only a small amount of computing time and almost a dozen different routines are involved. Each PE determines different eigenvalues bisecting disjunct intervals.

There are two straightforward ways to parallelize the solution step

- Each processing element (PE) may calculate a set of eigenvectors independently. This introduces replicated calculation of the Krylov sequence on each PE but requires only a final synchronisation to collect the contributions of the eigenvectors to the electron distribution. This method introduces different rounding errors and therefore different Krylov sequences in the tridiagonalisation and the eigenvector calculation step. Results show to be dependent on the numbers of PEs employed and are therefore not usable.
- It may be parallelized in a similar way as the tridiagonalisation. This requires less synchronisation than the tridiagonalisation, because α and β are reusable, though synchronisation is needed to calculate for each eigenvector the corresponding electron distribution. This variant produces reliable results independent of the number of PEs used.

Inverse iteration and final calculation of the electron distribution are overall parallelized.

3 Exclusive Usage of PVM

The CRAY T3D system is tightly coupled to the moderately parallel CRAY Y-MP/C90 vector systems [9]. This frontend typically uses time-sharing in a multi user environment, while T3D PEs are dedicated to one user.

Mapping the dynamic group model depicted in figure 1 on a massively parallel computer system like the T3D with a frontend, it appears natural to locate the master on the frontend and the groups on the T3D. This partitioning has some severe drawbacks:

1. No group functions between the T3D and the frontend are available. So the master is not able to broadcast to a whole group. Instead he has to send a message to the group leader who has to broadcast it to the rest of the group.
2. PEs on the T3D are dedicated while the frontend uses time-sharing. This leads to PE stalling while waiting to get some cycles on the frontend. Furthermore it may cause buffer overflow on the frontend if several groups are reporting results while the master is still kept waiting by the operating system.

While a solution to point 1 is straightforward, point 2 prevents significant speedups and may even cause fatal runtime errors.

Using only the T3D increases performance. Even with one PE exclusively used as master, t.m. 3 PEs in a group doing work, the T3D-only version is as fast as the T3D-YMP combination with 4 PEs in the group.

PVM is known to introduce a lot of overhead due to its way to handle messages. Indeed, introducing parallelism in the tridiagonalisation step employing 16 PEs increases run time by 40%. This is due to the fine grain parallelism in this part of the algorithm and the high communication costs for two reductions and the matrix vector multiplication Hu.

4 Combining PVM with SHMEM Routines

CRAY offers native communcation software for message passing, the SHMEM routines [1]. These routines provide simple one sided point-to-point communication (put, get) with low latencies ($\approx 2\mu s$) and high bandwidths (put: 120 MB/sec, get 60 MB/sec)[2] and group routines like reduction, broadcast and barriers[2]. Drawbacks of the routines are

- only data in static memory can be used for message passing.
- groups have to be specified as an active set.
- group routines need auxiliary arrays which have to be initialized.

An active set consist of
$$[pe\_root, log\_stride, number]$$
where pe_root is the lowest PE number in the group, log_stride is the stride ($\log_2$ based) between PEs, and number is the number of PEs involved. For example the set [7,1,3] consists of the PEs [7,9,11]. Thus, only simple regular patterns in the construction of groups are possible and flexibility is very restricted.

The need of auxiliary arrays in collective group functions causes situations, where a barrier is forced, there at least all group members do participate, before a shmem_barrier for this group is usable. Once initialized these arrays may be reused for the same group operation as long as group context does not change. This does not allow flexible dynamic group handling as needed.

Combining SHMEM routines with routines of a message passing system like PVM is feasible and overcomes drawbacks of both[3]. PVM simply allows dynamic group handling, while SHMEM routines provide low latencies and high bandwidth. Furthermore, exchanging PVM calls against SHMEM routines in hot spots only, reduces development costs.

Introducing SHMEM routines in the tridiagonalisation and the eigenvector calculation reduces runtime by 70%. The speedup[4] calculating a moderate sized problem using one static group is 2.5. The difference to the optimal value of 4.0

---

[2] There are more SHMEM routines, but they are not used in this context.

[3] MPI becomes an alternative, but was not available at the beginning of this project.

[4] 4 to 16 PEs. A single PE version of the program does not exist.

arises as well from the communication costs due to the Hu multiplication, as from serial parts as from load inbalance between group leader and members. The decomposition of matrix H has great impact on communication costs as will be discussed in section 5.

The dynamic group concept using 4 PEs proves to be effective. It reduces runtime by 18%, though changing group membership does imply a significant overhead. Data has to be sent to the new member, some indizes have to be recalculated by all members and three new synchronisations are necessary. This overhead counts for the difference to the expected 25% reduction. Using 8 PEs the overhead equals the winst and no runtime reduction is observed.

As expected, the algorithm scales with the number of distributions of sodium atoms considered concurrently. One distribution on 4 PEs takes the same time as two distributions on 8 PEs as well with as without dynamic groups.

5 Optimizing Matrix Layout

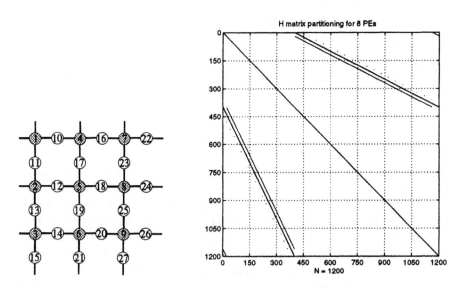

Fig. 3. Original numbering scheme and resulting structure of the H matrix and its partitioning on 8 PEs. Filled circles represent W atoms, empty circles O atoms.

One matrix layout is based on the numbering given in figure 3 together with a resulting partitioning of matrix H using 8 PEs. All W atoms are numbered first which allows simple access to the electronic density in each UHF cycle without index calculation. Furthermore it provides a balanced data distribution dividing N by the number of PEs. It shows strongly different communication patterns per PE varying from data exchange (receive and send) with 4 PEs as for PE0, to an exchange with only 1 PE as for PE3. This introduces communication inbalance, latencies and substantial data traffic. The communication pattern does depend on the number of PEs used and prevents predictable speedup.

Changing to the numbering depicted in figure 4 introduces a higher degree of data locality. This scheme preserves lattice structure on each PE. Data distribution is no longer balanced, as indicated. Each PE now has to exchange date with two PEs. The amount of data to be exchanged is $2\sqrt{N/3}$ words, equally for all PEs, and independent of the number of PEs involved.

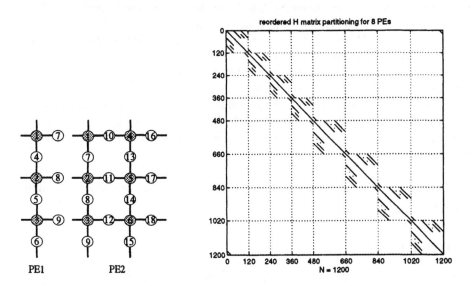

Fig. 4. In this work introduced numbering for 2 PEs, $N = 18$. Partitioning is parallel to y-axis. Filled circles represent W atoms, empty circles O atoms. On the right the resulting H matrix and its decomposition on 8 PEs is indicated.

Introducing this new numbering reduces communication costs and increases efficiency dramatically. Runtime is reduced to 70%, speedup and scalability are increased.

6 Conclusion

In this paper a parallel Lanczos algorithm to calculate the metal insulator transition of cubic tungsten bronzes Na_xWO_3 is presented. The bronzes are theoretically treated with an Anderson-Mott-Hubbard model resulting in a Hamiltonian matrix H that has to be solved in an unrestricted Hartree-Fock approach for different stochastic distributions x.

Two different approaches are employed to introduce parallelism. Coarse grain parallelisation is used to calculate different sodium distributions concurrently. This approach favors a master-slave paradigm where dynamic group membership, as depicted in figure 1, is necessary to reduce load inbalance between groups requiering different numbers of UHF iterations.

Fine grain parallelism is introduced within each UHF cycle, mainly in the tridiagonalisation and eigenvector calculation. In both parts a Krylov sequence is set up. Due to rounding errors in finite precision arithmetic, it is of great importance to use the same parallelisation strategies for both the tridiagonalisation as well as the eigenvector calculation.

Fine grain parallelism using PVM is not feasible. PVM introduced overhead increases runtime instead. Using SHMEM routines solely is not possible, either, because groups can not be set up flexible enough.

Combining PVM and SHMEM routines is possible, provides flexibility and reduces software development costs.

Still proper data distribution is essential to achieve effectivity. Even on the very fast T3D network an adequate numbering scheme reduces communication costs dramatically. An overall runtime reduction to 70% is achieved changing from the W-first numbering (figure 3) to the lattice orientated numbering shown in figure 4. At the same time it increases effectivity, speedup and scalability of the overall program.

Thanks to Hartmut Dücker for the PVM version and valuable hints and explanations.

References

1. R. Barriuso and A. Knies. Shmem user's guide for fortran. Cray Research, Inc., 1994.
2. K. Cameron, L. J. Clarke, and A. G. Smith. Cri/epcc mpi for cray t3d. http::/www.epcc.ed.ac.uk/t3dmpi/Product/Docs/mpi_t3d.ps.Z, 1995.
3. J. K. Cullum and R. K. Willoughby. *Lanczos Algorithms for Large Symmetric Eigenvalue Problems*. Birkhäuser, Vol. I Theory, Boston-Basel-Stuttgart, 1985.
4. J. K. Cullum and R. K. Willoughby. *Lanczos Algorithms for Large Symmetric Eigenvalue Problems*. Birkhäuser, Vol. II Programs, Boston-Basel-Stuttgart, 1985.
5. J. W. Demmel, M. T. Heath, and H. A. van der Vorst. Parallel numerical linear algebra. Technical report, LAPACK Working Note 60, UT CS-93-192, 1993.
6. J. P. Doumerc, M. Pouchard, P. Hagenmuller. In P. P. Edwards, and C. N. Rao, editors. *The Metallic and Nonmetallic States of Matter*, London, 1985. Taylor and Francis.
7. H. Dücker, Th. Koslowski, W. von Niessen, M. A. Tusch, and D.E. Logan. The metal-insulator transition in disordered tungsten bronzes - results of an anderson-mott-hubbard model. Accepted.
8. A. Geist, A. Beguelin, J. Dongarra, W. Jiang, R. Manchek, and V. Sunderam. *PVM: Parallel Virtual Machine - A Users' Guide and Tutorial for Networked Parallel Computing*. The MIT Press, Cambridge, Massachusetts, 1994.
9. W. Oed. The cray research massively parallel processor system cray t3d. ftp.cray.com:product_info/mpp, 1993.

PVM-Implementation of Sparse Approximate Inverse Preconditioners for Solving Large Sparse Linear Equations

Thomas Huckle

TU München, Institut für Informatik, Arcisstr. 21, 80290 München, Germany

1. Sparse Approximate Inverses and Linear Equations

We want to solve a system of linear equations $Ax = b$ in a parallel environment. Here, the $n \times n$-matrix A is large, sparse, unstructured, nonsymmetric, and ill-conditioned. The solution method should be robust, easy to parallelize, and applicable as a black box solver.

Direct solution methods like the Gaussian Elimination are not very effective in a parallel environment. This is caused by the sequential nature of the computation and solution of a triangular factorization $A = LR$ with lower and upper triangular matrices L and R. Therefore, we prefer an iterative solution method like GMRES, BiCGSTAB, or QMR (see [1]). In an iterative scheme we compute

$$x_0(= 0) \rightarrow x_1 \rightarrow x_2 \rightarrow ... \rightarrow \bar{x} = A^{-1}b \,,$$

using only matrix-vector multiplications and inner products of vectors in every iteration step. To display a very simple example, we can consider the splitting $A = M - K$, $Mx = Kx + b$, to derive the iteration

$$x_{j+1} = M^{-1}Kx_j + M^{-1}b \,.$$

This algorithm leads to convergent iterates if the spectral radius $\rho(M^{-1}K) < 1$.

For many important iterative methods the convergence depends heavily on the position of the eigenvalues of A. Therefore, the original system $Ax = b$ is often replaced by an equivalent system $MAx = Mb$ or the system $AMz = b$, $x = Mz$. Here, the matrix M is called a preconditioner and has to satisfy a few conditions:

- AM (or MA) should have a 'clustered' spectrum,
- M should be fast to compute in parallel,
- $M \times$Vektor should be fast to compute in parallel.

Often used preconditioners are

(i) Block-Jacobi-prec..: $M = inv(\text{block-diag}(A))$, is easy to parallelize, but in general with unsatisfactory convergence;

(ii) Polynom-prec..: $M = pol(A)$, with the same properties as (i);

(iii) Incomplete LU-decomposition of A, with $A \approx LU$ and defining $M = (LU)^{-1}$; but like Gaussian Elimination this is not very effective in a parallel environment.

A very promising approach is the choice of sparse approximate inverses for preconditioning, $M \approx A^{-1}$ and M sparse [8,3,2,6,5,7]. Then, in the basic iterative scheme only matrix-vector multiplications with M appear and it is not necessary to solve a linear system in M like in the incomplete LU-approach. Obviously, A^{-1} is a full matrix in general, and hence not for every sparse matrix A there will exist a good sparse approximate inverse matrix M.

We can compute such a matrix M by solving a minimization problem of the form $min\|AM - I\|$ for a given sparsity pattern for M. By choosing the Frobenius norm we arrive at an analytical problem that is very easy to solve. Furthermore, in view of

$$\min \|AM - I\|_F^2 = \sum_{k=1}^{n} \min \|AM_k - e_k\|^2$$

this minimization problem can be solved columnwise and is therefore embarrassingly parallel.

First we consider M with a prescribed sparsity pattern, e.g. $M = 0$, M a diagonal matrix, or M with the same sparsity pattern as A or A^T. We get the columnwise minimization problems $\|AM_k - e_k\|_2$, $k = 1, 2, ..., n$, with a prescribed sparsity pattern for the column vector M_k. Let us denote with J the index set of allowed entries in M_k, and the reduced vector of the nonzero entries by $\hat{M}_k := M_k(J)$. The corresponding submatrix of A is $A(:, J)$, and most of the rows of $A(:, J)$ will be zero in view of the sparsity of A. Let us denote the row indices of nonzero rows of $A(:, J)$ by I, and the corresponding submatrix by $\hat{A} = A(I, J)$, and the corresponding reduced vector by $\hat{e}_k = e_k(I)$. Hence, for the k-th column of M we have to solve the small least squares problem

$$\min \|\hat{A}\hat{M}_k - \hat{e}_k\| .$$

Mainly, there are three different approaches for solving this LS-problem. We can compute a QR-decomposition of $\hat{A}$ based on

(1) Householder matrices or
(2) the Gram Schmidt process,   or we can solve
(3) the normal equations $\hat{A}^T \hat{A} \hat{M}_k = \hat{A}^T \hat{e}_k$ iteratively using the preconditioned conjugate gradient algorithm.

For the general case it is not possible to prescribe a promising sparsity pattern without causing J and I to be very large. This would result in large LS-problems and a very expensive algorithm. Therefore, for a given index set J with optimal solution $M_k(J)$ we need a dynamic procedure to find new promising indices that should be added to J. Then, we have to update I and solve the enlarged LS-problem until the residual $r_k = AM_k - e_k$ is small enough or J gets too large.

In general the start sparsity pattern should be $J = \emptyset$. Only for matrices with nonzero diagonal entries we can set $J = \{k\}$ in the beginning for M_k.

We use a hierarchy of three different criteria for finding new promising indices for J and M_k. As a global a priori criterion for M we only allow indices that appear in $(A^T A)^m A^T$ for a given m, e.g. $m = 0, 1, 2, 3$, or 4. As a heuristic justification let us consider the equation

$$A^{-1} = (A^T A)^{-m-1}(A^T A)^m A^T ,$$

where the diagonal entries of $(A^T A)^{-m-1}$ are nonzero (moreover the maximal element of this matrix ia a diagonal entry), and therefore the sparsity pattern of $(A^T A)^m A^T$ is contained in the sparsity pattern of A^{-1} (Here we neglect possible cancellation). Similarly, by considering the Neumann series for A^{-1}, the sparsity pattern of $(I + A)^m$ or $(I + A + A^T)^m$ seem to be also a good a priori choice for the sparsity structure of A^{-1}. Such global criteria are very helpful for distributing the data to the corresponding processors in a parallel environment. For the maximal allowed index set J_{max} we get a row index set I_{max} and a submatrix $A(I_{max}, J_{max})$, which represents the part of A that is necessary for the corresponding processor to compute M_k. If one processor has to compute M_k for $k \in K$, then this processor needs only the submatrix of A that is related to the column indices $\cup_{k \in K} J_{max}$ and the row indices $\cup_{k \in K} I_{max}$.

Now let us assume that we have already computed an optimal solution M_k with residual r_k of the LS-problem relative to an index set J. As a second local a priori criterion we consider only indices j with $(r_k^T A e_j)^2 > 0$. We will see later that this condition guarantees that the new index set $J \cup \{j\}$ leads to a smaller residual r_k.

The final selection of new indices out of the remaining index set, after applying the a priori criteria, is ruled by

(a) a 1-dimensional minimization $\min_{\mu_j} \|A(M_k + \mu_j e_j) - e_k\|$,   or

(b) the full minimization problem $\min_{J \cup \{j\}} \|A \tilde{M}_k - e_k\|$.

In the case (a) we have to consider

$$\min_{\mu_j \in \mathbf{R}} \|r_k + \mu_j A e_j\|_2 = \min_{\mu_j} \|A(M_k + \mu_j e_j) - e_k\|_2 =: \rho_j .$$

For every j the solution is given by

$$\mu_j = -\frac{r_k^T A e_j}{\|A e_j\|_2^2} \quad \text{and} \quad \rho_j^2 = \|r_k\|_2^2 - \left(\frac{r_k^T A e_j}{\|A e_j\|_2}\right)^2 .$$

Hence, indices with $(r_k^T A e_j)^2 = 0$ lead to no improvement in the 1-D minimization. Now we can arrange the new possible indices j relative to the size of their corresponding residuals ρ_j.

In the case (b) we want to determine the optimal residual value σ_j that we get by minimizing $A \tilde{M}_k - e_k$ over the full index set $J \cup \{j\}$ with known index set J and a new index $j \notin J$. Surprisingly it is not too expensive to derive σ_j using this higher dimensional minimization. The additional costs for every j are mainly one additional product with the orthogonal matrix related to the old index set J. It holds [6]

$$\sigma_j^2 = \|r_k\|^2 - \frac{(r_k^T A e_j)^2}{\|A e_j\|^2 - \|Y^T \hat{A} \hat{e}_j\|^2} \ ,$$

with $\hat{A} = Q \begin{pmatrix} R \\ 0 \end{pmatrix} = YR$ and $Q = (Y \quad Z)$. Again we can order the possible new indices after the size of σ_j. Similarly to (a), indices with $(r_k^T A e_j)^2 = 0$ lead to no improvement in the residual and can be neglected.

Now we have a sorted list of possible new indices. Starting with the smallest ρ_j (resp. σ_j) we can add one or more new index to J and solve the enlarged LS-Problem. Numerical examples show that it saves operations if we add more than one new index per step. This has two reasons:

- First, we have no guarantee that with our criteria we really get a notable improvement in the residual by using the enlarged index set $J \cup \{j\}$ with smallest ρ_j or σ_j. In some examples we fall into a dead-end. We can reduce the likelyhood of such dead-ends by allowing more than one new index per step.
- Secondly, it is obvious that we save a lot of operations if we apply the index searching subroutine not so often. It may happen that in the final solution there will occur superfluous indices in M_k if we add more than one index per step. To avoid too many superfluous indices, we compute the mean value of the new residuals ρ_j (resp. σ_j) and remove all indices with larger residual than this mean value. After this step, we add at most p new indices to J with a prescribed parameter p. We stop this process if the reached residual r_k is smaller than a prescribed tolerance $\epsilon < 1$ or if $|J|$ is too large.

2. Computational Aspects

Iterative Solution

If we want to use an iterative solver (3), we can apply the conjugate gradient method on the normal equations $\hat{A}^T \hat{A} \hat{M}_k = \hat{A}^T \hat{e}_k$. No further evaluation is necessary, if we do not multiply both matrices. As preconditioner we can define $diag(\hat{A}^T \hat{A})$. The matrix×vector-multiplications in the cg-method can be done in sparse mode in two steps with $y = \hat{A}x$ and $z = \hat{A}^T y$, where x, y, and z are small dense vectors.

The advantage of this approach is, that we need no additional memory, no updates, and we need no old information if we want to solve the LS-problem for an enlarged index set J. Furthermore, sometimes we want to find factorized sparse approximate inverses that minimize for example

$$\|A M_1 ... M_l - I\|_F \ .$$

Hence, in every step for given $M_1, ..., M_{l-1}$ we have to compute the new approximate sparse inverse to $AM_1...M_{l-1}$. If we solve the resulting least-squares problems iteratively we can avoid the explicit product $AM_1...M_{l-1}$, that will be much denser than the original A.

The disadvantage of this iterative method is that we can not use old results connected with the index set J for solving the enlarged LS problem. This leads to a more expensive method than the QR-based solvers, especially if we add only one new index in every step. Hence, with this iterative solution method, one should add more than one new index per step, because then there are to solve fewer LS-problems.

Householder Orthogonalization

Now let us consider the use of an implicit QR-decomposition based on Householder matrices. First, we assume that we add only one new index in J per step. Then we begin with index sets $J_1 = \{j_1\}$ and I_1, and we have to compute the QR-decomposition of $\hat{A}_1 = A(I_1, j_1)$. In the Householder approach, we use one elementary Householder matrix $H_1 = I - 2q_1 q_1^T$, that transforms the matrix $\hat{A}_1$ via $H_1 \hat{A}_1 = \begin{pmatrix} R_1 \\ 0 \end{pmatrix}$ into upper triangular form.

In the second step we add one profitable new index j_2 to J, which leads to the new matrix

$$\hat{A}_2 = \begin{pmatrix} \hat{A}_1 & \hat{B}_1 \\ 0 & \hat{B}_2 \end{pmatrix},$$

where $\hat{B}_1$ is the part of the new column that is related to indices in I_1, while $\hat{B}_2$ is related to new indices that are only induced by the shadow of j_2. Now, we have to update the QR-decomposition. Therefore, we have to compute the QR-decomposition of

$$\left(\begin{array}{c|c} R_1 & H_1\hat{B}_1 \\ 0 & \\ \hline 0 & \hat{B}_2 \end{array} \right) = \begin{pmatrix} R_1 & \tilde{B}_1 \\ 0 & \tilde{B}_2 \end{pmatrix}.$$

We compute the new Householder vector q_2 related to the matrix $\tilde{B}_2$ with $H_2 \tilde{B}_2 = \begin{pmatrix} R_2 \\ 0 \end{pmatrix}$. This leads to the equation

$$\begin{pmatrix} 1 & 0 \\ 0 & H_2 \end{pmatrix} \begin{pmatrix} H_1 & 0 \\ 0 & I \end{pmatrix} \hat{A}_2 = \begin{pmatrix} R_1 & \tilde{B}_1 \\ 0 & R_2 \\ 0 & 0 \end{pmatrix}.$$

We can write this equation in a more convenient form by adding zeros in the vectors q_1 and q_2, to extend these vectors to the row length of $\hat{A}_2$. Then, we get

$$\tilde{H}_2 \tilde{H}_1 \hat{A}_2 = \begin{pmatrix} \tilde{R}_2 \\ 0 \end{pmatrix} \quad \text{with} \quad (\tilde{q}_1 \quad \tilde{q}_2) = \left(\begin{array}{c|c} q_1 & 0 \\ 0_{i_1} & q_2 \end{array} \right).$$

If we continue to add new indices to J, and to extend the vectors q_i and $\tilde{q}_i$ to the corresponding row length, then we get the matrix

$$\hat{A}_m = \begin{pmatrix} \hat{A}_1 & * & \cdots & \cdots & * \\ 0 & B_2 & * & & \vdots \\ 0 & 0 & B_3 & \ddots & \vdots \\ \vdots & \ddots & \ddots & \ddots & * \\ 0 & \cdots & 0 & 0 & B_m \end{pmatrix} \qquad (*)$$

and Householder vectors of the form

$$\tilde{H}_m \cdots \tilde{H}_1 \hat{A}_m = \begin{pmatrix} \tilde{R}_m \\ 0 \end{pmatrix}$$

with

$$\begin{pmatrix} \tilde{q}_1 & \tilde{q}_2 & \cdots & \tilde{q}_m \end{pmatrix} = \begin{pmatrix} q_1 & 0 & \cdots & 0 & 0 \\ 0_{i_1} & q_2 & \ddots & \vdots & \vdots \\ 0_{i_2} & 0_{i_2} & \ddots & 0 & 0 \\ \vdots & \vdots & \ddots & q_{m-1} & 0 \\ 0_{i_m} & 0_{i_m} & \cdots & 0_{i_m} & q_m \end{pmatrix}.$$

This matrix is a lower triangular matrix where additionally the last entries in every column are zero. Hence, we have to store only the short nonzero kernels q_l and the related lengths of these vectors. Then, every multiplication with H_l can be reduced to nontrivial arithmetic operations. In this way we can use the sparsity of A in the Householder matrices $\tilde{H}_l$.

To solve the LS- problem, we have to compute

$$Q^T \hat{e}_k = \tilde{H}_m \cdots \tilde{H}_1 \hat{e}_k .$$

We can update this vector for every new index j_{m+1} which takes only one product with the new Householder matrix $\tilde{H}_{m+1}$.

If we add more than one new index per step, we can derive the same sparsity structure in $\hat{A}_m$ and the Householder matrices $\tilde{H}_l$ if we partition the index set I in such a form that $I = I_1 \cup I_2 \cup \cdots I_s$ where I_l is the set of new row indices induced by the column index j_l.

This approach is numerically stable and can be used in connection with the index startegy (a). If we want to use the criterion (b) for choosing new indices then we need an explicit expression for the first columns Y of the orthogonal matrix Q. This leads to additional costs and Gram-Schmidt methods that automatically generate an explicit representation of Y are better suited for this case.

Gram-Schmidt Orthogonalization

Here, in the original form, we orthogonalize the new column Ae_j against all the previous orthogonalized columns. Again, let us assume that the index set

I is of the form $I = I_1 \cup \cdots \cup I_m$ with I_l related to a column index j_l. The QR-decomposition is of the form

$$\hat{A} = Q \begin{pmatrix} R \\ 0 \end{pmatrix} = (Y \quad Z) \begin{pmatrix} R \\ 0 \end{pmatrix} = YR \,.$$

The orthogonalization of the new column is evaluated by

$$r_j = Y^T(\hat{A}\hat{e}_j) \ , \quad q_j = \hat{A}\hat{e}_j - Y r_j \ , \quad r_{jj} = \|q_j\| \ , \quad q_j = q_j/r_{jj} \,.$$

Then, the new matrix R is given by

$$\left(\begin{array}{c|c} R_{j-1} & r_j \\ \hline 0 & r_{jj} \end{array} \right) \,,$$

and the matrix Y is built up by the vectors q_j, and of the form (*) with the same sparsity structure as $\hat{A}$.

Unfortunately, the Gram-Schmidt process is numerically unstable, and in many examples it will be necessary to use a stable generalization. A common more robust variant is the Modified Gram-Schmidt algorithm. In the k-th step, the k-th column of Q and the k-th row of R are determined. But, in contrast to the classical Gram-Schmidt procedure the inner products are computed not with $\hat{A}\hat{e}_j$ but with $\hat{A}\hat{e}_j - \tilde{Y}\tilde{R}\hat{e}_j$, where $\tilde{Y}$ and $\tilde{R}$ are all the previous evaluated columns and rows of Q and R. The increase of numerical stability causes a loss of sparsity: $(\hat{A} - \tilde{Y}\tilde{R})\hat{e}_j$ will be denser then $\hat{A}\hat{e}_j$. Hence, the evaluation of $Y^T(\hat{A} - \tilde{Y}\tilde{R})\hat{e}_j)$ will be more expensive than in the classical Gram-Schmidt algorithm with sparse $\hat{A}\hat{e}_j$.

For some classes of ill-conditioned matrices also the Modified Gram-Schmidt procedure gets numerically unstable. This occurs mainly for matrices with small, but very ill-conditioned submatrices $A(I, J)$. In this case, we have to use the Householder approach or some iterative refinement of the Gram-Schmidt algorithm, for example the method, introduced by Daniel, Gragg, Kaufmann, and Stewart [4]. Here, the columns q_k are refined iteratively to ensure the orthogonality against the previous columns. The main step in every iteration is similar to the original Gram-Schmidt algorithm.

Note, that with these Gram-Schmidt-like approaches Y will have the special sparsity pattern (*) if we order the index set I relative to the new columns of $\hat{A}$. For the LS-problem, we have to solve a linear equation with R and the right hand side $Y^T\hat{e}_j$, the j-th column of Q that is given explicitly. For the multiplication with Y^T and Y we can use the sparsity structure of Y, if we store one vector that contains the number of nonzero entries in every column of Y, resp. $\hat{A}$. In the same way, for evaluating σ_j we need one matrix-vector product $Y^T(\hat{A}\hat{e}_j)$, and can take advantage of the structure of Y and the sparsity of $\hat{A}\hat{e}_j$. Hence, for (b) this approach is very favourable, but to be numerically stable we need robust generalizations of the Gram-Schmidt orthogonalization.

PVM-Implementation

The master-slave model is very well suited for computing the columns of M. The master-process has to generate and start the slave processes, it has to distribute the columns and the related submatrices on this processes, and it has to collect the results of the slave processes. If we have p processors initialized under PVM, then every processor should compute $q = \lfloor n/p \rfloor$ columns M_k. Hence, the m-th slave process should compute M_k for $k = (m-1)q+1, (m-1)q+2, ..., mq$, and $m = 1, ..., p$.

The matrix A is stored in the Compressed Sparse Column format (CSC) in order to ensure fast acces to the columns of A. In a first step the master process computes a list of all column indices that occur in row i (CI), and sends this data to each slave process. The slave process uses the CSC-form of A for computing the inner products $r_k^T A e_j$, but additionaly it uses the CI-list to read off directly the indices j with $r_k^T A e_j \neq 0$. If the memory of a processor is not large enough to store the whole matrix A, we can use the global a priori criteria of Section 1. Then the smaller submatrix $A(I_{max}, J_{max})$ in CSC-format and the related part of the CI-list have to be sent to the slave process.

The slave processes apply the above described algorithm to compute their columns M_k and send their results back to the master process. The master process collects the results M_k of the slave processes and builts up the whole matrix M columnwise in CSC-format.

References

1. Barrett,R., Berry,M., Chan,T., Demmel,J., Donato,J., Dongarra,J., Eijkhout,V., Pozo,R., Romine,C., van der Vorst,H.: *Templates for the solution of linear systems: building blocks for iterative methods, SIAM, Philadelphia, 1994.*

2. Chow,E., Saad,Y.: Approximate Inverse Preconditioners for general sparse matrices, *Research Report UMSI 94/101, University of Minnesota Supercomputing Institute, Minneapolis, Minnesota, 1994.*

3. Cosgrove,J.D.F., Diaz,J.C., and Griewank,A.: Approximate inverse preconditioning for sparse linear systems, *Intl. J. Comp. Math. 44*, pp. 91-110, 1992.

4. Daniel,J.W., Gragg,W.B., Kaufman,L.C., and Stewart,G.W.: Reorthogonalization and stable algorithms for updating the Gram-Schmidt QR-factorization, *Mathematics of Computation, 30*, pp. 772-795, 1976.

5. Gould,N.I.M., Scott,J.A.: On approximate-inverse preconditioners, *Technical Report RAL 95-026, Rutherford Appleton Laboratory, Chilton, England, 1995.*

6. Grote,M., Huckle,T.: Parallel preconditioning with sparse approximate inverses, *SIAM J. Sci. Comput.*, (to appear).

7. Huckle,T.: Efficient Computation of Sparse Approximate Inverses, *TU München, Institut f. Informatik, TUM-19608; SFB-Report 342/04/96 A, 1996.*

8. Kolotilina,L.Yu., Yeremin,A.Yu.: Factorized sparse approximate inverse preconditionings I. Theory, *SIAM J. Mat. Anal. 14*, pp. 45-58, 1993.

Experience with PVM in an Industrial Environment

A. Blaszczyk[1], C. Trinitis[2]

[1] Asea Brown Boveri Corporate Research Heidelberg, Germany
[2] Institute of High Voltage Engineering, Technical University Munich, Germany
email: ab@decrc.abb.de, carsten@hsa.e-technik.tu-muenchen.de

Abstract: The paper presents experiences with operation of a parallel code used in ABB for 3D simulation of electric fields which is a part of design of high voltage equipment. The parallelization of this code is based on PVM communication software operated in heterogeneous workstation clusters including multiprocessor machines. The aspects of efficiency, reliability, cluster configuration and ease of use in a typical industrial environment are discussed.

1 Introduction

The hardware environment which is typical for CAD and development process in electrotechnical industry includes workstations connected via Ethernet or FDDI based network. Specialized high performance computers are relatively rare particularly in medium size but even in larger companies. The idea of clustering workstations using software like PVM [1] in order to increase the performance and improve utilization of hardware resources seems to be very attractive for a wide range of industrial users. However the most important barrier for creating clusters in industrial workstation networks is lack of applications.

In 1994 ABB Corporate Research started a project aimed at the parallelization of the in-house code POLOPT for 3D simulation of electric fields based on the boundary element method [2], [3]. This software is used by ABB companies in design of high voltage equipment like power transformers or gas insulated switchgear. The results of the project published in [4] show that a good efficiency of parallel computations can be achieved in a typical industrial hardware environment using PVM. In particular the computation times could be reduced for large problems from overnight to 1-3 hours and for smaller and medium size problems to ones of minutes.

The continuation of the project in 1995-96 has been aimed at making the parallel code available for designers. The efforts were focused on testing efficiency in a variety of hardware configurations including multiprocessor machines, reliability aspects, optimum cluster configuration and ease of use. In this paper after a short summary of the parallelization concept we present some results of these efforts.

2 Parallelization Concept

The details of the parallelization concept have been presented in [4]. The basic idea illustrated in Fig. 1 can be summarized as follows:
- The most time consuming part of the computation is the formulation and solution of a linear equation system which includes a dense and unsymmetric matrix.
- Each row of the matrix can be calculated independent from each other provided that the input data has been replicated on each processor. The parts of the matrix are never transferred through the network but are calculated and stored locally.
- The parallelization is based on a master-slave approach. The master determines the amount of work to be done by each processor (including itself) based on a Mandelbrot algorithm which implicitly takes into account the speed and current load of each processor.

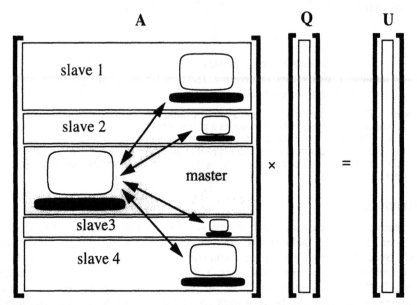

Fig. 1 Parallelization concept: each computer calculates a part of the matrix and stores it locally. The size of this part corresponds to the computer speed.

- The solution of the equation system is based on the iterative GMRES method. Its parallelization is straightforward since it can be limited only to matrix-vector-multiplication required during each iteration.
- The basic concept of parallelization is rather algebraic than topological (no domain decomposition); consequently the parallel efficiency does not depend on the geometry of the calculated problem but only on its size.

3 Efficiency

The parallel efficiency which can be achieved in a heterogeneous cluster of 8-10 workstations connected via Ethernet varies between 70 and 80% [4]. In many practical cases the efficiency is larger, even a superlinear speed up can be achieved. This effect is associated with dividing the large matrix (including often more than 100 MB) into many smaller parts which can be stored in the memory of a standard workstation instead of on the disk. Consequently the time consuming I/O operations can be avoided in parallel jobs and this leads to efficiencies larger than 100%.

During the last few years multiprocessor machines supporting distributed and shared memory architectures have been offered by many workstation manufacturers. The PVM based codes are able to run on both types of architectures without any modifications. We investigated the behavior of our application for IBM SP2 and SGI Power Challenge which represent both architectures and are currently the most frequently installed multiprocessor machines at ABB companies. Fig. 2 shows the relationships between computation time and number of processors for a medium size problem. Fig. 3 shows the corresponding relationship for parallel efficiency. As can be seen the Power Challenge processors are slightly faster than those of SP2, but on the other hand the efficiency of SP2 is better. Generally, both SP2 and Power Challenge are well suited for practical applications particularly for a number of processors up to 12 which is usually not exceeded in most industrial installations.

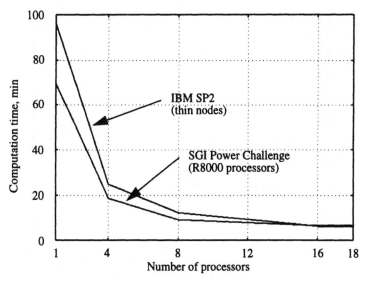

Fig. 2 Computation time for a medium size problem (matrix dimension 3500) on multiprocessor machines.

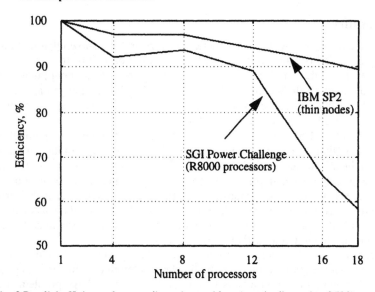

Fig. 3 Parallel efficiency for a medium size problem (matrix dimension 3500) on multiprocessor machines.

For a large number of processors the distributed memory architecture of SP2 shows much better efficiency than the shared memory of Power Challenge. It can be seen in Fig. 4 that the parallel efficiency for 57 processors is still 74%. This result confirms good scalability of the parallel code for distributed memory computers with fast communication between nodes. However, for designers of HV equipment this large number of processors is currently out of interest (or rather out of reach of their budget).

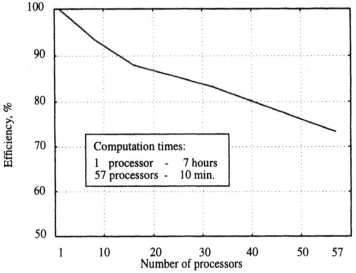

Fig. 4 Parallel efficiency for a large problem (matrix dimension 7000) on IBM SP2 (thin nodes, PVMe)

4 Reliability

The first experiences with the use of parallel version have shown that up to 50% of parallel jobs failed because of failures which occurred on one of the slaves or in the network. Many of these failures were associated with bad cluster configuration which could be avoided by dedicating workstations only for the parallel job and allocating appropriate resources (particularly enough disk space for storing the matrix, not mounted via NFS but locally on each workstation).

However, these first experiences turned out that the reliability of parallel jobs is much less than that of sequential ones and it must be improved. In particular the master task should be able to successfully finish the job even if one or more slaves have failed. Such a feature has been implemented based on the "timeout receive" function of PVM 3.3. After the loss of communication with a slave the master cancels it from the list of active slaves and continues the computations. Of course the cancellation of a slave leads also to the loss of the matrix part stored on it. This part must be recalculated by the master and other active slaves.

In Fig. 5 an example of a parallel job is shown during which 2 slave tasks failed. It can be seen that the parallel job could be finished although the efficiency is poor. The ability of finishing all started jobs is very important for very large problems (dimensions up to 20000) which usually take few hours of computations. A designer who starts such a job overnight expects the results the next morning. A good reliability of such calculations significantly increases the acceptance of parallel computations and is crucial for the integration of 3D field computations into short design cycles.

5 Cluster Configuration

The workstation clusters for parallel computations of electric fields have been arranged in 3 ABB companies: Hochspannungstechnik AG Zurich, Switzerland, Calor Emag Schaltanlagen Hanau, Germany, Power Transformers Ludvika, Sweden. In each of these companies approximately 30-100 workstations are installed and networked.

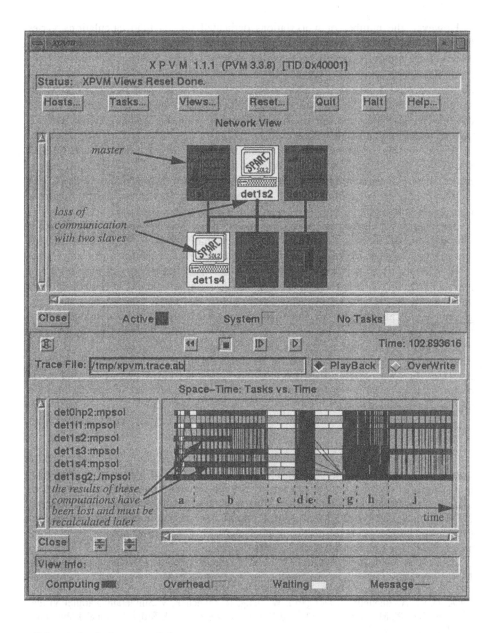

Solver tasks during computation:

a) replication of input data and initialization, **b)** matrix computation, **c)** timeout after loss of communication with slave det1s2, **d)** recalculation of matrix part lost on det1s2,
e) beginning of GMRES calculation, **f)** timeout after loss of communication with det1s4,
g) recalculation of the matrix part lost on det1s4, **h)** GMRES calculation,
f) field computation after solution of the basic equation system

Fig. 5 Parallel computation with failures on two workstations recorded with XPVM
(graphical console for PVM).

However, the number of workstations dedicated for 3D field calculations varies between 1 and 3. Other workstations belong to other groups or departments and are allocated for other tasks. This organization structure makes it difficult to find an optimum cluster configuration. Currently the user performing parallel computations manually has to write a hostfile including names and other data of all workstations which should be clustered. This procedure is not difficult but is beyond of competence of a designer who is rather interested in submitting a job without taking care where it is calculated and how to allocate resources.

An acceptable solution of this problem is offered by job management systems which can automatically write a hostfile and create a suitable cluster. Currently a project aimed at evaluating the ability of commercial job management systems (like CODINE or LSF) in handling parallel jobs for clusters of workstations including multiprocessor computers is being performed at ABB Corporate Research.

The automatic cluster configuration is essential not only for large jobs but also for jobs whose duration can be reduced from more than one hour to few minutes; this would allow almost interactive work with smaller or even medium size simulation tasks. The ability to perform the 3D simulation interactively is becoming more and more interesting for designers because of the growing efficiency of modelling software. For example, a designer using Pro/Engineer can easily change the parametric geometry and all changes are automatically inherited by the related finite element model. Since the creation of a new FE model can be done in few minutes the designer expects also short response times after pushing the "solver button". This quick response can be achieved by automatic creation of a cluster which would significantly increase the peak performance available to designers. This peak performance is required only from time to time but would allow to integrate 3D simulation into the day-to-day design process.

6 Conclusions

The parallel computations of electric fields based on PVM communication software can be performed with a good efficiency and reliability in an industrial environment including clusters of workstations and multiprocessor computers. If the aspects of cluster configuration and ease of use will be solved based on job management systems, PVM has a chance to become an integral part of simulation software used in the industrial design process. Therefore, from the viewpoint of industrial users a commercial support and software maintenance for PVM is essential.

References

[1] A. Geist, A. Beguelin, J. Dongarra, W. Jiang, R. Manchek, V. Sunderam: Parallel Virtual Machine. PVM 3 User's Guide and Reference Manual, Oak Ridge National Laboratory, Tennessee, May, 1994.

[2] POLOPT User's Guide (3D BEM solver for electrostatics with interfaces to FAM, I-DEAS and Pro/Engineer). ABB Corporate Research Heidelberg, 1996.

[3] Z. Andjelic, B. Krstajic, S. Milojkovic, A. Blaszczyk, H. Steinbigler, M. Wohlmuth: "Integral methods for the calculation of electric fields", Scientific Series of the International Bureau Research Center Juelich, 1992 (ISBN 3-89336-084-0).

[4] A. Blaszczyk, Z. Andjelic, P. Levin, A. Ustundag: "Parallel computation of electric fields in a heterogeneous workstation cluster", Lecture Notes in Computer Science Vol. 919, Springer Verlag, 1995 (proceedings of HPCN Europe Milan), pp. 606-611.

Taskers and General Resource Managers: PVM Supporting DCE Process Management

Graham E. Fagg[1], Kevin S. London[1] and Jack J. Dongarra[1,2]

[1] Department of Computer Science, University of Tennessee, Knoxville,
TN37996-1301
[2] Mathematical Sciences Section, Oak Ridge National Laboratory, Oak Ridge,
TN37831-6367

Abstract. We discuss the use of PVM as a system that supports General Process Management for DCEs. This system allows PVM to initialise MPI and other meta-computing systems. The impetus for such a system has come from the PVMPI project which required complex taskers and resource managers to be constructed. Such development is normally too time consuming for PVM users, due to the in-depth knowledge of PVM's internals, which are required to facilitate such systems correctly and reliably.

This project examines contemporary systems such as various MPIRUN systems and other general schedulers, and compares their requirements to the capabilities of the current PVM system. Current PVM internal operations are explained, and an experimental system based on the experience gained from the PVMPI project is demonstrated. Performance of some standardised plug-in allocation schemes that will be distributed with the new PVM 3.4 release are then demonstrated.

It is hoped that this project will provide users of dynamic meta-computing environments the "user controlled" flexibility that has previously been difficult to achieve without the user having to rely upon external third party scheduler and job control systems.

1 Introduction

PVM is one of a number of parallel distributed computing environments (DCEs) that were introduced to assist users wishing to create portable parallel applications [9]. The system has been in use since 1992 [1] and has grown in popularity, leading to a large body of knowledge and a substantial quantity of code accounting for many man-years of development.

The PVM system has proven itself to be very flexible in terms of both process and machine/node management, with user accessible library functions allowing complete user application control of the Virtual Machine and any other application executing upon it.

For many users, the default scheduling scheme and process control from the PVM console is sufficient. When more complex process control such as batch scheduling or run-time load balancing is required, the user is forced to either implement their own Resource Manager (RM) or rely upon third party software such as NQS, DQS, load-leveller, LSF or Far[2].

The implementation of a RM is non-trivial due to the state that it is required to maintain and the level of knowledge concerning the PVM internals that a user would need to produce a "correct" and reliable RM system.

The aim of this work is to illustrate a General Resource Manager (GRM) that is designed to be customised by a user so that do not have to concern themselves with PVM internals but instead can concentrate upon load-balancing , queueing and scheduling schemes that match their particular needs.

2 Requirements for Process Control and Resource Management

Many early cluster or meta-computing system users and administrators realised that the simple process allocation schemes offered on many single-user (fixed partition) MPP systems were barely adequate for the specialist machines and not complex enough to manage the dynamic newly emerging cluster environments with multi-user MPPs.

This lead to the early development of many Cluster Management Systems (CMS) which are typified by allowing users to submit jobs for execution under some form of centralised control. The standard features of such systems are:

Load Balancing: Jobs are sent to unloaded machines to improve performance and to minimise effects on other users.

Queueing: Jobs can be submitted to execute at a later time upon specialist or reserved resources. Usually multiple queues exist to separate long running, quick turn-around and interactive applications.

Experience has shown that many of these systems need to tailored to individual site needs, as well as support more complex features such as task migration which for non monolithic applications can be highly expensive in terms of additional resources required.

3 PVM 3.3

One of PVM's primary advantages over many other systems is its ability to interoperate across a diverse number of heterogeneous platforms with over 30 different types currently supported. To facilitate this portability, PVM utilises a daemon to interface between the user and any particular operating environment. The PVM user is thus insulated from the system dependent functions of each host architecture and is instead given a consistent core set of functions to manipulate a Virtual Machine that PVM creates. Through these core services users can monitor all aspects of the virtual machine and its applications.

3.1 Resource and Process Control

The PVM system consists of a number of hosts which each run a daemon *pvmd*. In the case of MPPs this daemon usually only runs on the front-end or service node and the daemon maintains a list of available nodes. Each daemon additionally holds a complete list of all hosts enrolled into the PVM system although only the master daemon is considered to hold the authoritative list.

The user library allows hosts to be added or deleted by sending either a *TM_ADDHOST* or *TM_DELHOST* message to the local daemon for the user application.

Each daemon maintains a list of the processes executing upon it (and its associated nodes), so there is no global list of processes thus reducing daemon to daemon messaging during fast process turn-over periods. Any process can start (*spawn*), signal or terminate (*kill*) any other process in the virtual machine by calling the user library which will send the appropriate message to the owning daemon. This daemon will then use local operating system services to perform the operation requested.

The task creation is performed by calling either pvm_spawn()/pvmfspawn() functions in a user application or the spawn command at the PVM console. This results in a *TM_SPAWN* request being sent to the local daemon. The local daemon would then decide on the allocation of resources using its own copy of the master host list, unless a *single* particular host was specified in the initial request. If two non-specific requests were sent from two different machines then the resulting process allocation would be determined by the previous requests that each particular daemon has already processed and would thus appear random. If an application needed to execute upon a set of machine $M_{1...N}$, then N separate spawn requests would need to sent, unless this set could be specified by a filter based on the architecture classes available.

3.2 Fault Tolerance

An important feature of PVM is that it contains user library hooks to monitor the Virtual Machine configuration and to **notify** changes by sending user level messages as they occur. This allows careful programmers to produce fault tolerant applications, although facilities such as check-pointing are not included in the current public release. They are however provided by several vendors and research projects such as Co-Check[8], which is based upon the Condor system[6].

3.3 Additional Daemons

In PVM 3.3 certain process and host management functions could be passed to specialist daemons to reduce the load upon the normal PVM daemons allowing them to remain more responsive. Three special daemons types were offered:

1. Hoster daemon which intercepts *addhost* messages and starts new *pvmds*.
2. Tasker daemon which is used to spawn tasks for the local *pvmds*.
3. Resource Manager daemon which catches almost all the management function requests between the user library and the *pvmds* allowing it to alter the look and feel of the whole Virtual Machine.

These daemons are activated by calling an appropriate register function. As the Resource Manager intercepts messages from the user processes directly, they need to be spawned after it has registered, else they would not be aware of its existence. A common solution to this problem would be to have the Resource Manager start the Virtual Machine directly.

4 PVMPI and specialised taskers

The PVMPI project[3] required PVM to start MPI-1[7] applications to allow inter-application communication using PVM across different MPLCOMM_WORLDs even if they were running under different MPI implementations. The usual spawn fork/exec mechanism was not suitable for this, as each different MPI implementation relied upon different techniques to obtain their initial operating environment, such as the number and identity of each process. Since the user API was to remain unaltered, a method had to be derived to allow the user to specify a host group or *cluster* within a single spawn instruction.

The solution was a specialist set of taskers with local host lists that understood how to initiate processes under each different implementation of MPI.

Thus the PVM system became a set of tasker managed clusters within the virtual machine as shown in figure 1, allowing PVM to replace all the non-standard MPIRUN schemes.

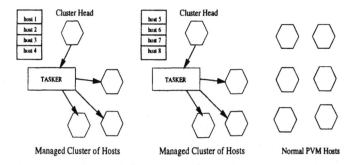

Fig. 1. Virtual Machine consisting of multiple host clusters and general hosts

Although the system suited the PVMPI projects needs, it could not be used as a general cluster management system for normal PVM applications as the processes started up by the taskers did not have the *correct* tids as excepted by the users library, due to the calling order within the PVM daemons. Other problems included inflexibility in the semantics of the current PVM API, made using a cluster only possible by spawning to the head of the cluster. Also manually starting up the taskers and giving them each a separate list was inconvenient and possibly error-prone. The solution was to keep the taskers (to avoid building their functionality into the already large PVM daemons) and migrate all the local host lists into a centralised Resource Manager which would pass spawn instructions to them when required.

5 General Resource Manager (GRM)

A Resource Manager can only be general if it meets most of the requirements listed in section two, including full user dynamic configuration. The GRM presented here mets these requirements by allowing users to slot in different modules for each of its the major functions.

The Resource Managers main functions are to control the addition and removal of hosts from the virtual machine, and the distribution of processes upon these hosts.

The GRM consists of seven main components as shown in figure 2. These being:

1. Relay mechanism to pass routine messages that are redirected to the resource manager from the user library back to the local PVM daemons. This handles the default process and host functions in the PVM system.
2. Host database: includes details about hosts such as associated processes, host clusters, current load, peak performance, memory free etc. This is used primarily for intelligent scheduling of job/spawn requests. Fault-tolerance of hosts is also handled here, as is automatic shutting down and bringing up of specific parts of the virtual machine.
3. Process database: includes application descriptions, requirements and run states. A list of processes constitutes a job.
4. Queue System: contains lists of jobs awaiting placement onto available hosts or specific named clusters.
5. Notification System: relays notification messages between daemons and user processes, using the data to update internal databases en-route.
6. Time Based Controller: A simple interface for scheduling job and host changes, such as removing clusters from the virtual machine during know busy interactive periods.
7. Check-point System: interface with the Co-Check system, that includes RM checkpoints so that if the system fails both the RM and the applications it is supporting either running or queued can be restarted from a known state.

5.1 Scheduler Operation

The Scheduler receives spawn requests and builds a host list that matches the requirements specified by the user request. If a request cannot be met immediately then the spawn details become a job which is queued. The choice of queue depends upon the priority of the request. A spawn request to a named set of hosts can now be completed in a single request instead of the N previously required. As the scheduler depends on a coherent database, change of state requests to the RM are processed first i.e. a spawn request would be delayed by a delete host or even a task exit message, hence the spawn request queue.

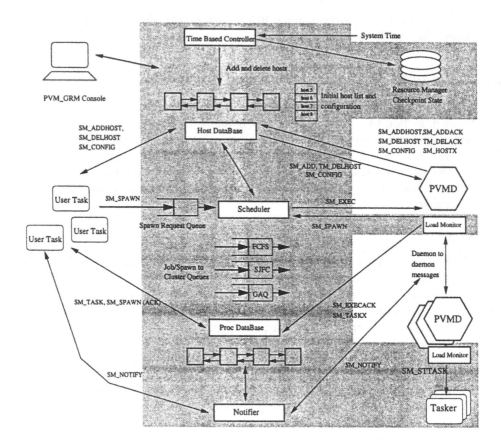

Fig. 2. Overall structure of the General Resource Manager (grey area) and its interaction with the rest of the Virtual Machine

Standard Scheduler The scheduler included with the standard distribution uses one of a set of best fit methods to allocate tasks to hosts if an open (non-specific) request is received. These rely on the load monitor system that has been added to the current PVM daemons, which on startup tests each host to measure their peak performance. The possible performance of each host is calculated as:

$$perf_{estimate} = \frac{perf_{peak}}{loading_{avg} + 1.0} \tag{1}$$

where $perf_{peak}$ is calculated as:

$$perf_{peak} = p_1 * int_{perf} + p_2 * float_{perf} + p_3 * double_{perf} + p_4 * memory\_bw_{perf} \tag{2}$$

and $loading_{avg}$ at time t is:

$$loading_{avg} = l_1 * loadavg_t + l_2 * loadavg_{t-\alpha} + l_3 * loadavg_{t-\beta} \qquad (3)$$

The values $p_{1...4}$ and $l_{1...3}$ are user configurable defaults. α and β depend upon the particular operating system implementation of the remote statics daemons used, and are included so that the scheduler can take into account previous loading to help predict the future loading.

$p_{1...4}$ can be varied per job request by adding them to the spawn command so that specific application characteristics such as high memory or networking bandwidth requirements can dominate process placement.

Four basic systems are included with standard distribution:

1. Round robin: similar to the original PVM spawn except it uses a single host table.
2. Spare cycles: uses least loaded machines, irrespective of their peak performance.
3. Low load performance: uses peak performance as calculated in equation (1), but allows only one PVM task per host.
4. High load performance: may run multiple jobs on each host (even if not MP) if it is expected to be faster than executing on lightly loaded machines. Upper limit to number of jobs per host is configurable.

Figure 3 shows the effects of each scheme on a large scale Monte Carlo chemistry simulation[10] in terms of execution time on a forty seven host heterogeneous network containing five separate architecture classes. It is interesting to note that although load average alone produces good results on homogeneous networks[4] more detailed information is required for optimal performance on heterogeneous systems.

Additional user level modules for scheduling can be included by either adding modules to the GRM source code which already includes software hooks, alternatively users can register the modules as separate PVM processes using the well defined interface supplied.

6   Conclusions

A GRM has been presented that will be distributed with PVM 3.4[5]. This resource manager is designed to be customised by the user and will provide support for job control and scheduling. The supplied scheduling schemes have been shown to provide improved performance using a number of user controlled criteria. The ability to checkpoint application queues and alter the default operation mechanisms makes the GRM highly reliable and flexible.

References

1. A. L. Beguelin, J. J. Dongarra, A. Geist, R. J. Manchek, and V. S. Sunderam. Heterogeneous Network Computing. *Sixth SIAM Conference on Parallel Processing*, 1993.

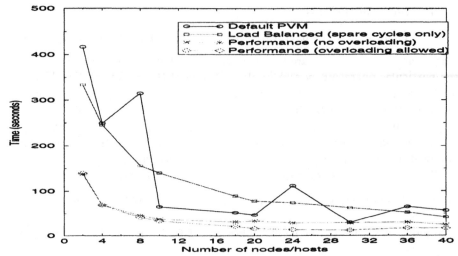

Fig. 3. default scheduling schemes performance on a heterogeneous network

2. Mark Baker and Geoffrey Fox. "MetaComputing: The Informal Supercomputer", Proc. of CRPC annual meeting, Argonne National Laboratory, Chicargo, pp. 26, 1996
3. Graham. E. Fagg and Jack J. Dongarra. PVMPI: An Intergration of the PVM and MPI Systems To appear: Calculateurs Parallhles, Paris, June 1996. Also: Department of Computer Science Technical Report, University of Tennessee at Knoxville, TN-37996, April 1996.
4. G.E. Fagg and S.A. Williams. Improved Program Performance using a Cluster of Workstations, *Parallel Algorithms and Applications*, Vol 7, pp. 233-236, 1995.
5. G. Geist, J. Kohl, R. Manchek, and P. Papadopoulos. New Features of PVM 3.4 and Beyond. Proceeding of *EuroPVM 95*, pp. 1-10, Hermes, Paris, 1995.
6. M. Litzkow, M. Livny, and M. Mutka. Condor – a hunter of idle workstations. In *Proceedings 8th IEEE International Conference on Distributed Computing Systems*, pp. 104-111, June 1988.
7. Message Passing Interface Forum. MPI: A Message-Passing Interface Standard. *International Journal of Supercomputer Applications*, 8(3/4), 1994. Special issue on MPI.
8. Georg Stellner and Jim Pruyne. Resource Management and Checkpointing for PVM Proceeding of *EuroPVM 95*, pp. 130-136, Hermes, Paris, 1995.
9. Louise Turcotte. "A Survey of Software Environments for Exploiting Networked Computing Resources", *MSSU-EIRS-ERC-93-2*, Enginerring Research Center, Mississippi State University, Febryray 1993.
10. Shirley A. Williams and Graham E. Fagg. "A Comparison of Developing Codes for Distributed and Parallel Architectures", Proc. of BCS PPSG *UK Parallel' 96*, pp. 110-118, Springer Verlag, London, 1996.

A PVM Implementation of a Portable Parallel Image Processing Library

Zoltan Juhasz[1]* and Danny Crookes[2]

[1] Department of Information Systems, University of Veszprem
Veszprem P.O.Box 158, H-8201 Hungary
E-mail: juhasz@elod.vein.hu
[2] Department of Computer Science, The Queen's University of Belfast
Belfast BT7 1NN, United Kingdom

Abstract. This paper presents a portable parallel image processing library, which provides a high-level transparent programming model for image processing application development. The library is implemented using the PVM message-passing environment in order to achieve maximum portability. The paper describes the layered software model developed to provide extensibility and to hide the details of parallelism and the idiosyncrasies of the various communication technologies. Implementation details of the image processing library and the abstract communications layer are described and we report on the performance of the library operations we achieved on Ethernet and ATM network based workstation clusters.

1 Introduction

In recent years workstation clusters have become a popular alternative to mainstream parallel computing. For some applications, networked workstations can provide near supercomputer performance with minimal additional investment. As a result, many scientific and high-performance applications have been ported to these systems. Computer vision and image processing (CVIP) applications are one of the potential candidates that can benefit from this architecture [3]. Unfortunately, the efficient implementation or porting of CVIP programs requires parallel programming expertise, which is a drawback for image processing application developers.

To make available the performance of workstation clusters, but at the same time hiding the parallelism, we have developed a portable, parallel image processing (IP) library, which provides the user with a high-level transparent programming model for implementing image processing algorithms.

The variety of message passing libraries (eg. PVM, MPI, p4) and communication technologies (eg. Ethernet, ATM, FDDI, HIPPI) commonly used with workstation clusters increases the difficulty of porting the application code. To

---

* Supported by the Hungarian National Science and Research Foundation (OTKA) under Grant No. F007345.

abstract out the different communication systems, we have developed an abstract image processing communications layer, which is used to implement the image processing library. This gives a layered software model, as shown in Fig. 1.

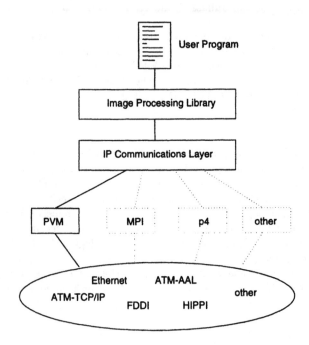

Fig. 1. Hierarchy of the image processing library

The purpose of this paper is to report on some aspects of the implementation of the IP communications layer using the PVM message-passing library, and to illustrate with timings the performance achieved on different systems for some fundamental image processing operations.

The structure of the paper is as follows. Section 2 gives a short introduction to the high-level image processing model. Section 3 describes the abstract communications layer and its use in implementing the IP library routines. Section 4 presents the strategy of the implementation of the IP communications layer and highlights performance-critical points of workstation clusters. Section 5 presents timings for the image distribution, histogram calculation and convolution operations achieved on the Ethernet and ATM-based workstation clusters. The paper ends with conclusions and directions for future work.

2 High-level Programming Model

The library provides, as a set of C data types and routines, image processing specific data structures and operations. Data structures include image, vector (1-D arrays, eg. histograms) and template types. The template (or mask, window, neighbourhood) is a 2-D array holding weights for neighbourhood operations such as convolution. The library operations are at the complete image level and are based on Image Algebra[5, 1] and related programming languages like Apply, Adapt[7] and Tulip[6]. The main operation classes of the library are:

- image I/O (eg. file input and output, image display and capture);
- point operations (eg. Img2 := Img1 > Thresh);
- neighbourhood operations (eg. convolution of an image with a template);
- global operations (eg. finding a histogram or calculating the Fourier Transform of an image).

The use of the library routines is demonstrated by the example program of Fig. 2. This program carries out a simple contrast transformation on an image read from file, then displays it after applying a smoothing operation.

```
IMAGE img1, img2, img3;
img1 = read_image("test.img");
img2 = linear_stretch(img1);
img3 = neighbourhood(img2, CONVOLVE, smoothing_templ);
display_image(img3);
```

Fig. 2. Example high-level image processing program

3 Portable Implementation of the Library Using the IP Communications Layer

The image processing library routines are implemented using the data-parallel (SPMD) programming model, where one of the processors will also serve as a controller. Each image is split into segments (plus an overlapping border area) and each segment is allocated to one processor. Obviously, several of the library operations require interprocessor communication. For instance, when an image is read from file, its segments have to be sent to the appropriate processors; or for convolution, the border area of the image segments first have to be updated; or in calculating the histogram of an image, the segment histograms have to be gathered and accumulated, and the global histogram then broadcast to all processors in the system.

All the necessary communications operations are provided in the image processing communications layer as a set of C functions. Examples of routines are:

```
int distribute_image(IMAGE img)
int receive_segment(IMAGE img)
VECTOR gather_histogram()
int broadcast_histogram(VECTOR hist)
int send_histogram(VECTOR hist)
```

```
VECTOR receive_histogram()
int swap_border(IMAGE img, TEMPLATE templ)
```

Using these routines, the programming of the image processing library operations becomes independent of the underlying communications mechanism, as it
is demonstrated in the followings.

1. Library routine for reading an image:

```
IMAGE read_image(char *filename)
{ IMAGE img;
  if controller task {
    read_file(img, filename);
    distribute_image(img);
  }
  else /* ordinary task */
    receive_segment(img);
  return(img);
}
```

2. Program for applying a neighbourhood operation F to an image:

```
IMAGE neighbourhood(IMAGE img, (*F)(), TEMPLATE templ)
{ IMAGE tmp;
  swap_border(img, templ);
  apply function F to all neighbourhoods in local segment
  return(tmp);
}
```

3. Implementation of the library routine for histogram calculation:

```
VECTOR find_histogram(IMAGE img)
{ VECTOR hist;
  hist = histogram(img);
  if controller task {
    hist = gather_histogram(); /* also accumulates */
    broadcast_histogram(hist);
  }
  else { /* ordinary task */
    send_histogram(hist);
    hist = receive_histogram();
  }
  return(hist);
}
```

4 PVM Implementation of the Communications Layer

To decide the best way to implement each of the communications layer routines, it is necessary to know the ratio of message startup time vs. transmission time. We model communication time to send a message of k words with the linear relationship $t_{comm}(k) = T_s + kT_w$, where T_s is the startup time for a communication and T_w is the time to transfer one word of data. With current networks T_s can typically be around 1000 times T_w[2]. Thus for small message lengths, the startup time dominates communication time. Consequently, the primary goal when implementing the IP communications layer is to minimise the number of messages during the execution of the algorithms.

4.1 Image Distribution

In our implementation images are distributed over the worker processors using the horizontal row partitioning method. This method lends itself to efficient implementation and enables us to balance the load easily for arbitrary number of processors. In contrast to such parallel systems where the communications startup cost is low and therefore small messages are used to achieve maximum communication pipelining (eg. each row of the image is a separate message), we use the largest possible message size in order to transfer each image segment possibly in one single message and thus minimise startup overhead.

Having an image I of size N to be distributed over p processors using packets of length K, $(1 \leq K \leq N/p)$, the total image distribution time is

$$t_{distr} = (p-1)\left[\frac{N}{Kp}(T_s + KT_w)\right] \quad,$$

which for the best case (when $K = N/p$) results in a startup overhead equal to $(p-1)T_s$. For large images this is a negligible proportion of the total transfer time, but for small images and a large number of processors this still can be the dominating factor.

4.2 Border Swap

The borders in our implementation are updated by sending and receiving a message of length $b\sqrt{N}$ for each border (b is the size of the border area in rows). Normally with nearest-neighbour communication, several such borders can be swapped simultaneously. On the shared Ethernet network, however, this communication operation will become serialised, resulting in an overall swap time of $t_{swap} = 4(p-1)\left(T_s + b\sqrt{N}T_w\right)$. Using shorter messages or different data distribution technique would increase communication time.

4.3 Histogram Gather and Broadcast

The histogram calculation involves two types of communication operations. The local segment histograms have to be gathered – and accumulated – to create the global, image level histogram. Then this global histogram has to be broadcast to all tasks involved.

We implemented both the gather and broadcast operations as a series of $(p-1)$ individual send-receive communication pairs. The gather-accumulate operation could be carried out by the pvm_reduce() function but in this implementation we would have liked to avoid the use of dynamic process groups. The histogram broadcast operation could also be implemented by either the pvm_bcast() or the pvm_mcast() function. Since not all parallel systems support these routines we have decided to provide a straightforward implementation, which when required could be replaced by a more efficient vendor-specific collective communication routine.

5 Performance Results

In this section we present the performance of the data distribution and two important image processing library operations, namely the histogram calculation and the convolution. The experiments have been carried out on the following two systems: (i) a 5-processor HP9000/715 workstation cluster using Ethernet-TCP/IP network and (ii) a 5-processor SGI cluster using ATM-TCP/IP network communication.

First, we have performed round-trip communication tests to obtain values of T_s and T_w. The ATM-TCP/IP network parameters are $T_s = 735\mu sec$ and $T_w = 0.065\mu sec$, while the Ethernet cluster parameters are $T_s = 727.5\mu sec$ and $T_w = 0.92\mu sec$. From these values it can be anticipated that the ATM network will perform better than the Ethernet network only when transmitting few very large messages.

The data distribution times measured on the two different clusters – in function of the number of workstations – are shown in Fig. 3. It is evident that for small image sizes the startup time of each processor communication dominates. For larger images, such as 512×512 and 1024×1024 (Ethernet) or 1024×1024 (ATM), this startup time is negligible and distribution time is basically independent of the cluster size.

The execution times for the histogram and convolution operations are shown in Fig. 4 and Fig. 5. During histogram calculation relatively small amount of computation is performed and short messages are exchanged. Since the computation time is smaller than T_s the total time is determined by the communication performance. Since startup time is similar for both networks, the two clusters perform almost identically for this operation as it can be seen in Fig. 4.

As shown in Fig. 5 the performance of the convolution operation is similar on the Ethernet and ATM-based systems. However, we have experienced communication problems (transfer times an order larger than expected occuring at

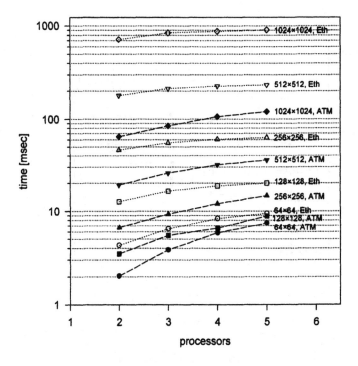

Fig. 3. Performance of the image distribution operation

random) with the ATM network, which decreased performance for small image sizes (see the convolution results for the 64x64 image size), therefore in such cases the use of the Ethernet-based cluster was preferable.

6 Conclusions

In this paper we presented a PVM-based parallel image processing library that provides a high-level portable programming environment for workstation cluster computing. A layered software model has been developed to hide parallelism and communication systems details from the user. An abstract image processing commmunications layer ensures that the source code does not have to be modified in case the library is implemented on a new architecture.

We have compared the performance of two clusters in executing the image processing library operations. Our results indicate that the viability of either networking technology depends on the size of the input images, and that fast networking technology, such as ATM cannot be exploited with current network protocols. New, 'lightweight' protocols are required to reduce the large communication startup time. Due to the layered design of our library, however, these network-level changes affect only the abstract communications layer, of whose

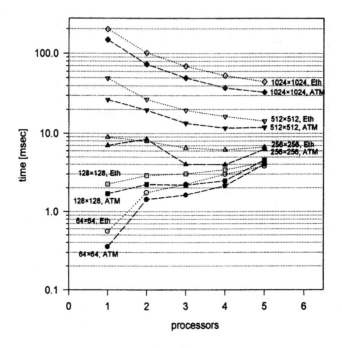

Fig. 4. Performance of the histogram operation

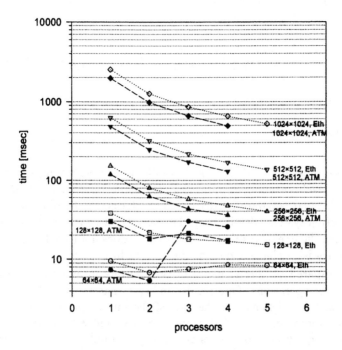

Fig. 5. Performance of the convolution operation

routines can be easily modified should improved new protocols become available.

The library can be extended with further routines by the user by keeping to the layered model described in the paper. It is hoped that this library can form the basis of a generally used parallel image processing environment and facilitate efficient and high-performance image processing/computer vision work.

References

1. D. Crookes, P. J. Morrow and P. J. McParland, IAL: a parallel image processing programming language, IEE Proceedings, Part I, Vol 137 No 3 (June 1990) pp 176-182.
2. Jack J. Dongarra and Tom Dunigan, Message-Passing Performance of Various Computers, Technical Report, ORNL, August 1995.
3. S. G. Dykes, X. Zhang, Y. Zhou and H. Yang, Computation and communication patterns of large-scale image convolutions on parallel architectures, in *Proc. 8th Int. Parallel Processing Symposium*, IEEE Press, April, 1994, pp. 926-931.
4. Al Geist et al., PVM User's Guide and Reference Manual, September 1994.
5. G. X. Ritter, J. N. Wilson and J. L. Davidson, Image Algebra: an overview, Computer Vision, Graphics and Image Processing, No. 49 (1990) pp 297-331.
6. J. A. Steele, An abstract machine approach to environments for image interpretation on transputers, PhD Thesis, The Queen's University of Belfast, 1994.
7. Jon A. Webb, Steps Toward Architecture-Independent Image Processing, *IEEE Computer*, February 1992, 21-31.

Porting of an Empirical Tight-Binding Molecular Dynamics Code on MIMD Platforms

M. Celino

ENEA, HPCN Project,
C.R. Casaccia sp100, C.P. 2400 - 00100 Roma A.D. (Italy)

Abstract. A Molecular Dynamics code, utilized for the study of atomistic models of metallic nanostructured materials, has been ported on MIMD platforms by means of the PVM message passing libraries. The nanostructured materials represent a challenging problem for the parallelization strategies due to their intrinsic dishomogeneity and to the slow relaxation toward the equilibrium configuration. The interaction potential, derived from the second moment approximation of a tight-binding Hamiltonian, the Parrinello-Rahman-Nosé and the VI order predictor-corrector Gear algorithms are implemented efficiently in the parallel code. The parallelization strategies utilized and the molecular dynamics code are described in detail. Benchmarks on several MIMD platforms allow performances evaluation and future improvements.

1 Introduction

To study the structural properties of metallic nanostructured materials [1] an empirical tight-binding Molecular Dynamics (MD) code [2, 3] has been utilized.

The nanostructured materials are polycristals whose crystal sizes are of the order (tipically 1 to 20) nanometers so that 50% or more of the solid consists of incoherent interfaces between crystal grains of different orientations [4, 5, 6]. In order to obtain stable configuration of realistic nanostructured materials, we have reproduced the experimental condensation process of several atomic clusters. In particular, we have chosen to reproduce the dynamics of Palladium nanophase materials since its asymptotic grain size in ball-milling treatment [7] is very small and hence realistic simulations may require a relatively small number of atoms. Three samples, constitued by three different types each of atomic cluster, have been simulated. In each sample there are about 15000 atoms. Thus we have büilt three starting configurations (monodisperse) of different grain size. Then we have started three simulations to investigate the simulated condensation at the temperature $T = 300 K$.

The modellization of nanostructured materials represents an interesting system from the point of view of parallelization strategies, because this kind of materials are highly dishomogeneous and because they need very long simulation runs to reach the equilibrium configuration.

To allow the study at fixed external temperature and pressure, allowing deformations to the simulation cell, the algorithms of Parrinello-Rahman-Nosé are implemented [8, 9, 10]. Furthermore, to obtain high accuracy in the calculation of the trajectories and good total energy conservation during the long run to reach the equilibrium configuration, the VI order Gear predictor-corrector is implemented [11, 12].

In the following we will describe the main features of the algorithms implemented (section 2), the parallelization strategy, the performances and the scalability obtained (section 3) and then the conclusions will be deduced and future improvements described (section 4).

2 Description of the algorithms

To better reproduce the metallic properties of the system we have utilized a many-body potential derived from the second moment approximation of a tight-binding Hamiltonian [2, 3]. In this scheme the ion-ion interaction is described as made up of an effective band term plus a short-range repulsive pair potential. The total cohesive energy of the system, composed by N atoms, is:

$$E = \sum_{i=1}^{N} \left(E_r^i + E_b^i \right) \tag{1}$$

where the repulsive part is assumed pairwise and described by a sum of Born-Mayer ion-ion repulsion

$$E_r^i = \sum_{j>i}^{N} A \exp\left[-p\left(\frac{r_{ij}}{r_o} - 1\right)\right] \tag{2}$$

and the attractive part can be written for an atom i:

$$E_b^i = -\left\{ \sum_{j=1}^{N} \xi^2 \exp\left[-2q\left(\frac{r_{ij}}{r_o} - 1\right)\right] \right\}^{1/2} \tag{3}$$

In these expressions: r_{ij} represents the distance between atoms i and j, r_o is the first-neighbors distance. The parameters ξ, q, A and p are determined by fitting the experimental values of cohesive energy, lattice parameter, bulk modulus and the shear elastic constants C_{44} and $C' = \frac{1}{2}(C_{11} - C_{12})$ and by taking the equilibrium condition into account.

It is clear that the interatomic force depends not only by the interacting atom but also by their neighbours inside the cutoff distance. To take correctly in account the embedding of each interacting atoms, a double loop of order N are needed. We will show that this causes an extra communication among all the processor during the dynamical loop.

With respect of the parallelization strategy the second important algorithm is represented by the Gear predictor-corrector for the integration of the differential equations describing the dynamical evolution of the system. The main characteristic of this approach is that it involves the knowledge of all the dynamical variables up to the fifth order derivative of the position coordinates. Thus the code must load 18 arrays of lenght N plus 3 arrays for the components of the forces. Furthermore, because both terms in the interaction energy depend upon the distance among the atoms, it is convenient to calculate and to store all the distances, because two loops of order N are needed to calculate the forces.

It is worth noting that we use the Verlet list to speed up the calculation of the distances among the atoms, implying a N^2 loop every n_t time-steps (where n_t depends on the diffusion constants of the material). We have to underline that, because we utilize from 2 to 8 processors, each processor must calculate the distance among 2000-7000 atoms, thus it is not useful to use other algorithms, such as the linked-cell algorithm [11], to save time in the evaluation of the distances values.

In the end the Parrinello-Rahman-Nosé algorithms induces modifications to the dynamical equations of the atoms that can be partitioned among the PEs in the same way of atomic coordinates. The dynamics of the simulation cell and of the Nosé parameter is replicated on all the nodes.

3   The parallelization scheme

3.1   Description of the code

We will show in the following that the best parallelization strategy developed so far allows a parallelization of 98% of our code [13]. We will show that, despite the needs of the communications overhead, the parallel version is more convenient (i.e. allows large speedup) than the scalar one. Furthermore, the parallel version allows easy portability on many parallel platforms.

The most simple and efficient parallelization technique for a MD code, with a low number of atoms, consists in distributing the data among the processors with the atomic decomposition technique [14, 15]. This method consists in assigning the atoms to the processors without caring about their physical position in the simulation box. The key point of the parallelization is to measure the cpu time needed for each algorithm. In our case the evaluation of the distances and the calculation of the repulsive and attractive forces is the 98% of the serial code.

As shown in the flux diagram (Figure 1), the code starts with the definition of the variables: every processor defines its local variables and allocates the same quantity of RAM. Then, while the processor nought reads the starting data from the disk, the other processors wait to receive the complete set of data. A broadcast instruction of processor nought will assure the distribution of the data to every other processor.

Now, each processor has all the informations to start the calculations but it does not know yet which atoms are assigned to it. Thus it is necessary a routine for an initial load balancing. With this routine, repeated by every processor, every node knows its group of atoms and the limits of each group assigned to the other processors. This routine assigns the atoms to the processors proportionally to their interactions with the other atoms, thus every node has about the same

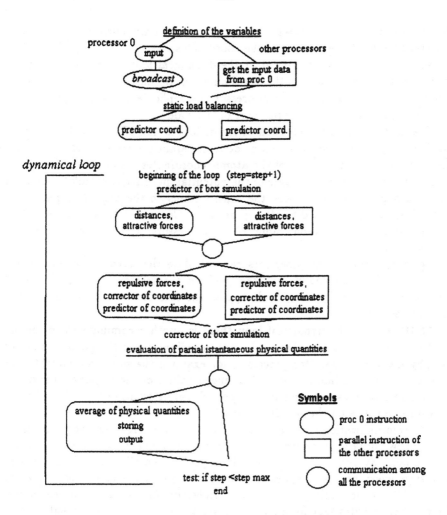

Fig. 1. Flux diagram of the parallel code. For simplicity the flux is splitted in two: on the left there is the processor nought flux diagram, while that pertaining to the other processors is on the right. The circles represent the communications routines where the data are exchanged among all the processors.

number of interaction forces to evaluate. This static load balancing can be enough for our scopes, because we want to simulate only the relaxation of the system forward the equilibrium configuration.

The next step, before the beginning of the dynamical loop, is a step of coordinates prediction. In the serial code, this routine was included in the dynamical loop. Since in a parallel code it is essential to communicate as less information as possible, as shown in the flux diagram, the processors exchange the coordinates no more than in two points of the code. At the end of the dynamical loop the code performs the coordinates correction and a predictor for the next step of the loop. In this way the processors exchange the coordinates only once during the job. After the predictor step, there is an overall communication to let all the processor know the new values of the coordinates.

Now every node is ready to start the dynamical loop. The first step is the predictor of the coordinates of the box simulation. Because it is only an operation among 3x3 matrix arrays, every processor can replicate the same operation as the others, and this forms a serial part of the code.

The next step is the evaluation of the distances and of the repulsive part of the forces. These two routines can be done in parallel: every processor performs the operation only on the set of the atomic coordinates assigned to it.

Due to the many-body nature of the potential, in order to calculate the total force, each processor also needs the contributions to the attractive part from atoms that can be assigned to other processors. Consequently, at this point, the code needs a communication among all the processors, as shown with a circle in the flux diagram.

Again in parallel, the processors can calculate the force on each atom assigned to them, and finally they can perform a corrector and a predictor of the coordinates. Finally, every node applies the corrector to the box simulation coordinates.

At the end of the dynamical loop there is another communication among all the processors. They exchange the new coordinates and the partial values of the physical quantities. One processor, for example the processor nought, will evaluate the total istantaneous and averaged values of the physical quantities. Furthermore it will store the data on the disk before printing the value of some relevant quantities on the standard output.

The communication scheme utilized is described in Figure 2: every processor sends its own data to every other processor thus implying communication of pairs of processors at a time. The best way is to perform parallel communications at each step: every processor sends its own data to another one, and then it waits for the data to be received from a third processor. In the case of p processors, the number of bytes sent from one processor to another is proportional to N/p for each iteration and the cycle must be repeated $(p-1)$ times. The time is only $p-1$ time the send and receive of N/p data, so

$$t_c \propto 2(p-1)\frac{N}{p} \simeq 2N \tag{4}$$

It is worth noting that the time for communication does not depend on p.

3.2   RAM occupancy

Each node has the same amount of RAM occupancy because the system is replicated on all the processors ($\simeq 3N * 8$ bytes). The arrays for the derivatives, till the fifth order ($\simeq 18N * 8/p$ bytes), the arrays for the forces ($\simeq 3N * 8/p$ bytes) and the matrix for the storage of the distances ($\simeq Nn * 8/p$ bytes where $n = 100$) are splitted among the processors. Furthermore the Verlet list is also splitted ($\simeq N * 4/p + Nn * 4/p$ bytes).

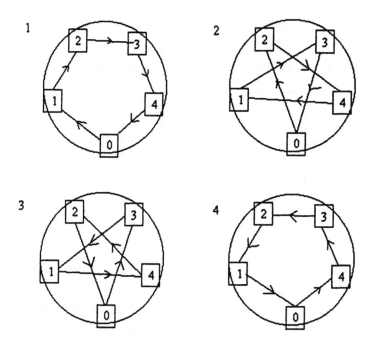

Fig. 2. Schematic of communication utilized in the code. In this example, 5 processors, in 4 steps, exchange their own data with all the others.

Thus the RAM occupancy on each node, for about 15000 atoms and 8 processors, is about 3.7 Mbytes.

Thus we can study more than 15000 atoms, but in this case we must change some algorithms. For example, for system with a great number of atoms the Verlet list is no more the more efficient way to calculate the distances, in fact it would be preferable to use the linked-cell algorithm.

3.3 Performances

In the following we report performances of our code on several parallel computers. We have benchmarked the parallel code timing the whole run on test cases. In the test cases we have adopted the parameters used in production runs, including the static load balancing and the distribution of the data performed by processor nought. Furthermore, the updating of the Verlet list is done every 50 time steps, this is a crucial parameter for the benchmark because it tunes the part of the code scaling in time with N^2.

We have tested the parallel version of the code on the following parallel computers: Ibm SP2 with High Performance Switch (HPS), cluster of workstations with ethernet and Cray T3D.

The IBM SP2 is present in the ENEA Frascati Research Center and the cluster of workstations is a cluster of SUN Spark20 that is in the ENEA Casaccia Research Center. The CRAY platform instead, it is kindly provided only for

Table 1. Speedup obtained on several parallel platforms

Processors	IBM SP2	Cl. work.	CRAY T3D
1	1.0	1.0	1.0
2	1.9	1.8	1.9
4	3.8	2.5	3.9
6	5.6	1.7	-
8	7.0	-	7.2

benchmark by the Edinburgh Parallel Computing Centre (EPCC, Edinburgh, Scotland).

As shown in Table 1, obviously the worst speed up is obtained running on the cluster of workstations. On the contrary, due to the high efficiency of the IBM HPS and the Cray inter-processors network, speed-up shows high efficiencies.

4 Conclusions and future improvements

We have ported and utilized efficiently an empirical tight-binding Molecular Dynamics code for the study of nanostructured materials. Furthermore, the Parrinello-Rahman-Nosé and the Gear predictor-corrector algorithms are implemented. The code was benchmarked on several MIMD platforms to verify the portability and the efficiency.

In the future, we will allow to the code to perform efficiently simulation runs also at different values of temperatures and pressures. In these cases, it can be important to implement an efficient dynamical load balancing routine to take in account for the large structural transformations that can happen in the system. The dynamical load balancing can be implemented according to a SPMD (Single Program Multiple Data) parallel scheme. The SPMD scheme is constitued by the following steps: 1) each processor performs the evaluation of the optimum number of interactions to be calculated, 2) each processor evaluates the best load balancing and 3) reorganize its own coordinates to send only its own information to the other processors that need of it.

Furthermore, the linked-cell algorithm, to calculate the distances among the atoms, will be implemented to allow the study of a greater system.

5 Acknowledgment

We acknowledge all the Computational Materials Science Group of ENEA, in particular Drs. V.Rosato, F.Cleri and G.D'Agostino and F.Pisacane for useful discussion and valuable suggestions.

We are in debt to Dr. O.Tomagnini (IBM ECSEC of Rome) for his kind and professional collaboration.

Furthermore, we would thank the Edinburgh Parallel Computing Centre to have given us the possibility to spend a period in the University of Edinburgh.

References

1. M.Celino, G.D'Agostino and V.Rosato, Nanostruc. Mat. 6 (1995) 751-754; Mater. Sci. and Eng. A204 (1995) 101-106.
2. V.Rosato, M.Guillope and B.Legrand, Phil. Mag. A, 59 (1989) 321-336.
3. F.Cleri and V.Rosato, Phys. Rev. B, 48 (1993) 22-32.
4. H.Gleiter, J. Appl. Cryst. 24, (1991), 79-90.
5. R.W.Siegel, "Nanophase materials: structure-property correlations", in "Material Interfaces", eds.: D.Wolf and S.Yip, Chapman-Hall, London 1992, pag. 431-460.
6. J.A.Eastman, M.R.Fitzsimmons and L.J.Thompson, Phil. Mag. B, 66 (1992) 667.
7. J.Eckert, J.C.Holzer, C.E.Krill and W.L.Johonson, J. Mat. Res. 7 (1992) 1980.
8. M.Parrinello and A.Rahman, Phys. Rev. Lett. 45 (1980) 1196-1199; J. Appl. Phys. 52 (1981) 289-297.
9. H.C.Andersen, J. Chem. Phys. 72 (1984) 2384-2393.
10. S.Nosé, Mol. Phys. 52 (1984) 255-268; J. Chem. Phys. 81 (1984) 511-519.
11. M.P.Allen and D.J.Tildesley, "Computer simulation of liquids", Clarendon Press, Oxford, 1987.
12. C.W.Gear, "Numerical initial value problems in ordinary differential equations". Prentice-Hall, Englewood Cliffs, NJ (1971).
13. M.Celino, Technical report ENEA n.205Q (1995).
14. D.M.Beazley, P.S.Lomdahl, N.Gronbech-Jensen, R.Giles and P.Tamayo, Comp. Phys. 3 (1995).
15. J.Plimpton, J. of Comp. Ph. 117 (1995) 1-19.

Automatic Test Pattern Generation with Optimal Load Balancing

H.-Ch. Dahmen, U. Gläser, H. T. Vierhaus

German National Research Center for Information Technology (GMD)
Institute for System Design Technology
Schloß Birlinghoven, D 53754 St. Augustin, Germany
email: [dahmen, glaeser, vierhaus]@gmd.de
Tel: (+49)2241-14-2875, Fax: (+49)2241-14-2242

Abstract. Test patterns are used to prove the correct functionality and absence of manufacturing faults after producing a chip. The automatic generation of those test patterns for sequential circuits is harder than NP-complete problems and therefore an interesting algorithm to parallelize.

In this paper we describe a parallel approach for automatic test pattern generation (ATPG) using PVM with optimal load balancing. The main advantage over existing approaches is a dynamic solution for partitioning the fault list and the search tree resulting in a very small overhead for communication without the need of any broadcasts and an optimal load balancing without idle times for the test pattern generators.

1 Introduction

The main objective of our present work is to develop a distributed algorithm for test pattern generation with the following features:
- reducing the run times for test generation for large circuits from days to hours
- executable on any network of workstations (including heterogeneous networks) and on the high performance parallel computer IBM SP2
- minimal overhead for communication
- optimal load balancing without idle times for the test pattern generators

To meet these features, we designed a communication scheme with "active servers" for test pattern generation and a "passive client" to organize the fault list and one server to handle fault simulation. In contrast to standard client server models, our "active servers" for test pattern generation send messages to the fault list handler client, requesting a fault to be treated. In this way there is no need for any broadcasting to synchronize the test pattern generators, synchronization is done by the fault list handler, distributing fault by fault after every request from one "active server".

This communication scheme with a dynamic dividing of the fault list and of the search tree is in contrast to the existing approaches [RaBa92], [AgAg93], [Krau94], [SiAg95] with mostly a static dividing of the fault list and of the decision tree. Since the communication overhead needed in our approach is quite low, we can use any standard network of UNIX workstations for test pattern generation even with slow connections between the workstations. For even faster test pattern generation we can use the IBM SP2 high performance parallel system (or any other parallel computer with UNIX nodes).

The rest of the paper is organized as follows: In section 2 we introduce briefly the problem of ATPG, in section 3 we provide details of the parallel sequential ATPG approach. In section 4 some implementation details are presented. Experimental results and comparison with the existing approaches are shown in section 5, followed by conclusions in section 6.

2 ATPG - The Problem

The rapid growth of integrated circuit technology has allowed VLSI devices to provide increased functionality while decreases in costs have allowed widespread use of integrated circuits in commercial applications. To handle issues of size and complexity most design processes assume a hierarchical design methodology and use automation in the design process. The increased complexity of such circuits makes testing a more crucial need to ensure correct operation and provide minimum standards of reliability. Testing these circuits for correct functionality and absence of manufacturing faults is currently one of the most pressing problems in industry.

The run time to compute test patterns for circuits are often days and weeks, therefore it is necessary to distribute the computation to cut down these long run times.

3 The Parallel ATPG Approach

In the following, our PARSET algorithm (Parallel Test Generation for Synchronous Sequential Circuits) is described in detail.

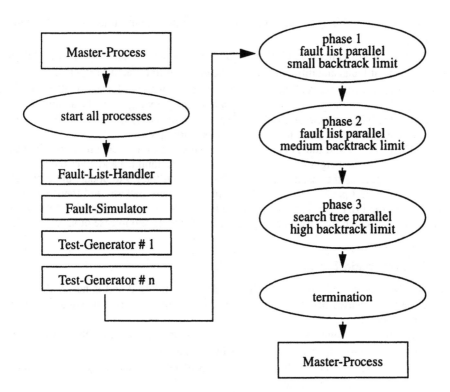

Fig. 1. PARSET Overview

3.1 The Algorithm Overview

PARSET aims at parallel test generation with each test generator handling a different fault at the same time (parallelization via the fault list) during the first two phases (figure. 1), during the third phase each test generator handles the same fault but a different part of the decision tree (parallelization via the search tree). Whenever a test generator finds a test sequence for a fault or a fault is proved to be redundant, this is reported to the fault list handler. In case a test sequence is found, this sequence is simulated by the fault simulator. In case of untestability the fault list is updated (during the third phase only if all the test pattern generators report untestability), and the fault simulator updates also its own fault list for untestable and covered faults.

We found out that in general it does not make sense to parallelize the fault simulator because this parallelization will not increase the efficiency of the approach significantly. The reason is the different complexity of test pattern generation (> NP complete) and fault simulation ($\in$ P).

3.2 The Processes

PARSET consists of four different processes:

- The master-process (MS): A single top level process handling subprocess incarnation and starvation.
- The fault-list-handler (FLH): A single process handling the fault list, a "passive client".
- The fault-simulator (FS): A single process handling fault simulation, a "normal server".
- The test-pattern-generators (TPG): A number of processes handling test generation for different faults and different parts of the decision-tree, "active servers".

The master process starts all the other processes and watches for termination. In big intervals it checks all the processes for being alive, thus allowing to handle abnormal termination of the TPG. In case of abnormal termination of one TPG it sends a message to the FLH to reset the fault the terminated TPG was working on.

The FLH, the FS and the TPG are based on the same executable program. The MS sets them up to do their right work. The FLH organize the fault list and gives the faults to the TPG and receives the results from the TPG and from the FS to update the fault list. The FS receives test patterns from the TPG, simulates these patterns and sends the covered faults to the FLH. The TPG work on one fault per time, during the third phase only on a part of the search tree. Results are reported to the FLH and in case of success to the FS.

3.3 The Message Types

The message passing scheme of PARSET is shown in figure 2 and is nearly the same during the three phases of test generation. There are eight different messages sent between the processes:

- Incarnation messages sent by the master to any subprocess and termination messages received from any subprocess by the master.
- A "fault message" is sent from the fault list handler to any test generator after a request from the test generator.
- The "pattern sequence message" is sent from a test generator to the fault simulator in case of success.

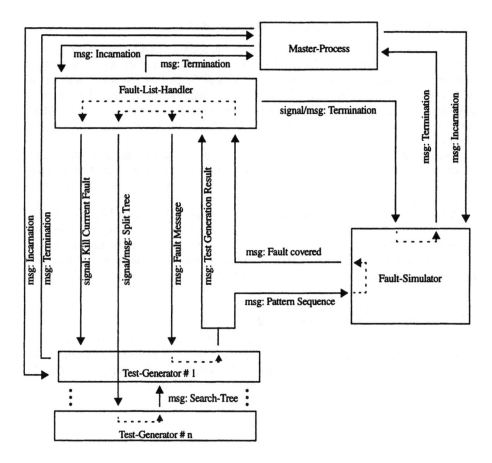

Fig. 2. The PARSET Message Passing Scheme

- The "test generation result message" is sent from any test generator to the fault list handler after finishing with one fault, also asking for the next fault is included.
- The "fault covered message" sent from the fault simulator to the fault list handler.
- A kill signal is sent from the fault list handler to any test generator if the fault which the test generator is actually dealing with is found to be covered by the pattern currently investigated by the fault simulator.

The kill signal is used to avoid redundant work in the test pattern generation process on one hand and to reduce the number of test patterns on the other hand. Although a test sequence for a covered fault is sometimes useful to cover some other faults, we decided to stop test generation in this case for reduced test generation time.

The following messages occur only during the phase for handling the search tree parallelism:

- The "split search tree signal" is send to a test generator from the fault list handler to start a splitting of the decision tree
- A "part of the search tree" is send from one test generator to another test generator after receiving the "split search tree signal"

3.4 Communication and Memory Management

The communication overhead is very low in our approach. Each process works on its own data structure copy. The message sizes for the communication are fixed to some few bytes with two exceptions: the pattern sequences and the parts of the search tree. For a pattern sequence, the message size depends on the length of the sequence, for the decision tree the message size depends on the position of the first optional decision, which is normally located close to the root of the tree. Thus the size of this messages is usually not larger than a few hundred bytes.

One limitation of this approach is that every process has to store the data structure of the circuit and for this reason there is redundancy in the main memory management. This redundancy cannot be avoided when working on a network of workstations with low communication overhead.

A typical communication diagram is shown in figure 3. Each black line between the processes indicates a communication message between them. The starting point and the end point of the lines also indicate the times when the message is sent and received. It is visible that the fault list handler (FLH) and the fault simulator (FS) have the highest communication effort, while the communication effort of the test generators (TPG) is relatively low.

3.5 Idle Times and Load Balancing

The master (MS) has only little work to do and is idle most of the time. For this reason we normally start this process on the same physically machine as the fault list handler. In figure 3 this is not correctly shown, because XPVM displays a sleeping process (UNIX sleep) like a working one, but work is only done at the black lines.

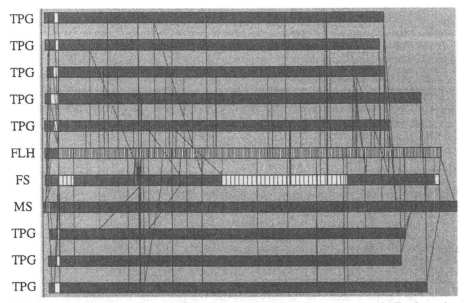

Fig. 3. An XPVM output displaying communication of processes and idle times in s208 circuit with 8 TPGs

Due to the dynamic handling of the fault list and the splitting of the search tree, the idle times of the test generators are very low. The amount of idle time depends on the time the process is waiting for a fault from the fault list handler and, in phase three, on the time the process is waiting for a part of the search tree. If the fault list handler is idle and a test generator finished its work sending a "test generator result" message, a new fault is immediately returned to the test generator and there is no idle time for the test generator. In some cases, when the fault list handler is communicating with the fault simulator, there is some small amount of idle time for the test generator, since the communication of the fault list handler with the fault simulator has higher priority than the communication with a test generator. This priority is given to the communication with the FS to avoid sending one fault to a TPG and then receiving a covered fault message from the FS for this fault resulting in a kill message to the TPG.

The time waiting for a part of the search tree is also very short, because the signal handling is performed immediately by the test generator that should send the part of the decision tree. We found in experiments that the idle times of the test generators are usually less than 1% of the overall time of a test generator resulting in an optimal load balancing for every TPG.

To avoid idle times by the TPG it is important to run the FLH and the FS physically on their own machines, so they can react very fast to each received message, especially in the case of many available TPGs. This results into idle times for the FLH and FS, which must be accepted to avoid idle times by a big number of TPG.

4 Implementation Issues

PARSET is implemented in C++ language with about 20.000 lines of code. For parallelization we use PVM [GeBe94] and UNIX signals [Brow94]. PARSET is based on the existing sequential algorithm FOGBUSTER [GlVi95]. For sending and receiving the existing C++ objects we added member functions like send and receive to our classes. To handle the different processes we built a number of process objects in the master process with member functions for execution.

5 Experimental Results

We tested the software on the ISCAS '89 benchmark [BrBr89] circuits using the stuck-at fault model with the reset option. The name of the circuits stands for the number of signals in this circuits, but another important fact is the sequential depth of the circuits. For the test generation processes we used SUN IPC and SUN Sparc 1 workstations plugged to the GMD Ethernet network and the IBM SP2 at GMD with 34 nodes, using standard Ethernet communication between the nodes.

We present results for most of the circuits up to the s526 and also one result for the s1423, for which ATPG is proved to be extremely difficult.

In table 1 we show complete results for a number of circuits received on the SP2. The 'Time for TPG' is the maximum of the times used in every TPG for test pattern generation. The 'Killed Time' is the sum of all times in test pattern generation which are aborted by a kill signal. 'PVM Time for Commun.' is the maximum of time used in the TPG for sending / receiving messages. The small amount of communication time shows the efficiency of our communication scheme.

In table 2 we show a comparison of our results with the parallel GENTEST. Note the different hardware for PARSET and GENTEST and that the cpu time for GENTEST is not including idle times.

Table 1. ISCAS '89 benchmark results for PARSET on the SP2

Circuit Name	Fault Coverage	Efficiency	Redundant Faults	Aborted Faults	Max Time for TPG in s	No of CPU's	Sum of Killed Time in s	PVM Time for Commun. in s
s298	89.73	95.27	16	14	18	4	<1	<1
s344	97.00	100.00	10	0	7	4	<1	<1
s349	97.08	99.42	8	2	4	4	<1	<1
s382	96.16	97.95	7	8	481	20	138	2
s386	81.77	99.74	69	1	8	4	<1	<1
s400	94.47	98.08	15	8	330	10	42	<1
s420	60.69	97.44	158	11	18	8	20	<1
s444	86.70	98.50	55	7	227	20	2,745	2
s526	80.04	84.10	22	86	12,967	8	5,184	12
s1423	43.23	44.72	25	862	90,000	32	17,312	415

Table 2. ISCAS '89 benchmark results for PARSET in comparison with others approaches

Circuit Name	PARSET SUN/Ethernet SUN SPARC IPX / 1+			PARSET IBM SP2 IBM SP2			GENTEST parallel SUN SPARC 2		
	Fault Coverage Efficiency	Time for TPG (s)	No CPU	Fault Coverage Efficiency	Time for TPG (s)	No CPU	Fault Coverage Efficiency	Time for TPG (s)	No CPU
s298	95.21	83	8	95.27	18	4	93.3	193	4
s344				100	7	4	95.0	105	4
s349	99.42	80	8	99.42	4	4	95.7	44	4
s382				97.95	481	20	100	47	12
s400	92.31	4406	8	98.08	330	10	93.6	648	12
s420	97.21	113	8	97.44	18	8	99.5	63	12
s444	91.20	11,117	8	98.50	227	20	97.5	1,320	12
s526	86.88	31,444	8	84.10	12,967	8	81.2	3,641	12
s820	96.12	2666	8				85.9	1,763	12
s832	96.09	4793	8				79.9	2,309	12
s953	82.85	1057	8				99.9	3	4
s1196	100	25	8				100	7	4
s1423				44.72	90,000	32			

The results for "GENTEST par" where taken from [AgAg93]

6 Conclusion and Further Work

This paper presents a new dynamic approach to parallel sequential test generation. The new PARSET algorithm allows for efficient test generation for synchronous sequential circuits. Idle times of the test generation processes are very low, due to a dynamic fault list handling and search space partitioning. Communication overhead is very low. The software is implemented using the PVM package and thus can work on an existing standard network of workstations and also on high performance parallel computers like IBM SP2. Experimental results for the ISCAS '89 benchmark circuits are encouraging. In further work we will mainly concentrate on extensions of this method. In detail we will concentrate on the ordering of the fault list, because its seems to be a very important factor for efficient test generation.

7 References

[AgAg93] P. Agrawal, V. D. Agrawal, J. Villoldo: *Sequential Circuit Test Generation on a Distributed System*, Proc. DAC '93, pp. 107-111

[AgTo93] M. J. Aguado, E. de la Torre, et al.: *Distributed Implementation of an ATPG System using Dynamic Fault Allocation*, Proc. ITC '93, pp. 409-418

[BrBr89] F. Brglez, F. Bryant, D. Kozminski: *Combinational Profiles of Sequential Benchmark Circuits*, Proc. 1989 Int. Symp. Circ. and Systems

[Brow94] C. Brown: *UNIX Distributed Programming*, Prentice Hall International (UK) Ltd.

[FuSh83] H. Fujiwara, T. Shimono: *On the Accelaration of Test Generation Algorithms*, IEEE Trans. on Computers (C-32), pp. 1137-1144, 1983

[GeBe94] A. Geist, A. Beguelin, et al.: *PVM: Parallel Virtual Machine - A Users Guide and Tutorial for Networked Parallel Computing*, 1994, avaiable via ftp from netlib2.cs.utk.edu

[GlVi95] U. Glaeser, H. T. Vierhaus: *FOGBUSTER: An Efficient Approach to Sequential Test Generation*, Proc. EURODAC '95

[GlVi96] U. Glaeser, H. T. Vierhaus: *Mixed Level Test Generation for Synchronous Sequential Circuits using the FOGBUSTER-Algorithm*, Accepted for Transactions on CAD

[GlVi92] U. Glaeser, H. T. Vierhaus: *MILEF: An efficient approach to mixed level automatic test pattern generation*, Proc. EURODAC '92, pp. 318-321

[GoKa91] N. Gouders, R. Kaibel: *Advanced Techniques for Sequential Test Generation*, Proc. 2.nd European Test Conference, Munich, 1991

[KlKa93] R. H. Klenke, L. Kaufman, et al.: *Workstation Based Parallel Test Generation*, Proc. ITC '93, pp. 419-428

[Krau94] Peter A. Krauss, *Applying a New Search Space Partitioning Method to Parallel Test Generation for Sequential Circuits*, SFB-Bericht 342/09/94 A (TUM-I9415), Institut für Informatik, Technische Universität München, 1994

[Micz83] A. Miczo: *The Sequential ATPG: A Theoretical Limit*, Proc. DAC '83, pp. 143-147

[NiPa91] T. Niermann, J. H. Patel: *HITEC: A test generation package for sequential circuits*, Proc. EDAC '92, pp. 214-218

[RaBa92] B. Ramkumar, P. Banerjee: *Portable Parallel Test Generation for Sequential Circuits*, Proc. ICCAD '92, pp. 220-223

[SiAg95] J. Sienicki, M. Bushnell, P. Agrawal, V. Agrawal: *An Adaptive Distributed Algorithm for Sequential Circuit Test Generation*, Proc. IEEE EURO-DAC '95, pp. 236-241

Parallel Crystal Growth for Chip Placement

Evgenia Hirik and **Iaakov Exman**

Institute of Computer Science
The Hebrew University of Jerusalem
Jerusalem - 91904 - Israel
Email: ianex@cs.huji.ac.il

Abstract. *Parallel Crystal Growth*, a new parallel extension of Simulated Annealing (SA), is presented. Sub-problems - the crystals - are solved in parallel, and iteratively merged into bigger crystals. The essential novelty is that processors spend time in a specific sub-problem, only as long as it is worthwhile, by *locally computed merging criteria*. Intra crystal search, is not repeated. Only inter-crystal exploration is done after merging. Whole problem termination is similarly determined. This leads to controlled speedup super-linearity, as demonstrated by experimental results with the PVM implementation.

1 INTRODUCTION

A problem is divided into a number of sub-problems. Each sub-problem is called a crystal, keeping with the physical analogy of SA algorithms. An initial crystal is assigned to just one processor. Along the time, crystals agglomerate into bigger crystals, and the corresponding processors, merge into clusters of processors which jointly solve the bigger crystal. The number of processors solving a sub-problem is kept proportional to the sub-problem size.

Simulated Annealing [2] has well known advantages (e.g. it is not trapped in local minima), but is very time-consuming. Recently, several SA parallelization attempts have been made [1], [4], [5]. These algorithms suffer from overheads, affecting diverse stages of the calculation. Parallel Crystal Growth tries to consistently overcome overheads throughout the calculation.

2 PARALLEL CRYSTAL GROWTH

The overall **Parallel Crystal Growth Algorithm** is:

1. **loop: forall clusters** - until termination condition is reached
   (a) **loop: forall processors in a cluster**
       i. run the SA algorithm, each with its own Temperature regime.
       ii. at the end of the outer SA loop, broadcast solution within the cluster.
       iii. periodically check whether a received solution is better than one's own; in the positive case, update solution.

(b) **merge crystals**
  i. test for merging condition;
  ii. if positive: merge with best crystal mate and coalesce corresponding clusters.

One starts with crystals with no less than a minimal size. As soon as no significant improvements are achievable, ready sub-problems are merged. At any time there may be running sub-problems of different sizes.

Each processor independently manages its own temperature, to increase the probability that processors - even those in the same cluster - go along different search paths.

3   WHEN AND WHICH CRYSTALS TO MERGE?

The question of when and *which* crystals to merge is central to our algorithm. One looks for a crystal, among the current ones, whose Acceptance Rate (AR)

$$AR = \frac{number\_of\_accepted\_moves}{number\_of\_attempted\_moves} \qquad (1)$$

becomes lower than a threshold. Once this crystal is located, it is merged with the crystal with the minimal Acceptance Rate among the remaining ones.

For the local acceptance rate threshold ART, we propose the following:

$$ART = K * \alpha^m * (1 - p) \qquad (2)$$

p is the crystal's fractionality - i.e. the number of modules in the crystal divided by the total number of modules, for the problem of modules placement in a chip. The $(1-p)$ term expresses the fact that smaller crystals need less time to explore their sub-problem space, since the smaller the relative crystal size, the bigger the probability of a module to be moved from its current place.

α is the factor by which the temperature decreases in each SA outer loop iteration; m is the number of temperature changes from the beginning of the optimization until the previous merging. The term α^m takes into account the regime in which the examined cluster starts the SA, without explicitly mentioning temperature T values (since T may vary from processor to processor). This is important for two reasons: first, as the value of T is smaller, more time is needed for the cluster to cover all search possibilities; second, we are not interested in reaching the frozen state with more than one cluster.

K is a constant that has to do with the ratio between the problem size and the number of the clusters at the beginning of the optimization. It allows one to compensate crystal over-fragmentation, by fast initial merging.

4   TERMINATION CRITERIA

Overall termination is determined by considering what can be done during the last stage of the algorithm, i.e. after the last merging:

1. interchange of modules that have never been in the same cluster - i.e. they were in *different* clusters before the last merging.
2. correction of eventual distortions made during the initial partition of the problem into crystals, by giving some extra time to reposition modules attached to the *same cluster* since the initial division.

Formally, the optimization terminates, similarly to the merging criterion, when a probability to exchange modules, falls below a final threshold f:

$$AR * PX < f \qquad (3)$$

AR is the acceptance rate and PX is the probability to choose a pair of modules that meets the above demands. If P_{diff} is the probability to choose modules that have never been in the same cluster (till the last merging) and P_{same} is the probability to choose modules that were allocated by the initial division to the same cluster, then:

$$PX = P_{diff} + P_{same} \qquad (4)$$

Assuming, as it is very reasonable, that the initial division allocates an identical number of modules to each cluster, P_{same} is given by:

$$P_{same} = \frac{1}{n_1} \qquad (5)$$

where n_1 is the initial number of clusters. P_{diff} will be obtained below indirectly.

4.1 High quality Solutions

High quality solutions are obtained when both demands of eq. (4) are satisfied. It is easier to calculate PX through its negation:

$$PX = 1 - \neg PX \qquad (6)$$

$\neg PX$ can be formulated as the probability to choose a pair of modules, for which there exists a merging i (besides the last one) such that before that merging, the modules were in different clusters and those clusters were chosen for the i_{th} merging. Formally,

$$\neg PX = \sum_{i=1}^{Number\_of\_mergings-1} Prob_i \qquad (7)$$

where $Prob_i$ is the probability that a pair of modules were in different clusters before the ith merging and these clusters were chosen for the ith merging. $Prob_i$ is given by:

$$Prob_i = \sum_{j=1}^{n_i} \sum_{k=1,(k!=j)}^{n_i} P_{jk}^{i'} P_{jk}^{i''} \qquad (8)$$

where $P_{jk}^{i'}$ is the probability that the chosen modules were in the jth and k^{th} clusters before i^{th} merging; $P_{jk}^{i''}$ is the probability that the acceptance rate of the jth cluster became lower than a threshold and the kth cluster has the minimal acceptance rate at that time (j_{th} and k_{th} clusters were chosen for i^{th} merging); n_i is the number of clusters before the ith merging.

Using p_l^i to designate the fractionality of the lth cluster before the ith merging, $P_{jk}^{i'}$ and $P_{jk}^{i''}$ are respectively defined as:

$$P_{jk}^{i'} = p_j^i * p_k^i \tag{9}$$

$$P_{jk}^{i''} = p_j^i * \frac{1}{n_i - 1} \tag{10}$$

Inserting these definitions into eq. (6) we finally obtain the expression for PX:

$$PX = 1 - \left(\sum_{i=1}^{Number\_of\_mergings-1} \frac{1}{n_i - 1} \sum_{j=1}^{n_i} p_j^i \sum_{k=1,(k!=j)}^{n_i} P_{jk}^{i'} \right) \tag{11}$$

4.2 Super-linear Speedups

Super-linear speedups are obtained by totally or partially relaxing some of the demands for high-quality solutions.

Since in eq. (4), the second term P_{same} only reflects a "correction" for eventual distortions, it is relaxed first. Thus $PX_{speedup}$, the probability to choose a pair of modules which were in different clusters until the last merging, is:

$$PX_{speedup} = PX - P_{same} \tag{12}$$

or actually subtracting P_{same} from eq. (11)

$$PX_{speedup} = 1 - \left(\left(\sum_{i=1}^{Number\_of\_mergings-1} \frac{1}{n_i - 1} \sum_{j=1}^{n_i} p_j^i \sum_{k=1,(k!=j)}^{n_i} P_{jk}^{i'} \right) + \frac{1}{n_1} \right) \tag{13}$$

4.3 Trade-offs between High-quality and Super-linearity

It is clear that the expression for $PX_{speedup}$ in eq. (13) does not stand for the ultimate achievable speedup. One can push it further, while still keeping an acceptable quality, as hinted in the following paragraphs.

Some algebraic manipulation together with reasonable approximations lead to the next formula:

$$PX'_{speedup} = 1 - \left(\sum_{i=1}^{Number\_of\_mergings} \frac{1}{n_i} \sum_{j=1}^{n_i} p_j^i \sum_{k=1,(k!=j)}^{n_i} P_{jk}^{i'} \right) \tag{14}$$

which obeys $PX_{speedup} < PX'_{speedup} < PX$ within an extended range of number of processors.

On the basis of eq. (14), one can obtain superlinearity (with solution quality above 80%, as shown experimentally in section 6), by assigning to p_j^i (for all j,i) its upper bound value 1. This termination criterion is:

$$PX = 1 - \left(\sum_{i=1}^{Number\_of\_mergings} \frac{1}{n_i} \sum_{j=1}^{n_i} \sum_{k=1,(k!=j)}^{n_i} P_{jk}^{i'} \right) \qquad (15)$$

Thus, we see that one can control quality vs. speedup trade-offs by tuning the values of p_j^i in the interval [actual high quality value of p_j^i, 1].

5 CRYSTALS FOR CHIP PLACEMENT

The special considerations concerning the initial division of a problem into crystals, and consequences along the optimization, are presented in the context of chip placement. These problems divide very naturally into sub-problems.

In the chip placement problem (see ref. [6]), a chip is represented by a rectangle with rows and columns of square cells. A NetList of modules, with given functionalities, has to be placed into chip cells, to minimize wiring length. Connection wires pass through horizontal and vertical channels. Channels are *not* uniformly distributed throughout the chip. The cost function is:

$$C = \sum_{v=1}^{Nv} (c(v) + p(v)) + \sum_{h=1}^{Nh} (c(h) + p(h)) \qquad (16)$$

where v and h are vertical and horizontal channel indices; Nv and Nh is the respective number of those channels; c(l) is the number of connections that cross channel l, and $p(l) = (c(l) - thr) ** 2$ if $c(l) > thr$ - a penalty for channel l, above a predefined threshold thr.

The length of the SA inner loop is set equal to the number of the cells on the chip, which implies a good probability that modules of each cell are considered at least once during each outer loop iteration.

5.1 Initial Partition into Crystals

Different clusters have to place disjoint subgroups of modules M in non-overlapping areas S of the chip. No collision occurs during merging of crystals, which is just the union of respective modules and areas. Each processor in a cluster executes a sequential SA over its (M,S) space. The length of the Markov chain generated by the cluster, equals the number of cells in S.

From the very beginning *crystals partition* should obey the following: placement of modules in the same cluster is done before merging of these modules into a bigger cluster. The remaining work after merging, should focus on interchanges across the boundary of the just merged crystals.

This property means high intra-cluster and low inter-cluster connectivity among modules. Identifying modules with nodes of a graph and their connections with the graph edges, we are looking for a partition of graph's nodes into N disjoint subsets such that the number of interconnections between subgraphs is minimal and the node distribution among subgraphs is as uniform as possible.

First, assuming N processors, N disjoint groups of modules are obtained. We use an heuristic Graph Partitioning algorithm [3] for this purpose.

Next, modules are assigned to cells. Two more definitions are needed. A value that reflects the cell's remoteness from the chip channels:

$$cell\_remoteness = \sum_{v=1}^{Nv} D(v) + \sum_{h=1}^{Nh} D(h) \qquad (17)$$

where $D(l)$ is the distance (in square cells) from the current cell to channel l. Lower distances, mean that the cell is less remote, thus more accessible.

A level of accessibility for each module:

$$module\_accessibility = \frac{\text{number of the module's connections}}{\text{total number of connections}} \qquad (18)$$

We loop through the modules' groups and in each iteration pick up from each group a module with the highest accessibility among those modules that haven't been allocated yet. If the chosen module cannot be positioned into the cells that are already allocated to the module's group, we find among the neighboring cells, that with the highest cell accessibility and position the module in this cell; from now on this cell is considered allocated to the group.

Remaining empty cells are divided among the groups, according to the group membership of most of the cell's neighbors.

5.2 Modules Attraction in Disjoint Crystals

Finally, the existence of connections between groups of modules in disjoint crystals cannot be neglected. Additional terms are put in the cost function to represent them.

One introduces dummy modules in each of the surrounding crystals and artificially attract them towards the crystal in question. The computational contribution to the cost function is of the same form of regular modules. Dummy modules are positioned in the "baricenter" of the cluster that they represent, i.e. the first cell allocated to that cluster (with the highest accessibility value). The SA algorithm executed by each cluster is not allowed to move these modules.

6 RESULTS AND DISCUSSION

6.1 Simulation

The crystal growth algorithm in all its aspects was implemented with PVM and run on a cluster of Pentium BSDI computers linked by ethernet. The cluster either runs with one process in each processor or simulates bigger clusters. Three of the chips employed for performance measurement are characterized in Table 1.

The temperature regime is regulated by $\alpha=0.95$ and the frozen state was attained when for 4 consecutive Markov chains the acceptance rate fell below 0.03.

chip #	chip size	# of modules	# of v channels	# of h channels
7	10x10	150	6	7
21	20x20	580	15	9
26	25x25	943	13	17

Table 1. *Sample chips for experimental runs.*

Results for a super-linear case, with the termination criterion in eq. (15), are seen in figures 1 and 2. Quality is measured by cost function values relative to sequential results taken with a score of 100. It is found that speedups grow significantly with the increasing number of processors. For instance, for chip21 speedup reaches values up to 9.22 and 21.6 with respectively 6 and 8 processors. Nevertheless, the quality of solutions for the same runs does not fall below 85%.

Results for a quasi-linear case, the termination criterion in eq. (11), are seen in figures 3 and 4. For instance, the speedup with 6 processors on the same chip21 is only 4.52 but its solution quality is 95.27%.

It should be remarked that the graph partitioning algorithm to get the initial division into clusters takes a negligible time, relative to the whole crystal growth.

6.2 Discussion

Parallel Crystal Growth, a new parallel extension of Simulated Annealing, has been demonstrated. Its main goal is to overcome overheads displayed by many of the known parallel SA algorithms.

Two versions of termination criteria are proposed: one obtains a high quality solution with quasi-linear speedup, while the other impressive speedups with good enough quality of the solution. Moreover, speedup vs. quality trade-offs are clear. By varying p_j^i, the fractionality of the jth cluster before the ith merging, on the interval [actual high quality value of p_j^i, 1], one improves either speedup or solution quality at the expense of the other. The precise implications and limits of these expressions are the subject of further investigation.

The algorithm was implemented for the chip placement problem. Experimental results match well the expectations.

References

1. Boissin N. and Lutton J.L. - "A Parallel Simulated Annealing Algorithm" - *Parallel Computing* **19**, 859-872, (1993).

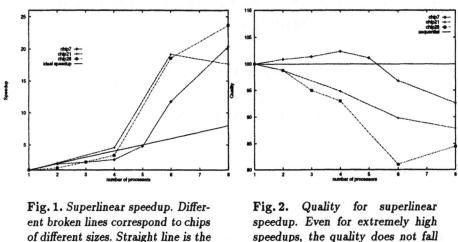

Fig. 1. *Superlinear speedup. Different broken lines correspond to chips of different sizes. Straight line is the linear speedup.*

Fig. 2. *Quality for superlinear speedup. Even for extremely high speedups, the quality does not fall below 80%.*

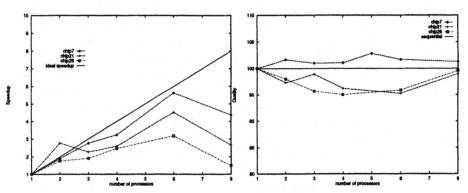

Fig. 3. *Quasi-linear speedup.*

Fig. 4. *High-quality for quasi-linear speedup. All results above 95%.*

2. Darema, F., Kirkpatrick, S. and Norton, V.A. - "Parallel algorithms for chip placement by simulated annealing" - IBM J. Res. Develop. **31**, 391-402 (1987).
3. Kernighan, B.W. and Lin, S. - "An efficient heuristic procedure for partitioning graphs" - Bell System Tech. J. - **49**, (2) 291-308 (1970).
4. Lee Kyung-Ceun and Lee Soo-Young - "Asynchronous Communication of Multiple Markov Chains in Parallel Simulated Annealing" - Int. Conf. on Parallel Processing, 169-176, (1992).
5. Nabhan, T.M. and Zomaya, A.Y. - "A Parallel Simulated Annealing Algorithm with Low Communication Overhead" - IEEE Trans. Parallel and Dist. Systems, **6**, 1226-1233 (1995).
6. Sechen, C. - "VLSI Placement and Clobal Routing Using Simulated Annealing" Kluwer Academic Publishers - Boston - 1988.

Efficient Message Passing on Shared Memory Multiprocessors

Massimo Bernaschi

IBM STS&S
P.le G.Pastore, 6 – 00144 Rome
e-mail: massimo@vnet.ibm.com

1 Introduction

For many years there has been a never ending discussion among defenders of the shared memory paradigm and supporters of the message passing approach. From a practical viewpoint, message passing is usually considered *cheaper* since it can be used, through packages like PVM [1], on inexpensive computing resources like networks of workstations, whereas shared memory has been available, in the past, only in large and expensive multiprocessor systems like Cray's. Recently the scenario has been changing: powerful multiprocessors RISC workstations running UNIX-like operating systems are offered by many vendors at very reasonable prices. These systems have good chances to become the typical *entry level* to perform parallel processing. However, two major issues must be addressed to exploit at its best this opportunity: the lack of a widely accepted (real or de-facto) standard for programming and the relatively poor level of the *sustained* performances of such systems. The latter is much more severe than the former because of a combination of *intrinsic* limitations like the hierarchical structure of memory subsystems (which causes long latencies) or the complexity of the memory-processor interconnections that need to support concurrent transactions and split the bandwidth available among all the processors[2].

In the present work we describe an efficient message passing mechanism for the shared memory environment which minimizes the operations required to the memory subsystem. We have included this mechanism in an experimental version of PVMe (the IBM implementation of the PVM package) that has been used on a PowerPC based SMP. In such way PVM message passing has become competitive in terms of performance with the additional advantage that many *already* existing PVM applications run more efficiently without additional effort.

2 Single-Copy Message Passing

Usually, in a shared memory multiprocessor controlled by the UNIX operating system the exchange of a message between two independent tasks (here we do not consider threads, so a "task" corresponds to a UNIX process) requires, at least, two memory copy operations. This is because of the restrictions imposed in the access to the virtual address space of another process by the operating system for security reasons. As a consequence, a message is sent either by using operating system services like UNIX-domain sockets and message queues, or at *user* level by means of shared segments. Luckily, to share a segment, tasks need to interact with the operating system just at initialization time; after that, the access is completely transparent although, at times, other system facilities (like *semaphores*) are required to control it. In any case messages which belong to a private data area must be copied by the sender in the shared segment and then copied again by the receiver in *its* private target area.

Our idea is to eliminate one of these two memory copy operations whenever it is possible. In other words, instead of copying the entire message twice, the tasks exchange few basic info and then just one of the peers copies the data. In practice distinct scenarios need to be considered because the necessary information are not always available at the beginning of a transfer. When the sender is ready to transmit but the target task has not yet invoked the corresponding receive primitive, the sender does not know the final destination of the message. In this situation it can provide just the following minimal set of information:

- its identity
- the position of its private data area in the *global* virtual address space of the system
- the offset of the message within this area
- the length of the message
- the message identifier

When the receiver is ready it can complete the transfer by copying the message directly from the private data area of the sender to the final destination (which is in *its* private data area). In case the *receive* primitive is invoked before the *send* one, the final destination of the message is immediately available (along with other info) to the sender so it can copy directly the contents of the message and the receiver does not need to perform any memory copy operation.

However this is an optimal solution just if the receive primitive is *no-blocking* because the receiver has the chance to do other, hopefully useful, work during the memory copy operation. With a *blocking* operation it is better to have the receiver busy in copying the message because otherwise it would be idle during the transfer.

In any case the fundamental requirement is to access the private data area of another process. But, as we mentioned above, operating systems usually do not offer such possibility and to overcome this obstacle we have employed two techniques.

In our first experiments we tried to use *real* addresses instead of virtual ones. The basic scheme was the following. The task which did not make the copy (it could be the sender or the receiver according to the previous discussion) invoked a special routine we wrote and got the real address of its buffer (actually if the buffer was longer than the page size or was not page aligned the translation had to be done for each single page). Then it advertised the real address(es) to the active peer task which did the same translation for its own buffer. Once the two real addresses were available, that task entered in *kernel* mode. In the PowerPC architecture[6] the virtual-to-real translation mechanism can be turned off independently for data and instructions. So the task disabled data translation and made the copy using real addresses inside the kernel domain. Data translation was re-enabled at the end of the operation just before switching back to *user* mode.

Although the *kernel* mode memory copy worked smoothly the performance was not satisfactory. For reasons that are still unclear it was much slower than an equivalent (i.e. same length) operation in user mode with virtual addresses so it was not worthy using it and we have given up to this method.

Our second, and currently in use, technique is based on an extended and enhanced version of the *shmat()* and *shmdt()* system calls.

The standard *shmat()* allows to attach in the virtual address space segments either created by means of the *shmget()* primitive or corresponding to *mapped* files. Conversely the novel *shmat()* accepts as argument the identifier of any type of segment including those containing the tasks' private data area. These identifiers are not usually available to the tasks but another new system call allows to retrieve them starting from any address which belongs to the private segment. Each task right after the initialization (i.e. within the *pvm_mytid()*) communicates the identifier to the PVMe daemon.

To send a message, a process attaches the shared memory segment of the receiver (each PVMe task allocates a shared segment at initialization time [3]) and writes there the position of the message within its private data segment, the message length, its task identifier and the message tag. The receiver reads these info, attaches the sender's private segment (whose identifier is obtained the first time by means of a query to the daemon) copies the message directly in its own private data area and finally detaches that segment. The attach/detach operations are almost negligible in terms of time since our *shmat()* is one order of magnitude faster than the standard one. Such improvement is possible because the routines can not be invoked directly by the user task. This restriction solves any potential security issue allowing to by-pass the extensive (and expensive) checks that the standard *shmat()* is required to do.

Such simple scheme becomes a bit more complex because the sender may attempt to modify the contents of the data area corresponding a message before the receiver has completed the copy operation. To avoid the drawback we have carried out a sort of *copy-on-write* solution. The set of information mentioned above (the address of the message, its length,...) plus few others (like the process identifier of the receiver) is saved in a global data structure by the sender and a signal handler is installed to trap protection violations (the handler is installed just once since it is the same for any message). Then, before the end of the operation, the data area corresponding to the message is marked as read-only using the *mprotect()* system call. After that, any attempt to modify the read-only area forces the architecture to generate an exception. The operating system is invoked to manage it and what it does is to send a signal (SIGSEGV) to the offending task. Inside the handler the sender checks whether the address which has caused the exception is within the boundaries of one of the messages it has sent. If the check is negative, it means that the violation is a really illegal access. In that case there are no alternatives: the handler is deinstalled and the signal re-raised. As a result the default action is taken: the operating system ends the process writing its memory image in a file called *core*. On the other hand, a positive check means that the violation is recoverable if the original contents of the data area are saved before modifying them. To do that in a consistent way the sender adopts the following strategy:

1. attaches again the receiver's PVMe segment
2. checks whether the receiver is copying the message at that time. To allow this control the receiver, before starting the copy operation, sets a variable in the shared segment equal to the address of the source. If the copy is in progress the sender waits till it completes and then goes directly to step 3. Otherwise it:
   − suspends the receiver by means of a SIGSTOP signal
   − copies the entire message in a part of that segment already reserved for the purpose.
   − updates the receiver's message queue so it refers to the new area for that particular message
   − restarts the receiver by means of a SIGCONT signal
3. detaches the segment
4. changes the access type of the data area to read/write

Clearly, even if the recovery mechanism is not activated, the permissions on the data areas which constitute messages must be reset equal to the original read-write mode as soon as the contents have been successfully copied. This could be done easily by the owner (i.e. by the sender itself) but the receiver should notify it explicitly and there is no effective way of doing that asynchronously. The solution we have found is to give the responsibility of such duty directly to the receiver. In other words the receiver, just before detaching the sender's segment, changes the protection bits of the pages it has copied. Unfortunately, the *mprotect()* system call does not work in this context, but we have been able to write another new system call which solves the problem. The trick here is to access the page table without using the kernel services. This is not trivial but, at least, the architecture we use (PowerPC) allows to retrieve easily crucial information like the location, in real memory, of the page table itself.

All the potential troubles we have just described disappear if the sender may execute immediately the memory copy operation. As we have already stressed, this is not just possible but also preferable when a non-blocking receive is started before the matching send operation.

3 Hardware Platform

The development and the tests have been performed on an IBM J30 system[4]. The J30 is an SMP which supports up to 8 PowerPC 601 CPUs. The organization is similar to many others bus based SMP's, with the important exception that a *data crossbar* (i.e. a non-blocking switch) is used to carry data transfers. Cache consistency is guaranteed directly in hardware by means of a snooping mechanism which implements the well-known MESI protocol [5]. In particular the so called *intervention*, that is data movement between first level caches, is supported, along with the standard *copy-back* procedure, to allow multiple concurrent cache-to-cache transfers.

Second level physical caches are available to supplement the on chip caches of the 601's. The benefit that second level caches provide strongly depends on the application type. In general they are extremely useful for commercial workloads which usually produce high miss rates whereas for engineering/scientific applications they give just a small advantage.

Main memory is organized in *banks* where each back is a 32 byte wide unit comprised of 4 DIMMs (Dual in Line Memory Modules). Each memory bank is independent and multiple memory accesses may be occurring at the same time on a memory board. Memory data is transferred to and from the data crossbar in parallel through a memory board interface that is 288 bits wide. This value corresponds to 32 bytes of data plus 4 bytes used by the ECC mechanism. Since a data transfer takes three cycles, the maximum theoretical bandwidth between the memory and crossbar is 800 Mbytes/second at 75 Mhz. However the bandwidth as seen by a processor in a *memcpy()* operation of a large block is typically much lower because the complete memory hierarchy (registers, first level cache, second level cache, main memory) must be crossed twice. The actual results are presented in table 1.

Number of bytes	Mbytes/sec
128	120.697784
512	128.560443
1024	128.864804
4096	70.660741
8192	62.583319
16384	47.627846
32768	43.863026
65536	35.426844
131072	28.618803

Table 1. Performance of the memory copy operation

The PowerPC architecture defines a *weakly order* memory order, which allows the hardware to re-order load and store operations. Special hardware support plus explicit synchronization primitives can be used to enforce an ordered storage model at single processor level. On system-wide basis, the software is responsible to ensure ordering between processors and possibly I/O subsystem.

4 Results

The "single-copy" mechanism is employed in PVMe for all messages bigger than the page size (which is 4096 bytes in the PowerPC architecture). The changes required in the PVMe code have been few and quite simple since most of the complexity is hidden in the system calls we have described in section 2.

To verify the real advantage offered to PVM codes, we have run both simple synthetic benchmarks and some real-world applications.

Looking at figure 1 it is readily apparent the significant improvement that the new mechanism offers compared to a classical "double-copy" approach. As soon as the threshold of 4096 byte is reached, the bandwidth sharply increases and becomes about three times higher. For messages such that the cache effects are

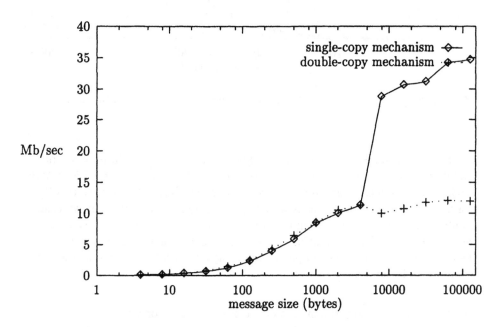

Fig. 1. Ping-Pong bandwidth. The message is left unchanged

negligible (i.e. when the message is bigger than the cache size) these results are nicely consistent with the data reported in table 1. An element which deserves special attention is the copy-on-write trick used to avoid message corruption. To measure its overhead, we have stressed it by using a test program which attempts, on purpose, to modify the contents of a message right after the sending. The results are reported in figure 2 and they show how there is still a good advantage with respect to the original "double-copy" mechanism.

About the applications, one of the best testcase we have found is the simulation of the immune system reaction to the introduction of a generic entity (antigen)

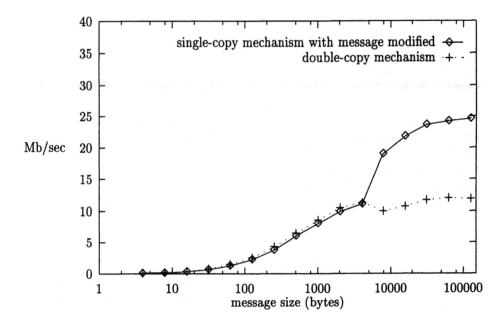

Fig. 2. Ping-Pong bandwidth. The message is immediately modified by the sender

which is potentially dangerous for the host organism. The code is based on a model developed by F. Celada and P. Seiden[7] which is implemented as a sort of *Extended Cellular Automata* [8] with non deterministic dynamics. In other words, the behavior of the cells and their interactions are ruled by stocastic events. Even if most of the computations are local to each site the movement of the cells is a non-local operation. As a consequence, for the sake of simplicity, the parallel version has been implemented using a typical master-workers scheme where at each iteration:

- the master distributes a subset of the sites to each worker
- master and workers carry out the local part of the simulation on their subset of sites
- the workers send back the results to the master
- the master carries out the non-local part of the computation (i.e. cells movement)

With this pattern the code is specially suitable for the exploitation of the single-copy mechanism because neither the master nor the workers need to modify immediately the contents of the data areas after the sending of the messages. The results shown in table 2 confirm this remark.

Number of processors	double-copy speed-up	single-copy speed-up
2	1.78	1.92
4	3.37	3.79

Table 2. Speedup of the immune system simulator with 256 sites and 1024 distinct *receptor* types

5 Conclusions

We have described a non-conventional scheme for the message passing on SMP systems. It may represent a possible solution to the performance issues of existing or new PVM applications on such systems and it has the important feature of being completely transparent to the users which are not required to change their codes. We like to stress that the same mechanism could be used for other message passing libraries (e.g. MPI) and that, although with some effort, we expect it to be portable to other SMP platforms as well.

References

1. V.S. Sunderam, G.A. Geist, J. Dongarra and R. Manchek,
   "The PVM concurrent computing system: evolution, experiences and trends",
   Parallel Computing, vol. 20, n. 4, April 1994.
2. J.D. McCalpin,
   "Memory Bandwidth and Machine Balance in Current High Performance Computers"
   IEEE Computer Architecture Technical Committee Newsletter, December 1995.
3. M. Bernaschi and G. Richelli,
   "PVMe: an enhanced implementation of PVM for the IBM 9076 SP2"
   Proc. of *HPCN Europe 1995*, Bob Hertzberger and Giuseppe Serazzi Editors, Springer, 1995.
4. J.O. Nicholson,
   "The RISC System/6000 SMP System",
   Proc. of *CompCon 95*, San Francisco, March 1995.
5. J. R. Goodman, "Using cache memory to reduce processor-memory traffic",
   Proc. of *International Symposium on Computer Architecture*, 1983
6. S. Weiss and J. E. Smith,
   "Power and PowerPC (Principles, Architecture, Implementation)",
   Morgan Kaufmann Publishers, 1994.
7. P. Seiden, F. Celada "A computer model of cellular interactions in the Immune System",
   Immunology Today 13, 2 1992, 55-62.
8. S. Wolfram,
   "Universality and Complexity in Cellular Automata",
   Physica 10 D 1-35, 1984, 91-125.

Development of PVM Code for a Low Latency Switch Based Interconnect

A. Touhafi , W. Brissinck , E.F. Dirkx
atouhafi-wfbrissi-efdirkx@info.vub.ac.be

V.U.B. TW-INFO Pleinlaan 2, B-1050 Brussel Belgium

Abstract. Message passing as a means to distributed parallel computing is becoming more and more attractive for communication intensive
applications. In this type of applications where the interchanged messages are rather short a small end-to-end latency is required. To achieve
a small end-to-end latency it is necessary to have a fast switch based
network with improved software. PVM as message passing tool in combination with an ethernet do not fulfill these requirements. By writing
a device driver optimized for small messages and by porting the PVM
routines to a Low Latency Switch Based Interconnect, better results can
be achieved.

1 Introduction

This work is related to a larger co-design case study, which is performed at the
department of parallel computing systems at the Free University of Brussels.
The case study deals with the design, study and implementation of a parallel
discrete event simulator (PDES). The PDES uses PVM as message passing tool.
Examination of the used simulation models [1] [2] and testing of the written code
proves that the PDES is very communication intensive.The size of the exchanged
packets is about 200 bytes. For this type of applications,standard PVM in combination with an ethernet based multiprocessor is not satisfying through the long
introduced latency.

Many study's have shown that switch based high-speed local area networks
can support low latency data transfer [3]. However a finetuning of the software
and hardware components is necessary. Our goal is to reduce the end to end communication latency which is affected by the hardware overhead and the software
overhead. Therefore a study of the introduced software and hardware overhead
is important.

The software overhead is composed of three parts [4]:

- The interactions with the host operating system.
- The device driver.
- The higher layer protocols.

In our case a smaller latency can be achieved by improving the device driver and
some PVM routines.

The hardware overhead is mainly composed of:

- The used transmission medium.
- The busstructure.
- The troughput and latency of the switch.
- The adapter efficiency and intelligence.

Performant low-cost switches and relatively good adapters are developed by many manufacturers.

In this paper we will discuss the way we've ported PVM to a switch based low latency LAN. First an overview of the system is given afterwards we will focus on the developed device driver and the ported PVM routines.

2 System Overview

The used hardware platform is a multiprocessor built up of Pentium PC's (120 MHz). All machines are interconnected by a 10Mb/s ethernet and a low latency, fast,switch based interconnect [5]. The latter are gaining popularity because of their low cost and good performance. However, the overhead introduced by the bus architecture and the network adapter [6] are typical for this type of LAN.

The network adapter is based on the IEEE 1355 [7] standardized DS-links [8]. The DS-links provide point-to-point communication between devices. Each connected pair of DS-links (1 input-link and 1 output-link) implements a full-duplex, asynchronous, flow-controlled connection. The links can operate at a speed of up to 100 Mbps. To maximize the bus-throughput, the adapter has been designed for machines supporting a PCI interface.

The switch is a 32 way Asynchronous Packet Switch'[5]. It connects 32 high bandwidth serial communication links to each other via a 32 by 32 way nonblocking crossbar switch. The links operate concurrently. Each link can operate at up to 100 Mbps. The switch uses wormhole routing in order to minimize the latency in message transmission. When the 32 links are active,the message delay introduced by the switch is approximately 12 microseconds for 64 byte messages.A maximum throughput of 50 Mbps can be achieved for this messages [9].

3 The Device Driver

Linux [10] is used as operating system for the Pentium PC's cluster. Interactions between the operating system and the user-application can cause throughput reduction and long latency because of copy actions between user and kernel space [11]. Also the protocol used by the device driver to handle the calling processes and the end-to-end transmission protocol play an important role in the introduced latency and throughput reduction. For this reason, the implemented device driver tries to execute direct copy's from user space to the network adapter. Nevertheless, the device driver provides input and output buffers. In case a process wants to send a message through a link which is not free, it's data will be copied in a FIFO buffer associated to that output link. Once the data is copied, the calling process can resume it's operation. This in opposit to a TCP/IP connection where the process will be put in a 'sleep' state as long as the connection is not established. The consequence is that the processor is IDLE for 80 percent and even more when communication intensive applications like the PDES are run.

The device driver foresees the following routines to the user:

- link_open: This routine will open a file descriptor.
- link_close: Opposite of link_open.
- link_send: The send routine is made very simple which is favourable for the throughput and the latency. This is possible due to the small size of PDU's. There is no need for packetization and fault handling since this is done by the network adapter for packets smaller then 4 Kb. The link_send routine will first check if the destination link is free. In case it is free, the data will be send directly to the network adapter. Otherwise the data is copied in a data-buffer at kernel-space until the output link becomes free.
- link_recv: The receive function of the device driver has a notion of the TID of processes defined by PVM. This way the header could be reduced to a minimum of 5 bytes.

  The receive function works in several steps:
  - First it checks if the data with the correct TID has already arrived. This is done by checking the *arrived-list* in which all the arrived messages are registered. To each link an *arrived-list* is associated.
  - If the data is available it is send to the calling process.
  - If no data is available in the *link-buffer-space* with the right TID, the process is put in a sleep state and is registered in the *link-wait-list* in case of a blocking receive. In case of a nonblocking receive the process continues it's operation once the *arrived-list* is checked.
  - If some data arrives at a specified link, the *link-wait-list* is checked for sleeping processes. The sleeping process with a correct TID will be woken and the data is written directly to the user data space.
  - If there is no process with the right TID, waiting for the received data, the data is written in the *kernel-buffer-space* and the arrival is registered in the *arrived-list*.

  The receive function can be accessed as long as there is:
  - No process reading data from a *link-buffer*.
  - No process reading data directly from a link.
  - No data (comming from a link) is being written to the *link-buffer-space*.
- link_select:This function allows the calling function to find out if the driver is ready to perform link_send or link_recv.

4. Porting PVM to the Hardware Platform

The changes deal with the direct communication between tasks located on different machines. The communication task-deamon, deamon-deamon and task-deamon-deamon-task are handled by the standard PVM routines. This makes that only a few routines had to be changed. The communication between two deamons is established through the TCP/IP (ethernet) network, while the task-task communication goes through the low-latency network. The performance improvement is achieved because we do not use TCP/IP to exchange data. Instead the data is handled directly by the device driver who tries to minimize buffering and interactions with the adapter.

The changed PVM routines are:

- pvm_initsend: The data will be always left in place, this in order to minimize the copy-actions.
- pvm_send: The following pseudo-code shows the structure of the pvm_send routine.

> **if**
>> (task_to_task-communication) and
>> (receiving_host different_from local_host) and
>> (local_host connected_to_LowLatencyLAN) and
>> (receiving_host connected_to_LowLatencyLAN)
>> **then** info=our_send
>> **else** info=standard_pvm_send

This approach is chosen for the ease of portability to new PVM releases. The actions that must be executed are: including the our_send routine, renaming pvm_send to standard_pvm_send and replacing pvm_send by the pseudo-code. The our_send routine, first maps the TID to the appropriate link number by using the Hosttable. Then the databuffer is located and directly send from userspace to the network adapter.

- pvm_recv: The implementation of pvm_recv is similar to that of pvm_send. The pseudo-code shows it's implementation.

> **if**
>> (task_to_task-communication) and
>> (sending_host different_from local_host) and
>> (local_host connected_to_LowLatencyLAN) and
>> (receiving_host connected_to_LowLatencyLAN)
>> **then** bufid=our_recv
>> **else** bufid=standard_pvm_recv

The our_recv routine can be used as a blocking- and a non-blocking receive dependent on the use of the link_recv routine.

- pvm_pk*: The packing routines are also an important source of latency. Improvements are possible in case the multiprocessor is homogeneous. In that case an application specific packing routine that makes an appropriate bit-stream can be used. A possible implementation is to use a preprocessor conditional that can be set or reset. We have defined the HOMOGEN_MACHINE conditional. If the HOMOGEN_MACHINE option is set during compilation of the code, the pvm_pk* routines are replaced by the faster packing routine.

– pvm_mcast: A lot of improvement can be achieved if the network adapter supports multicasting. At this stage, the network adapter does not provide multicasting support. But it will be implemented in the near future. This is possible since the interfacing part of the network adapter is implemented with reconfigurable logic (FPGA) [12].

5 Conclusion

Communication intensive applications using small packets, loose a lot of performance when executed on a multiprocessor interconnected by an ethernet, with PVM as message passing tool. The meassured end-to-end latency for messages of 64 bytes is almost 2 ms. Better performances can be achieved when a low latency switch based interconnect is used in combination with ported PVM routines. The end-to-end message delay introduced by the software for short packets is summarized:

Table 1. meassured end-to-end message delay in micro-seconds.

16 bits	32 bits	64 bits
114	172	290

Despite the optimized device driver and PVM-routines, the latency introduced by the software is still a large part of the total end-to-end latency. Better results could be achieved if the network adapter could bypass the kernel when reading and writing userdata.

The number of routines that must be ported can be kept small by using a combination of two networks. In our case, all the PVM control functions uses the ethernet while the task-task communication is executed using the fast switch based interconnect.

References

1. Parallel Discrete Event Simulation, R.M. Fujimoto, communication of the ACM octobre 1990.
2. Distributed Discrete Event Simulation, Misra J., ACM comput.surv.18,1 March 1986.
3. Low-Latency Communication Over ATM Networks Using Active Messages. IEEE Micro Volume 15 nr.1 p46-p53.
4. Distributed Network Computing over Local ATM networks, IEEE journal on Selected Areas in Communication Special Issue of ATM LANs, 1995.
5. Asynchronous Packet Switch ST-C104, SGS-THOMSON engineering data.

6. Applying programmable logic technology for generic network adaptor design and implementation, A.Touhafi, W.Brissinck,E.F. Dirkx, 8th IEEE Lan/Man Workshop proceedings.

7. IEEE Draft Std P1355 Standard for Heterogeneous Interconnect (HIC) (Low Cost Low Latency Scalable Serial Interconnect for Parallel System Construction), IEEE Standards Department 445 Hoes Lane, P.O. BOX 1331 Piscataway, NJ08855-1331, USA.

8. Networks, routers and transputers, Chapters 5 and 6, M.D.May, P.W.Thompson, P.H.Welch.

9. Networks, routers and transputers, Chapter9, M.D.May, P.W.Thompson, P.H.Welch.

10. Linux Kernel Hackers Guide , Michael K. Johnson, johnsonm@sunsite.unc.edu.

11. ATM on Linux, Werner Almesberger, Laboratoire de Reseaux de Communication, werner.almesberger@lrc.di.epfl.ch august 1995.

12. The Programmable Logic Data Book, Xilinx, 1994.

PSPVM: Implementing PVM on a High-Speed Interconnect for Workstation Clusters

Joachim M. Blum, Thomas M. Warschko, and Walter F. Tichy

University of Karlsruhe, Dept. of Informatics
Postfach 6980, 76128 Karlsruhe, Germany
email:{blum,warschko,tichy}@ira.uka.de

Abstract. PSPVM in an implementation of the PVM package on top of ParaStations high-speed interconnent for workstation clusters. The ParaStation system uses user level communication for message exchange and removes the operating system from the critical path of message transmission. ParaStations user interface consists of a user-level socket emulation. Thus, we need only minor changes to the standard PVM package to get it running on the ParaStation system.
Throughput of the PSPVM is increased eight times and latency is reduced by a factor of four compared to regular PVM. The remaining latency is mainly (88%) caused by the PVM package itself. The underlying sockets are so fast $(25\mu s)$ that the PVM package is the limiting factor. PSPVM offers nearly the raw performance of the network to the user and is object-code compatible to regular PVM. As a consequence, we achieve an application speed-up of four to six over traditional PVM using regular ethernet on a cluster of workstations.

1 Introduction

The PVM package is available on a wide range of parallel machines. These machines cover dedicated parallel machines as well as networks of workstations. On many dedicated parallel machines the supplied PVM uses proprietary send/receive calls to speed up communication. On networks of workstations all communication is usually done via the Unix socket interface. Socket communication relies on operating system functionality and common network protocols and is therefore limited in its speed. As long as these implementations were based on slow sockets the time lost in the rest of the PVM package (e. g. slow list implementation, packing/unpacking) was not the limiting factor. In this paper we present user level sockets on top of a high-speed interconnect where the speed is so fast that the PVM package limits the performance of the whole system.

2 Related Work

The public domain PVM package supports various multiprocessor environments. All of these are either based on shared memory multiprocessors (SunMP, SGI) or on message-passing multiprocessors (iPSC/860, Paragon, CM5). Common to all is that there is a daemon running on a host computer. This daemon organizes the communication with the other systems in the virtual machine and allocates/deallocates compute nodes on the local parallel machine.

A hardware configuration comparable to the ParaStation is the IBM SP2. It is built by bundling together regular RS/6000 workstations in a 19-inch rack and connecting them with a high-speed communication switch. Each node is controlled by its own instance of the AIX operating system. The interface communicates in user space or in system space. Only one process per node can communicate in user space concurrently. IBM has implemented a specially adopted PVM version for the SP2 which is called PVMe[BR92]. This reimplementation is missing the powerful dynamic task creation of regular PVM. Crays PVM version for the T3D is also missing dynamic spawning of new tasks. Both implementation use special hardware features to speed up communication.

3   The ParaStation System

The ParaStation system is based on the retargeted MPP network of Triton/1 [PWTH93, HWTP93]. The goal is to offer MPP-like communication performance while supporting a standard, but efficient programming interface such as UNIX sockets. The ParaStation network provides a high data rate, low latency, scalable topologies, flow control at link level, minimized protocols, and reliable data transmission. It is dedicated to parallel applications and is not intended as a replacement for a common LAN. These properties allow the use of specialized network features, optimized point-to-point protocols, and controlling the network at user-level without operating system interaction.

The ParaStation system library provides multiple logical communication channels on one physical link. Multiple channels are essential to set up a multiuser/multiprogramming environment which is needed to support various interfaces such as Unix sockets. Protocol optimization is done by minimizing protocol headers and eliminating buffering whenever possible. Within the ParaStation network protocol, operating system interaction is completely eliminated, removing it from the critical path of data transmission. The functionality to support a multiuser environment is realized at user-level in the ParaStation system library.

3.1   ParaStation System Library

During normal operation the ParaStation system library interfaces directly to the hardware without any interaction with the operating system (see figure 1). The device driver is only needed during system and application startup.

Since the hardware consists of a retargeted MPP network, many protocol features are already implemented in hardware. The gap between hardware capabilities and user requirements is bridged within the ParaStation system library (see figure 2) which consists of three building blocks: the hardware abstraction layer, the central system layer, and an application layer which uses the standardized user interface (sockets).

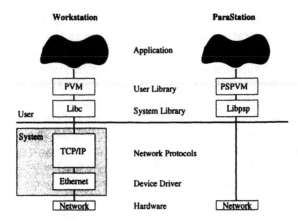

Fig. 1. Difference between traditional network interfacing and the ParaStation solution

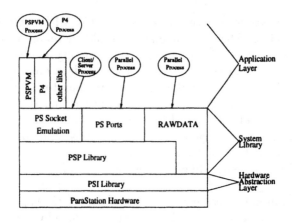

Fig. 2. ParaStation system library

Hardware Abstraction Layer. This layer provides an abstraction of the underlying hardware. It is normally used only by the ParaStation system layer. The implemented functionality of this layer consists of highly optimized send/receive calls, status information calls, and an initialization call to map communication buffers into user space.

Since messages at this level are addressed to nodes rather than individual communication channels, message headers simply contain the address of the target node, the number of data words contained in the packet, and the data itself. While sending a message, data is copied directly from user-space memory to the interface board and the receiving function does the same thing vice versa, eliminating all intermediate buffering. This layer provides a true zero-copy protocol.

System Layer (Ports). The system layer provides the necessary abstraction (multiple communication channels) between the basic hardware abstraction layer capabilities and a multiuser, multiprogramming environment. We reassembled operating system functionality at user level to meet our primary design goal of efficiency.

To support individual communication channels (called *ports* in ParaStation) on top of the node-addressed hardware protocol, the system layer adds information about the sending and receiving *port* in each packet. This concept is sufficient to support multiple processes by using different port-ids for different processes. For reasons of efficiency, semaphores necessary to ensure mutual exclusion while sending and receiving messages, are implemented at user level. Deadlock-free communication while sending large messages which cannot be buffered by the hardware is ensured by a combination of sending and receiving message fragments. Prerequisite for sending a message fragment is that the network will accept it. Otherwise incoming messages are processed to prevent the network from blocking.

The resulting implementation of these concepts contains no system call on the critical path of communication. Furthermore we tried to provide a zero-copy behavior (no buffering) whenever possible. To prevent deadlock situations, sometimes a single buffering is necessary. This technique leads to high bandwidths and low latencies.

Socket Interface. The socket interface provides an emulation of the standard UNIX socket interface (TCP and UDP connections), so applications using socket communication can be ported to the ParaStation system with little effort. Porting an application is as easy as adding a prefix to all socket calls. For connections outside a ParaStation cluster, regular operating system calls are used automatically. The interface can even handle file access when using `read/write` calls instead of `send/recv`. `Send/recv` calls which can be satisfied within the ParaStation-cluster do not need any interaction with the operating system.

Application Layer. ParaStation implementations of parallel programming environments such as PVM [BDG$^+$93](see section 4), P4 [BL92], TCGMSG [Har91], and MPICH [GL] use ParaStation sockets for high-speed communication. This approach allows us to easily port, maintain, and update these packages. We can use the standard workstation distribution rather than reimplementing the functionality on our own.

The structure of the ParaStation system library provides well known interfaces (UNIX sockets, PVM) to guarantee as much portability between different systems as possible as well as low-latency, maximum-throughput interfaces (raw-data port, hardware layer) to get maximum performance out of the hardware.

3.2   ParaStation Hardware

The ParaStation hardware [WBT96] consists of the retargeted and reengeneered MPP-network of Triton/1 [PWTH93, HWTP93]. Data transport is done via a table-based, self-routing packet switching method which uses virtual cut-through routing. Every node is equipped with its own routing table and with four buffers. The buffering decouples the network from local processing. The size of the packet can vary in the range from 4 to 508 bytes. Packets are delivered in order and no packets are lost. The network topology is either a two-dimensional toroidal mesh or a bidirectional ring. For both topologies a deadlock-free routing scheme is provided.

The ParaStation hardware is a PCI interface card. PCI was chosen because it meets our throughput requirements and it is available in several systems of different vendors (Intel-based systems, Digital's Alpha stations, IBM's PowerPCs, and Sun's UltraSparcs). The first implementation of the ParaStation system runs on Digital Alpha workstations with Digital UNIX (OSF/1); ports to other platforms are under way. At the time of this writing the first implementation on PCs running Linux is tested and shows excellent communication results.

4 PSPVM: ParaStation PVM

PSPVM is based on the ParaStation sockets. These sockets have the same functionality and same semantics as regular Unix sockets. The only two visible differences to the regular sockets are a prefix *pss* to each sockets call, and the higher speed of these sockets. Due to these sockets the changes to the standard library are small and an update to new versions of PVM is an easy task. We can use all advantages of the regular PVM package and communicate in user space without interaction with the operating system. Multiple processes are allowed on each node. The loopback of the ParaStation sockets is specially optimized (80MByte/s) and is mostly limited by the memcpy speed of the system.

An effect of a slow communication subsystem is that many programmers use *PvmRouteDefault* which sends a messages via the daemon on the same node which in turn contacts the daemon on the destination node, and delivers the message finally to the destination process. This routing scheme causes at least two process switches and two additional process-to-process latencies. Fast sockets reduce communication time so much that process switching and message transaction in the daemons are no longer a negligible part of the communication latency. To speed up communication, users should use the already provided *PvmRouteDirect* which establishes direct communication channels to the destination process.

The ParaStation system library supports most of the functionality of PVM and other message passing environments such as MPI and P4. A current project adjusts many programming environments to the ParaStation by using the ParaStation system library as a kernel. This approach enables us to only reimplement the environment-specific functionality and offer the whole speed of the hardware to the user.

5 Benchmark Results

The evaluation of PSPVM is done at two important levels. First, the latency and throughput of the ParaStation implementation is compared to the standard implementation. Second, an application benchmark is run on both implementations to compare the effect of an efficient communication subsystem to user applications. The tested ParaStation cluster consists of eight 21064A Alpha workstations (275MHz, 64MB memory). Our tests are all done with the same program object code, just linked with the other library.

5.1 Latency and Throughput of Sockets and PVM

Coarse-grained parallel programs depend only on throughput of the communication system, whereas fine-grained parallel programs also depend on low latency. The main reason that fine-grained parallel computation is inefficient on workstation clusters is that the latency of the communication subsystem is on the order of milliseconds. To measure the throughput und latency of the different communication subsystems we used a simple pairwise exchange benchmark. Both processes send and receive messages at the same time. This scenario is common in real applications. In the PVM versions of the program, packing and unpacking of the data is included. To get a measurement of the communication time and not of the packing, we used *PvmDataInPlace* buffer allocation.

Message	Throughput (in MByte/s)				Latency (in μs)			
	Standard		ParaStation		Standard		ParaStation	
size	PVM	sockets	PSPVM	sockets	PVM	sockets	PSPVM	sockets
4	0,01	0.01	0,04	0.31	733	566	200	25
16	0,04	0.05	0,17	1.21	748	631	191	26
64	0,16	0.19	0,61	3.53	764	655	204	36
500	0,69	0.70	3,16	8.34	1480	1411	316	119
2000	0,80	1.07	5,75	8.68	5073	3717	712	460
16000	0,79	0.89	6,41	8.73	41273	35816	5109	3663
64000	0,8	0.83	6,02	8.55	159682	153977	21214	14966

The PVM package adds about 200 μs to the latency caused by the underlying communication subsystem. This loss is not the dominating factor when built on top of regular sockets (loss of 29%), but it limits the performance when the system is built on top of ParaStation's user level sockets (loss of about 800%!!). Fine-grained parallel programs use many small messages and so these latencies dominate the performance of the whole application. On the other hand, the throughput of PVM is about 8 times higher on ParaStation. This improvement is achieved by just replacing the system calls by user level calls which use an additional PCI interface card. Therefore PSPVM has a substantial advantage even for coarse-grained programs. These numbers show that work must be done on the PVM system itself to speed up the message handling.

5.2 ScaLAPACK

The second application benchmark, *xslu*, taken from ScaLAPACK[1][CDD+95] is an equation solver for dense systems. Numerical applications are usually built on top of standardized libraries, so using these libraries as benchmarks is straightforward. ScaLAPACK is available for several platforms and presented results are directly comparable to other systems.

ScaLAPACK on ParaStation with PSPVM								
Problem	1 workstation		2 workstations		4 workstations		8 workstations	
size (n)	Runtime [s]	MFlop	Runtime [s]	MFlop	Runtime [s]	MFlop	Runtime [s]	MFlop
1000	5.0	134	3.36	199	2.95	226	2.74	244
2000	34.4	155	20.8	257	13.6	394	9.80	545
3000	109	165	62.3	289	39.2	459	27.9	647
4000			138	309	84.0	508	54.6	782
5000					152	547	96.4	865
6000					251	573	157	920
7000							234	978
8000							334	1022
ScaLAPACK on Ethernet with PVM								
Ethernet	n=3000	165	n=4000	232	n=6000	320	n=8000	261

The above table confirms scalability of performance in term of both problem size and number of processors. The efficiency of the two, four, and eight processor clusters are 94%, 87%, and 77% respectively. Remarkable is that we get more than a Gigaflop on an 8-processor cluster. These are real measured performance figures and not theoretically calculated numbers. The last line shows the maximum performance one gets using ScaLAPACK configured with standard PVM (Ethernet). The best performance in this scenario is reached at a problem size of n=6000 on a 4-processor cluster. Using even more processors results in a drastic performance loss due to bandwidth limitation on the Ethernet. For ParaStation, in contrast, we see no close limitation when scaling to larger configurations. And it is even possible to improve the ParaStation performance by using a better interface than PVM.

6 Conclusion and Future Work

In this paper we have presented an efficient way of using PVM for fine-grained parallel programs on workstation clusters just by replacing the regular socket by user level sockets which use a parallel communication interface card. This change results in a much better performance of the whole system. We have experienced a speedup of three to four over regular PVM on a number of applications. It is also shown that the PVM library itself is now the limiting factor. To eliminate this bottleneck we plan to redirect PVM calls to the ParaStation system library, where most of the PVM functionality is already implemented. This reimplementation will be object-code compatible to regular PVM and will deliver the native

---

[1] Scalable Linear Algebra Package.

performance of the ParaStation environment. Most additional work will be done in an efficient implementation of allocating/deallocating PVM-specific buffers. Due to the PCI interface we are not limited to the DEC Alpha workstations and we are working on a port to Linux on PC/Alpha and ports to other platforms such as Windows NT on PC and Alphas are scheduled[2].

On the network side we target a 100MByte/s application-to-application throughput in a new-generation ParaStation board with fiber-optic links.

References

[BDG+93] A. Beguelin, J. Dongarra, Al Geist, W. Jiang, R. Manchek, and V. Sunderam. *PVM 3 User's Guide and Reference Manual*. ORNL/TM-12187, Oak Ridge National Lab., 1993.

[BL92] Ralph Buttler and Ewing Lusk. *User's Guide to the p4 Parallel Programmimg System*. ANL-92/17, Argonne National Laboratory, October 1992.

[BR92] Massimo Bernaschi and Giorgio Richelli. PVMe: an enhanced implementation of PVM for the IBM 9076 SP2. In *ISCA*, 1992.

[CDD+95] J. Choi, J. Demmel, I. Dhillon, J. Dongarra, S. Ostrouchov, A. Petitet, K. Stanley, D. Walker, and R. C. Whaley. Scalapck: A portable linear algrbra library for distributed memory computers – design issues and performance. Technical Report UT CS-95-283, LAPACK Working Note #95, University of Tennesee, 1995.

[GL] William Gropp and Ewing Lusk. *User's Guide for* mpich, *a Portable Implementation of MPI*. Argonne National Laboratory.

[Har91] R. J. Harrison. Portable tools and applications for parallel computers. *International Journal on Quantum Chem.*, 40:847–863, 1991.

[HWTP93] Christian G. Herter, Thomas M. Warschko, Walter F. Tichy, and Michael Philippsen. Triton/1: A massively-parallel mixed-mode computer designed to support high level languages. In *7th International Parallel Processing Symposium, Proc. of 2nd Workshop on Heterogeneous Processing*, pages 65–70, Newport Beach, CA, April 13–16, 1993.

[PWTH93] Michael Philippsen, Thomas M. Warschko, Walter F. Tichy, and Christian G. Herter. Project Triton: Towards improved programmability of parallel machines. In *26th Hawaii International Conference on System Sciences*, volume I, pages 192–201, Wailea, Maui, Hawaii, January 4–8, 1993.

[WBT96] Thomas M. Warschko, Joachim M. Blum, and Walter F. Tichy. The ParaPC/ParaStation project: Efficient parallel computing by workstation clusters. Technical report, University of Karlsruhe, Department of Informatics, March 96.

---

[2] The information about the supported systems and many other useful information are available at http://wwwipd.ira.uka.de/parastation

Implementing the ALWAN Communication and Data Distribution Library Using PVM

Guido Hächler and Helmar Burkhart

Informatics Department
University of Basel, Switzerland
{haechler, burkhart}@ifi.unibas.ch

Abstract. ALWAN is a parallel coordination language and programming environment developed at the University of Basel. The design goals of ALWAN are to increase the programmability of parallel applications, enable performance portability, support the reuse of software components, and mixed-language programming. In this paper we summarize the language and describe the code generation for PVM-based environments. We report on performance measurements on IBM SP2, INTEL PARAGON, and CRAY T3D.

1 Introduction

Parallelism has been studied for quite a while and different paradigms, languages, and tools have been developed. PVM has certainly made an important contribution to our field because it was the first message passing library that offered portability. But its terminology is system-oriented (e.g. packing and unpacking of buffers) and important aspects such as data distributions need to be programmed by hand. The goal of our research is to increase the programmability of parallel applications by preserving as much performance as possible for different parallel architectures (performance portability). This paper describes ALWAN, a new skeleton programming language and its code generation for PVM-based systems.

Program development within the ALWAN environment is outlined in Fig. 1 to which the Roman numerals refer. A coordination description (I) written in ALWAN is transformed into a source code skeleton (IV) using the ALWAN compiler (III). Predefined ALWAN modules, where frequently used topologies or routines are collected, may be imported and re-used (II). Procedures declared as EXTERNAL define the interface to code written in other (sequential) languages (V). The high-level parallel coordination constructs are translated into ALWAN library (VI) calls with appropriate parameters. This library is implemented for various machines, interfacing to the virtual machine layers available on the given platform. Finally, all code parts are compiled (VII) and linked to form an executable program (VIII). Porting to a different platform only requires a recompilation on the target machine, thus replacing the ALWAN library with the appropriate new one.

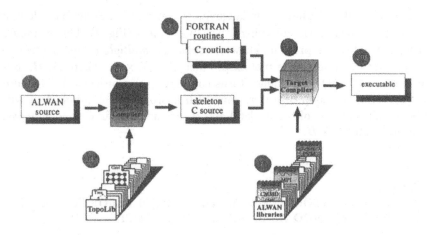

Fig. 1. Overview of the ALWAN environment

2   The ALWAN Language

ALWAN is a parallel coordination language for programming skeletons of parallel algorithms at a high level of abstraction. The language with its MODULA-2 like syntax and its strong typing concept, encourages structured parallel programming. Only few new concepts have been added reflecting the necessities of parallel programming:

Process topologies are the core of ALWAN programs. A topology is an entity consisting of a data and a statement section. During runtime, the topology behaves like a parallel procedure (the procedure body is executed in parallel on every node), working on different local data (SPMD). Statements outside a topology are applied to a single set of data (SPSD). The inherit mechanism allows to use *topology templates*, which describe the logical arrangement of processes and their interprocess communication paths for a given process topology.

Communication is explicit and is given by assignments. The assignment statement in combination with the *@location* construct (added either on the left or right-hand side) indicates a communication. If the location specifier denotes a single communication partner (*direction*), a point-to-point (direct) communication is expressed. Location specifiers denoting a set of processes (*group*) indicate collective communications (broadcast, scatter, gather, ...). The *active* statement enables the selection of subgroups of processes which initiate a communication.

Data distribution is described by partitioning and mapping constructs. Indices of arrays can be split in various manners (*block-wise, cyclic, ...*).

ALWAN supports both a local view and a global view of distributed data. Thus, index computations which are necessary within PVM programs are hidden from the programmer.

For a detailed description of the language we refer to [1]. Some basic language features are shown in the following program fragment (Fig. 2). The example is a matrix multiplication program known as *broadcast-multiply-rotate* algorithm [2]. The program computes the matrix product $C = A \times B$ (where A, B, and C are $n \times n$ matrices) on a square torus complex ($p \times p$ processes). The core of the algorithm is a loop of p iterations with a broadcast of subblock A within rows, a multiplication of subblocks A and B (accumulated in C) and an upwards rotation of subblock B.

```
1   TYPE
2     PartMatrix(n,p:LONGINT) = ARRAY [0..n-1],[0..n-1] OF LONGREAL
3       PARTITIONED AS BLOCK(n CDIV p),BLOCK (n CDIV p);
4     ...
5
6   TOPOLOGY Fox(A,B: PartMatrix; VAR C: PartMatrix; n,p:LONGINT);
7   INHERIT TopoLib.Torus(p,p);            (* Inherit torus properties *)
8   VAR                                     (* Private variables *)
9     i: INTEGER;                           (* Iteration counter
10    a: PART OF PartMatrix(n,p);           (* Local buffer*)
11  BEGIN
12    InitBlock($C, LENGTH($C,1));          (* External procedure 1*)
13    FOR i := 0 TO p-1 DO                   (* Systolic steps *)
14      ACTIVE (row_id+i) MOD p = col_id DO (* Sender selection*)
15        a@row:=$A                         (* Broadcast *)
16      END;
17      multiply(a, $B, $C, LENGTH($C,1));  (* External procedure 2 *)
18      $B := $B@south                      (* Rotate *)
19    END
20  END Fox;
21    ...
```

Fig. 2. ALWAN fragment of the broadcast-multiply-rotate algorithm

The process topology is built by a parallel procedure (**TOPOLOGY**, line 6) that **INHERIT**s torus attributes (**row_id, col_id, row, south, ...**) from the topology library (line 7). In an object-based way, torus-specific elements are accessible within this block. The matrices are declared as block-wise partitioned arrays (lines 2 and 3) using the **PARTITIONED** attribute; the necessary additional matrix buffer is declared using the **PART OF** notation (line 10). The variable a denotes a two dimensional array, which size corresponds to the size of a local subblock of a partitioned array. As ALWAN is intended to be a coordination language, external routines can be called within an ALWAN program (lines 12 and 17), which enables the reuse of sequential software components. The systolic steps clearly reflect the algorithmic structure. Line 14 defines the selection of the processes which broadcasts its matrix block to all row neighbours (the **@row** construct represents this process group). **$A** is a notation to refer to a local subblock of matrix A. In the rotation step (line 18) the local subblocks of B are overwritten by the corresponding subblocks of the process in the south.

3   The ALWAN Compiler

ALWAN programs are translated to ANSI C code. As the language enforces a strong typing concept many possible errors can be caught at compile-time. Additional code is inserted to perform run-time checks and to produce debugging, timing and tracing information (specified by compiler switches). The parallel language constructs are translated into a series of statements including calls to the ALWAN library. To illustrate this part, we describe the translation scheme of a direct communication statement. The ALWAN fragment in Fig. 3 (left side) specifies a data transfer where all processes of the first column of a torus topology fetch the first column vector of the matrix s of their eastern neighbour and store it in the last column vector of its local matrix d. The generated code is outlined on the right side of the figure.

ALWAN code

Generated code

```
TYPE
   T = RECORD
      id: INTEGER;
      pos: ARRAY [0..2] OF REAL;
   END;

VAR
   d,s: ARRAY [0..9],[0..9] OF T;

ACTIVE col_id = 0 DO
   d[.., 9]:= s[..,0]@east
END;
```

```
1   typedef struct {
2      short int id;
3      float pos[3];
4   } T;
5
6   T d[10][10], s[10][10];
7   ...
8   _active=eval(col_id == 0, my_pid);
9   _passive=eval(col_id == 0, pid(west));
10  _dRange=makeDesc(.., 9);
11  _sRange=makeDesc(.., 0);
12  _elemDesc=makeDesc(T);
13  ...
14  alw_move(d, s, _active, _passive, east
15     _dRange, _sRange,_elemDesc, ...);
```

Fig. 3. Translation scheme for a direct communication statement

The communication statement defines three groups of processes, the *active* processes (first column of torus), which are receiving data, the *passive* processes (second column of torus) from which data is fetched and all other processes, which are not involved in this communication. On line 8 the flag _active is set on all processes which belong to the first column. A process sets the _passive flag if it is a communication partner of an active process (line 9). Thus, the process has to reevaluate the active condition replacing its pid by the pid of the process for which it is the specified communication partner (in this case the pid of its western process). Processes may have set the _active and the _passive flag. ALWAN allows the assignment of subranges of arrays, therefore range descriptors must be generated (line 10, 11). To enable execution in an inhomogeneous environment, a description of the array element type is also required (line 12). A call of ALWAN library routine **alw_move** concludes the series of generated statements (line 14,15).

4   The ALWAN Library

As shown in the previous section, ALWAN statements are translated into appropriate library calls. The ALWAN library handles:

- Process management (creation, synchronisation, switching between SPMD and SPSD, ...)
- Communication (direct, collective, reduction functions, ...)
- Data management (dynamic allocation, distribution, collection, index transformation, ...)
- Input and output
- Various checks, timers, and trace generation

The library is split into two parts: an architecture independent part and a part with routines which vary on different virtual machines. Currently the library is fully implemented for PVM and MPI; a limited implementation exists for the PARAGON native NX2 communication library and for the virtual-shared memory environment of the Convex Exemplar. We pick up the example from the previous section and outline the implementation of the **alw_move** routine for PVM environments:

```
1   int alw_move (void *dest, void *src, int active, int passive, DIRECTION dir,
2                 R_DESC dRange, R_DESC sRange, E_DESC elemDesc, ...)
3   {
4     if (passive) {
5        alw_pack (src, sRange, elemDesc);
6        pvm_send(dir.send, channel++);
7     }
8     if (active) {
9        pvm_recv (dir.recv, channel++);
10       alw_unpack(dest, dRange, elemDesc);
11    }
12    return 0;
13  }
```

Fig. 4. PVM implementation of alw_move library routine

The pseudo-code in Fig. 4 shows a simplified PVM implementation. As both the **active** and **passive** flag may be set and **dest** and **src** may denote overlapping storage areas (e.g. the rotation step in the matrix multiplication algorithm, line 18 in Fig. 2), care must be taken, not to overwrite storage by unpacking received data (line 10) before data which must be sent is packed (line 5). The routines **alw_pack/alw_unpack** pack/unpack data into/from PVM message buffers, according to the range and element type descriptors. The latter is described by an array of tokens, which are interpreted to call the appropriate **pvm_[u]pkXXX** routines. The tokens contain information about type, size, array lengths and offsets to the record base address. The EBNF grammar in Fig. 5 specifies the token stream allowed.

TypeDesc   ::= SimpleDesc | ArrayDesc | RecordDesc.
SimpleDesc ::= **SIMPLE** size (**BYTE_D** | **INTEGER_D** | **REAL_D** | ...).
ArrayDesc ::= **ARRAY** size length TypeDesc.
RecordDesc ::= **RECORD** size fieldCount { offset TypeDesc }.

Fig. 5. Grammar of token stream describing data types

For the record type in our previous example (Fig. 3) the following token stream is generated by the routine **makeDesc** (line 12): { **RECORD, 16, 2, 0, SIMPLE, 2, INTEGER_D, 4, ARRAY, 12, 3, SIMPLE, 4, REAL_D** }. These tokens describe a record of size 16 bytes containing 2 fields, the first starts with offset 0 and is a simple type (2 byte integer). The second starts with offset 4 (assuming that floats are 4 byte aligned) and is an array with 3 elements of type float (array size = 12 bytes).

If the library is implemented for a homogeneous environment, the communication routines can be optimized using the **pvm_psend** / **pvm_precv** routines, but in the case of overlapping storage areas a buffering mechanism may be necessary. If a range descriptor specifies non-contiguous storage areas, data compaction and expansion is required. Some PVM implementations allow further optimizations, such as the usage of the CRAY T3D-specific **pvm_fastsend** / **pvm_fastrecv** communication routines.

5   Results

As ALWAN offers higher level constructs, we need to analyze the resulting performance and potential overhead. A first source of loss is the interpretation of the token stream which describes the data type to be transferred. Table 1 shows the execution times for the transfer of an array of records for various parallel architectures. The record type is pick up from the example in Fig. 3 (line 1-4) . The first row lists the times for a native pvm implementation, where

Table 1. Transfer times for 1024 records

Architecture	PARAGON	CRAY T3D	SP2	WS cluster
PVM native	31.0 ms	514.6 ms	7.6 ms	100.8 ms
ALWAN	40.2 ms	540.8 ms	9.4 ms	110.9 ms
ALWAN optmized	2.6 ms	0.97 ms	1.5 ms	44.7 ms

pvm_[u]pkshort(&p→id,1,1) and **pvm_[u]pkfloat(p→pos,3,1)** are called 1024 times. The data is transferred between two processes back and forth several times and the average transfer time is taken. The second row shows the time of the corresponding ALWAN implementation, while the third row shows an ALWAN implementation where an optimized library is chosen, which takes advantage of the homogeneity of the architecture. This optimization could also be implemented in the native pvm version, but would require changes in the source code, while the ALWAN implementation only needs a substitution of the library.

The second test case investigates the transfer of an array of integers (4 bytes each) with different array lengths. Figure 6 summarizes the relative losses of an ALWAN implementation versus a native pvm version (**pvm_[u]pkint(a,len,1)**). The relative overhead introduced depends on architecture and array length. An interesting case is the ALWAN implementation for the CRAY T3D. A library implementation using the regular *pack* and *send* (*receive* and *unpack*) mechanism caused relative losses over 50% for small messages. Replacing these routines with the T3D-specific **pvm_fastsend** / **pvm_fastrecv** routines caused a dramatic improvement, even shorter communication times could be observed. As the message length is limited for these routines, larger messages require several calls, which leads to a performance decrease. The ALWAN library therefore transparently uses the different variants depending on the message length.

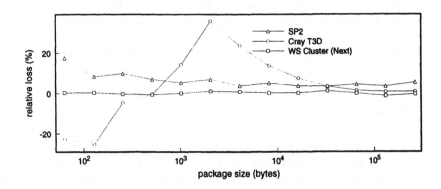

Fig. 6. Relative losses of ALWAN vs. native PVM implementation

As programs written in ALWAN are not bound to a specific virtual machine model, the best library version can be chosen for a given platform. Figure 7 shows a comparison of ALWAN library implementations on the PARAGON.

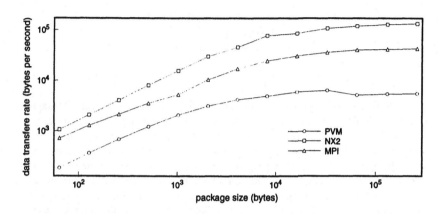

Fig. 7. Comparison of ALWAN library variants on PARAGON

6 Conclusions

ALWAN is a new approach for the development of parallel software. Software engineering concepts such as programmability, portability, and reusability favour the usage of such an approach. We have shown that the performance is close to the performance of message passing libraries. A public domain version and a user manual of ALWAN are available.

Acknowledgements

This research work has been supported by the Swiss National Science Foundation (SPP IF 5003-034357, SPP ICS 5003-45361) and Swiss Federal Office of Education and Science (ESPRIT HPCN project 21028). Robert Frank is a co-designer and implementator of ALWAN . We gratefully acknowledge help of other laboratory members: N. Fang, W. Kuhn, E. Lederer, G. Prétôt, K. Ruchti-Crowley, and B. Westermann.

References

1. Helmar Burkhart, Robert Frank, Guido Hächler, Peter Ohnacker, and Gérald Prétôt. ALWAN Programmer's Manual. Technical Report (in preparation 1995).
2. Geoffrey C. Fox, Mark A. Johnson, Gregory A. Lyzenga ,Steve W. Otto, John K. Salmon, and David W. Walker, Solving Problems on Concurrent Processors Vol. 1, Prentice-Hall International, 1988.

Transformation of Functional Programs into Data Flow Graphs Implemented with PVM

Ralf Ebner, Alexander Pfaffinger

Technische Universität München
Institut für Informatik
D–80290 München

Abstract. We present an implementation of the functional language FASAN for automatic coarse-grain program parallelization on workstation clusters. It is designed primarily for recursive numerical algorithms with distributed tree-like data structures and it exploits the maximal inherent parallelism of a program. Based on the stream and data flow semantics of the language, the compiler generates C procedures for building the data flow graph as dynamic data structure. FASAN schedulers evaluate the function nodes in parallel, and provide for all necessary communication using the PVM library. The new concept of "wrapper streams" for tree data structures avoids superfluous synchronization.

1 Introduction

Modern numerical algorithms concerning partial differential equations are often based on the technique of recursive domain decomposition or sub-structuring [2]. Instead of arrays, trees are the natural and suitable data structures for implementations of those recursive algorithms. However, their parallelization can be tedious when using conventional parallel programming concepts like pure message passing. Moreover, the organization of the message passing parts in the program may depend on the architecture and topology of the parallel system, so an efficient solution may not be portable in an easy way to other systems.

Functional languages allow a more abstract way of programming than explicit parallelizing techniques. It is a *declarative* programming style: We describe *what*, but not *how* to compute. So the programmer does not have to bother whether the program is executed sequentially or in parallel. A compiler could determine the call graph dependencies, divide the functional program into threads, and generate code in an appropriate thread language, like e. g. Cilk [3].

With our approach, the compiler generates procedures that construct the data flow graph dynamically, and a runtime library is responsible for the program execution through evaluation of the functions associated with the graph nodes. Therefore, we present FASAN (Functions Applied to Streams in Algorithms of Numerical Computing, [7], [8]), a functional auto-parallelizing language. It focuses on recursive functions and distributed tree-like data structures. This is a basic conceptual difference to the functional data flow language SISAL [6], which is based primarily on iterations and arrays, and evaluates recursion and trees only sequentially. It also differs from parallel languages like PCN [10], where a potentially parallel execution of recursive calls must be stated explicitly. FASAN is intended for coarse-grain parallelization on workstation clusters, whereas most parallel functional language proposals focus on fine-grain parallelism for shared memory computers [11].

2 Functional Programming in FASAN

The functional programming concept is inspired by the mathematical function notion, like $y = \sin(\frac{\pi}{2}x)$, where sin is applied without any side effect to x. In general, single-assignment variables and the lack of references (in form of aliases or pointers) avoid an implicit mutable computation state, enable automatic parallelization, and make programs easier to understand.

The aim of the FASAN system is to provide the programmer with a tool for producing a coarse-grain parallel and portable version of his application on a higher level of programming. The determinacy of a FASAN program relieves him from the typical parallel programming difficulties: proper synchronization, elimination of deadlocks, and description of complex communication structures or data distribution, as they arise for example with sparse grid algorithms [2].

FASAN describes a parallel functional layer between sequential modules and the overall framework that may also comprise database queries, visualization etc. Normally, it organizes the numerical computation part of an application, and is based on sequential functions written in conventional languages like C or Fortran. Thus, elaborated algorithms of standard numerical libraries may be included.

Functional programs are often considered less efficient than procedural ones, because their runtime system has to allocate all memory dynamically. It often copies memory blocks because it must not modify them selectively. On the other hand, a functional language does not force the programmer to code an explicit execution sequence of commands, but leaves the compiler or scheduler more freedom in finding the potential parallelism and an appropriate evaluation order of functions. So better program performance can be achieved through less idle time and less communication amount. A prominent example is SISAL, whose optimized compiler made the weather code run six times faster than the Fortran implementation did [1].

3 Function Definitions for Data Flow Graphs

A FASAN program consists in a set of function definitions. A function represents (part of) a data flow graph. Every function called in the definition body is represented by a graph node. The edges of the graph are the input, local, and result parameters interpreted as data streams. In fact, all functions take parameters in form of data streams. The single-assignment semantics requires exactly one value definition for each input or local stream in the function body, possibly controlled by conditionals. The FASAN system uses the stream semantics of the language for generating both concurrent threads and pipelines of the individual function nodes in the graph.

Tree structures in FASAN are specified to have asynchronous behavior. A tree is built as a *wrapper stream*. Whenever such a wrapper stream is decomposed, its constituent streams connect directly the sending and receiving function nodes and subtrees. They avoid superfluous synchronization induced by the recursive call hierarchy that ordinary tree data structures would cause.

In order to illustrate the conversion of function calls into data flow subgraphs and the benefit of wrapper streams, we take as an example the 2-dimensional quadrature of a function defined on a rectangular area. We follow an algorithm presented in [5] that is based on recursive domain decomposition. The underlying hierarchical basis does not require the absolute function value, but the value surplus for every newly added support point with respect to its four neighbors.

In the function ar_quad, the values of the four boundaries of the square are supplied in N, S, W, and E. M denotes the west-east separator line in the middle of the area. Figure 1 shows the FASAN function and the corresponding data flow graph (the test for the termination condition is omitted).

```
function ar_quad (N, S, W, E): i;
begin
   ⟨ Wn, w, Ws ⟩ := W;
   ⟨ En, e, Es ⟩ := E;
   im, M := sep_line (N, S, w, e);
   i1 := ar_quad (N, M, Wn, En);
   i2 := ar_quad (M, S, Ws, Es);
   i := im + i1 + i2;
end
```

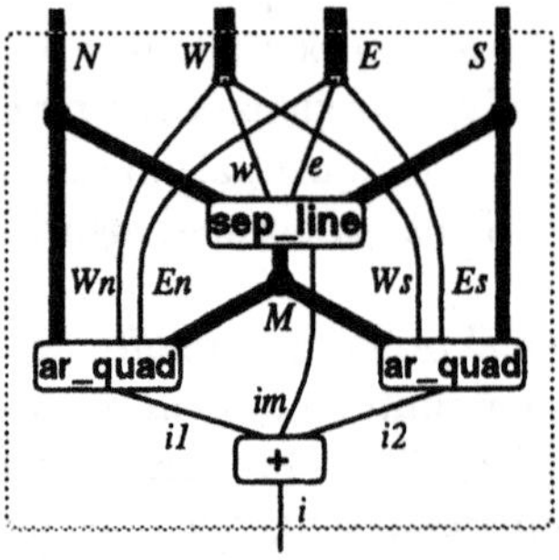

Figure 1. The function ar_quad and its data flow graph.

During the evaluation of the function, the recursive calls to ar_quad are replaced by identical data flow graphs. W (E respectively) is a wrapper stream that the ⟨⟩-operator decomposes into the northern subtree Wn (En), the central point w (e), and the southern subtree Ws (Es). The subtrees are passed directly to the recursive ar_quad call.

In contrast M is passed entirely to the ar_quad calls. With ordinary tree structures, these recursive calls would have to be delayed until the whole tree structure of M is computed. But since M is organized as wrapper stream, the ar_quad calls can immediately use those parts of M that are already computed. Given enough processors, the asynchronously working function nodes lead to maximal parallelism of the program: Concurrent work of sep_line and the two ar_quad calls is possible in every recursion depth, in spite of the syntactic dependence via M [8].

Beside the increased number of function nodes that can potentially work in parallel, wrapper streams enable a sensible automatic data distribution: They care for data to remain local on a host whenever possible, since direct connection streams need not transmit data over the network if sender and receiver nodes are located on the same host. This concept works well together with *tail recursive* functions (i.e. the recursion is the outermost function call), like for example in the more elaborated numerical quadrature algorithm aqho [4]. Aqho organizes the points of a sparse grid without boundary for the function evaluations in a tree structure (analogous to M in ar_quad). It traverses and refines this structure adaptively according to both local and global error estimations, adding gradually further points of support for a more exact function approximation. These repeated traversals can be expressed in FASAN as follows:

```
function iterate (tree):  ret;
begin
  new_tree, estimation := traverse (tree);
  if isstopped (estimation)
    then ret := result (new_tree);
    else ret := recycle iterate (new_tree); fi
end
```

Any such tail recursive function can be interpreted as iteration. This fact is reflected in the generated data flow graph: Instead of unfolding the recursive calls as a pipeline of **traverse** nodes (that's what the compiler actually produces without the keyword **recycle**, see Fig. 2, left), we can reuse the already existent graph by redirecting the output stream **new_tree** of **traverse** to the old input streams of **iterate** (Fig. 2, right) [8]. Note the direct connection streams (dotted in Fig. 2) that are drawn through decomposition of wrapper streams.

4   Implementation of the Data Flow Graphs with PVM

We now focus on the program execution based on the graph generating and sequential evaluation procedures compiled out of a FASAN program. On every host, a scheduler, the core of the FASAN runtime system (RTS), runs in a PVM process. It selects the functions to be executed from an *active set* queue in FIFO order. A so-called *synchronous* function or an external function call is put into the active set if all its input streams contain data. An *asynchronous* function (explicitly stated as such in the FASAN program) can be activated already with data on only one input stream.

A start program, the master PVM process, spawns the RTS processes on all hosts. Then it sends information about the successfully spawned processes to all and a function call to one of them. A subsequent message containing input data triggers unfolding of the data flow graph and distribution over the RTS processes (see Fig. 3). Finally, the output data is available for the start program. The

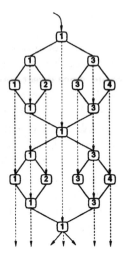 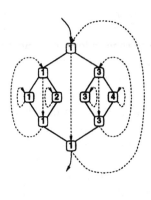

Figure 2. Basic structure of the aqho data flow graph, unfolded for pipeline use (left), and as cyclic memory-saving graph (right). The node numbers give a straight-forward distribution scheme on four hosts.

master process is free to perform arbitrary computation on the input and output data, including data base queries, visualization etc.

As we already mentioned, there are two ways to evaluate a function node: The sequential execution on the stack results in depth recursion and is economic with respect to memory usage. The second possibility is the unfolding procedure for building a data flow subgraph. This method enhances potential parallelism, analogous to breadth recursion. Unfortunately, for a cascade recursion like ar_quad, memory consumption increases exponentially with the recursion depth. Therefore, the user of a FASAN program can limit the unfolding procedures to the upper nodes of the call tree, either by giving explicitly the maximal call depth, or by specifying a limit for an integer *location parameter*, which is calculated as attribute during the execution of the unfolding procedure for every function call annotated like:

```
i2 := ar_quad(M, S, Ws, Es) on next_host;
```

The function next_host is a so-called *location function*, a programmer supplied load distribution scheme. It must calculate the number of an appropriate destination host for the migration of the function node. For this purpose, it may use the location parameter of the calling function, the actual and maximal host index, and dynamic information about system load and computation time. The location functions may also use knowledge of the architecture and topology of the parallel system, while the FASAN program itself remains architecture independent. The compiled FASAN program is based on the PVM library. All code for node migration and inter-process communication via streams between two function nodes is compiled automatically.

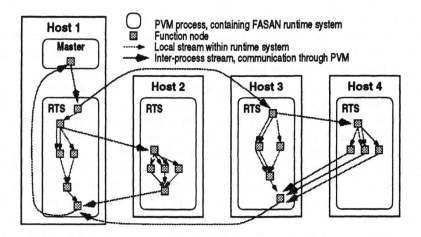

Figure 3. A snapshot of a (partially unfolded) data flow graph distributed over RTS processes.

The RTS library defines a protocol on top of PVM for packing and sending as well as receiving and unpacking stream and data element structures, function nodes, and system signals. When sending a function node, its input and output streams are blocked (Fig. 4). Their identifiers are sent together with the node to the destination RTS, which, in turn, creates stream stubs and sends data requests back to the originating RTS. Only when these requests (for input streams) or data (for output streams) arrive, the blocked streams know their corresponding destination stubs and can begin sending data.

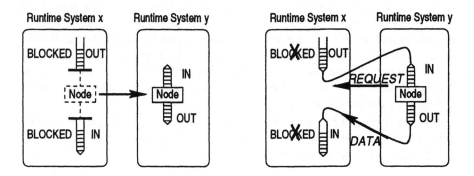

Figure 4. The migration of a function node (left), and the resulting inter-process streams (right).

Currently, migration of a function node is only possible during the unfolding procedure. For general dynamic load balancing, function nodes should be able to migrate later on as well. An extension of the communication protocol would then be necessary, with additional message types for requesting the blocking of remote stream stubs, and for acknowledgments of such requests.

Messages containing only streams occur when a wrapper stream is decomposed. The decomposition yields identifiers of the compound streams, together with sequence numbers of the actually needed data elements in the streams. If these streams are located in another RTS and not yet connected to the consumer, a new stream stub is created, and a data request message is sent to the producing RTS. Once such a direct stream connection has been established, the producer can immediately send all created data on this stream to the consumer. So no more data requests are needed from the second stream element on. The sequence numbers (which the stream identifiers arriving at the consumer carry) allow filtering all the data elements out of the direct stream that the control flow through conditionals has eliminated.

In order to adapt the communication rate and the package size of the PVM messages to the network bandwidth, the RTS processes can be told to accumulate messages until a given total number of bytes is reached, so that the network is not flooded with too many small packages.

5   Runtime Experiments and Observations

Several interesting observations have been made with the implementation of FASAN: For example, writing a (partially) generated data flow graph to a file and loading it in later computations can accelerate the startup phase. Also, the decision whether to unfold a function call as data flow graph or to execute it sequentially is a crucial question: Too many unfolded functions make up a gigantic graph, increasing the number of nodes exponentially with the call depth. Too small a data flow graph generates "fat" sequential nodes that decrease potential parallelism and disable communication between hosts for long periods. Besides, there is another tradeoff between short delay on inter-process streams and the overall number of data packages: We will have to explore heuristics that are needed for the decision on the amount of stream data to be accumulated before sending it to the receiving process.

We give results of the adaptive computation of $\int_{-1}^{1} \int_{-1}^{1} |xy + 0.05|^{-3.5} \, dx \, dy$ with the aqho program on a cluster of up to 16 workstations of type HP 9000/720 during moderate user load in the following table:

#hosts	ε	runtime	active time	fixed time speedup
1	10^{-5}	51.48 s	49.39 s	0.96
2	10^{-6}	59.20 s	82.84 s	1.40
4	10^{-8}	75.11 s	168.16 s	2.24
8	10^{-10}	87.43 s	330.21 s	3.78
16	10^{-12}	99.10 s	511.73 s	5.16

The error limit ε is chosen as small as possible so that the machines just need not swap memory pages. We used the simple static distribution scheme sketched in Fig. 2. The values for the *fixed time speedup* (sum of actual computation periods of all processes divided by the application runtime) are acceptable, since aqho is an adaptive algorithm. Note that the sequential program would not have been able to compute $\varepsilon < 10^{-7}$ due to the limited memory of a single machine.

6   Applications and Conclusion

The described FASAN programs for adaptive numerical quadrature have demonstrated the usefulness of the automatic generation of complex communication relations. At the moment, sub-structuring finite element codes [9] and the new adaptive sparse grid technique for the solution of elliptic differential equations [2] are being implemented with FASAN. Since the system can use the same code both for the sequential and the parallel program version, FASAN reduces the time to develop the parallel application considerably.

References

1. G. S. Almasi and A. Gottlieb. *Highly Parallel Computing.* Benjamin Cummings Publishing Company, Redwood City (CA) et al., 1994.
2. R. Balder. *Adaptive Verfahren für elliptische und parabolische Differentialgleichungen auf dünnen Gittern.* Dissertation, Technische Universität München, 1994.
3. R. Blumofe, C. Joerg, B. Kuszmaul, C. Leiserson, K. Randall, and Y. Zhou. Cilk: An Efficient Multithread Runtime System. In *Proceedings of the 5th ACM SIPLAN Symposium on Principles and Practice of Parallel Programming PPoPP, Santa Barbara, CA, July 19-21, 1995*, New York, 1995. ACM Press.
4. T. Bonk. A New Algorithm for Multi-Dimensional Adaptive Numerical Quadrature. In W. Hackbush, editor, *Proceedings of the 9th GAMM Seminar, Kiel, January 22-24, 1993*, Braunschweig, 1993. Vieweg Verlag.
5. T. Bonk. *Ein rekursiver Algorithmus zur adaptiven numerischen Quadratur mehrdimensionaler Funktionen.* Dissertation, Technische Universität München, 1995.
6. D. C. Cann. SISAL 1.2: A Brief Introduction and Tutorial. Technical report, Lawrence Livermore National Laboratory, Livermore (CA), 1992.
7. R. Ebner. Neuimplementierung einer funktionalen Sprache zur Parallelisierung numerischer Algorithmen. Diplomarbeit, Technische Universität München, 1994.
8. R. Ebner, A. Pfaffinger, and C. Zenger. FASAN — eine funktionale Agenten-Sprache zur Parallelisierung von Algorithmen in der Numerik. In W. Mackens, editor, *Software Engineering im Scientific Computing, Tagungsband zur SESC 1995, Kiel*, Braunschweig, to appear 1996. Vieweg Verlag.
9. R. Hüttl and M. Schneider. Parallel Adaptive Numerical Simulation. SFB-Bericht 342/01/94 A, Technische Universität München, 1994.
10. I. Foster, R. Olson, and S. Tuecke. Productive Parallel Programming: the PCN Approach. *Scientific Programming*, 1:51-66, 1992.
11. W. S. Martins. Parallel Implementations of Functional Languages. In *Proceedings of the 4th Int. Workshop on the Parallel Implementation of Functional Languages, Aachen*, 1992.

Computer-Assisted Generation of PVM/C++ Programs Using CAP

Benoit A. Gennart, Joaquín Tárraga Giménez and Roger D. Hersch

Ecole Polytechnique Fédérale de Lausanne, EPFL

gigaview@di.epfl.ch

Abstract. Parallelizing an algorithm consists of dividing the computation into a set of sequential operations, assigning the operations to threads, synchronizing the execution of threads, specifying the data transfer requirements between threads and mapping the threads onto processors. With current software technology, writing a parallel program executing the parallelized algorithm involves mixing sequential code with calls to a communication library such as PVM, both for communication and synchronization. This contribution introduces CAP (Computer-Aided Parallelization), a language extension to C++, from which C++/PVM programs are automatically generated. CAP allows to specify (1) the threads in a parallel program, (2) the messages exchanged between threads, and (3) the ordering of sequential operations required to complete a parallel task. All CAP operations (sequential and parallel) have a single input and a single output, and no shared variables. CAP separates completely the computation description from the communication and synchronization specification. From the CAP specification, a MPMD (multiple program multiple data) program is generated that executes on the various processing elements of the parallel machine. This contribution illustrates the features of the CAP parallel programming extension to C++. We demonstrate the expressive power of CAP and the performance of CAP-specified applications.

1 Introduction

Designing a parallel application consists of dividing a computation into sequential *operations*, assigning the operations to threads, synchronizing the execution of the threads, defining the data transfer requirements between the threads, and mapping the threads onto processors. *Implementing* a parallel program corresponding to the design specification involves mixing sequential code with calls to a communication library such as PVM, both for communication and synchronization. While designing the parallel program (i.e. dividing a problem into sequential operations) is an interesting task which can be left to the application programmer, implementing the parallel program is a time-consuming and error-prone effort, which should be automated. Moreover, debugging a parallel program remains difficult.

Portable parallel libraries such as PVM have become widely available, making it much easier to write portable parallel programs. Such libraries allow to write complex parallel programs with a small set of functions for spawning and identifying thread, and for packing, sending, receiving and unpacking messages. The simplicity of a library such as PVM makes it easy to learn and to port to new architectures.

This contribution proposes a coordination language for specifying a parallel program design. The Computer-Aided Parallelization (CAP) framework presented in this contribution is based on decomposing high-level operations such as 2-D and 3-D image reconstruction, database queries, or mathematical computations into a set of sequential suboperations with clearly defined input and output data. The application programmer uses the CAP language to specify the scheduling of sequential suboperations required to complete a given parallel operation, and assigns each suboperation to a execution thread. The CAP language is a C++ extension. The CAP preprocessor translates the CAP specification into a set of concurrent programs communicating through communication libraries such as MPI, PVM, and TCP/IP, or communicating through shared memory. The concurrent programs are run on the various processing elements of a parallel architecture. The CAP methodology targets coarse to medium grain parallelism.

CAP specifications produce programs that are deadlock free by construction and can be debugged using conventional debuggers. CAP has been developed in the context of multiprocessor multidisk storage servers. Simple arguments [4] show that careful dataflow control in such architectures is necessary to achieve the best performance. CAP gives the program designer control over the dataflow across threads in a parallel architecture. Both shared- and distributed-memory architectures are supported by CAP.

This contribution illustrates the features of the CAP parallel programming extension to C++. We demonstrate the expressive power of CAP and the performance of CAP-specified applications. Section 2 explains CAP methodology through an example. Section 3 shows expressive power of CAP and the performance of its PVM implementation.

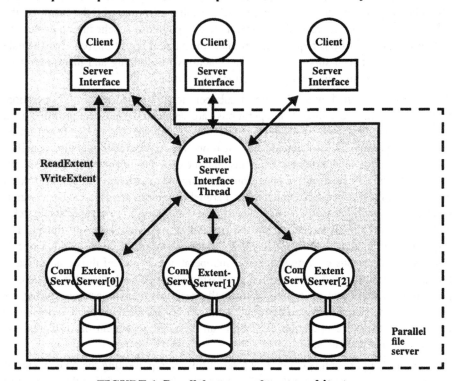

FIGURE 1. Parallel server software architecture

2 CAP overview

The abstract *parallel server* we consider consists of a set of *threads*. The clients connecting to the server are also modelled as execution threads. Figure 1 represents a parallel server with 3 disks and 3 processors. Each server processor executes two threads, one for disk accesses (ExtentServer[*]) and one for data processing (Compute-Server[*]). In our model, the client registers to the parallel server interface before starting any computation. The client then connects directly to the server's internal threads (ExtentServer[*] and ComputeServer[*]) to perform parallel computations. The complete server environment consists of the server threads and the client(s) threads. Threads may (but need not) share a common address space.

Operations are defined by a single input, a single output, and the computation that generates the output from the input. Input and output of operations are called *tokens*. Communication occurs only when the output token of an operation in transferred to the input of another operation. The basic mechanism for transferring the output token of an operation to the input of another operation is called *redirection*.

Using this terminology, *parallelizing* an operation consists of (a) dividing the operation in suboperations, each of them with one input and one output (the output of a suboperation is an intermediate result of the operation) ; (b) dividing the operation input token into several input subtokens to be fed to the suboperations ; (c) providing the *scheduling* of suboperations required to achieve the result of the original operation ; (d) assigning suboperations to threads in the parallel machine ; (e) merging the result of the suboperations to get the result of the parallelized operation. A schedule indicates the ordering of suboperations required to complete the parallel operation, based on the data dependencies between suboperations. The functions for dividing and merging values are called *partitioning* and *merging* functions respectively.

An operation specified as a schedule of suboperations is called a *parallel* operation. Other operations are called *leaf* operations, are sequential, and are specified in a sequential language such as C++. A leaf operation computes its output based on its input. No communication occurs during a leaf operation. A parallel operation specifies the assignment of suboperations to threads, and the data dependencies between suboperations. When two consecutive operations are assigned to different threads, the tokens are redirected from one thread to the other. Parallel operations are described as communication and synchronization between leaf operations. Threads are organized hierarchically. A set of threads executing on the parallel server is a *composite* thread. The sequential server threads are *leaf* threads. Composite threads execute parallel operations, and leaf threads execute sequential suboperations. Each leaf thread is statically mapped to an OS thread running on a processing element. Each thread is capable of performing a set of leaf operations, and usually performs many operations during the course of its existence. For load balancing purposes, tokens are dynamically directed to threads, and therefore operation execution can be made dependent on instantaneous server load characteristics.

Each parallel application is defined by a set of tokens, representing the objects known to the application, a set of partitioning and merging functions defining the division of tokens into subtokens, a set of sequential operations, and a set of parallel operations. Section 2.1 describes the tokens and partitioning/merging functions for the scalar-product operation. Section 2.2 is a formal description of the threads active in the parallel server. Section 2.3 describes the specification of the parallel scalar-product operation.

2.1 Tokens and partitioning/merging functions

A token consists of application-specific data, and a header indicating which operation must be performed on the data, as well as the context of the operation. Tokens are akin to C++ data structures with a limited set of types : the C++ predefined types, a generic array type, a generic list type, or other token types. This guarantees that the CAP preprocessor knows how to send tokens across any type of network.

```
token VectorPairT {
    int First ;
    int Size ;
    int ComputeServerIndex ;
    ArrayT<double> Left ;
    ArrayT<double> Right ;
} ;
```

```
token ResultT {
    double ScalarProduct ;
} ;
```

PROGRAM 1. Scalar product tokens

In the case of the scalar product operation, there are two token types : the *VectorPairT* token, and the *ResultT* token. The *VectorPairT* token consists of two arrays of floating point values, with the first index and size being *First* and *Size* respectively. The *ComputeServerIndex* filed of the *VectorPairT* indicates on which *ComputeServer* the scalar product must be computed. The *ResultT* token consists of a single floating point value called *ScalarProduct*. There is one partitioning function called *SplitVectorPair*, that divides a *VectorPairT* in several *VectorPairT* slices, and one merging function called *AddResult*, that accumulates the results of the scalar product operations applied to vector slices.

```
process ParallelServerT {
subprocesses :
    ExtentServerT      ExtentServer[2] ;
    ComputeServerT     ComputeServer[2] ;
    ClientT            Client ;
operations :
    ScalarProduct    in VectorPairT Input
                     out ResultT Output ;
} ;
```

```
process ComputeServerT {
operations:
    SliceScalarProduct
        in VectorPairT Input
        out ResultT Output ;
} ;
```

PROGRAM 2. Parallel server thread hierarchy

2.2 Thread hierarchy

When connecting to the parallel server, the client requests from the parallel server the list of server threads, the set of operations each thread can perform, and which input parameters must be provided for each operation. Program 2 is a formal representation of the parallel-server threads. The formal representation is captured by the CAP *process* construct. The *process* construct consists of the keyword *process*, the process name, an optional list of subprocesses, and a list of operations. In this example, the parallel server is called *ParallelServerT*. It contains two *ExtentServer* threads, two *Compute-Server* threads, and one *Client* thread. The set of threads making up the server (*process ParallelServerT*) can itself be seen as a high-level *composite* thread : the thread description is hierarchical. In the thread hierarchy, non-composite threads are referred to as leaf threads. Leaf threads perform sequential operations. Composite threads perform parallel operations. The *ParallelServerT* composite thread can perform one operation : the scalar product one a vector pair. The input and output tokens for the *ScalarProduct* operation are *VectorPairT* and *ResultT* tokens.

Threads run an infinite loop that receives tokens. The thread decodes the CAP header and executes the required operation on the data. The context information contained in the header both the operation that must be started after the current operation, and the address space in which the new operation will run. Communication between threads is asynchronous. Each thread has a mailbox for tokens. A token is kept in the mailbox until the thread is ready to execute it.

2.3 CAP parallel operation

As an example of operation parallelization, consider a scalar product of two vectors. The input token consists of two vectors of real numbers. The output token consists of a single real number, and the computation characterizing the operation is $SP(V_1, V_2) = \sum_{i=1}^{n} V_1(i) \cdot V_2(i)$. A parallelization of this computation would consists of dividing the input vectors in a number of slices sent to the threads in the parallel machine, computing the scalar product on each of the slices, and adding the slices' scalar products to find the vectors' scalar product. In this case, the suboperations are scalar product computations, the partitioning function selects slices in each of the two input vectors, and the merging function adds the slices' scalar product to the vectors' scalar product.

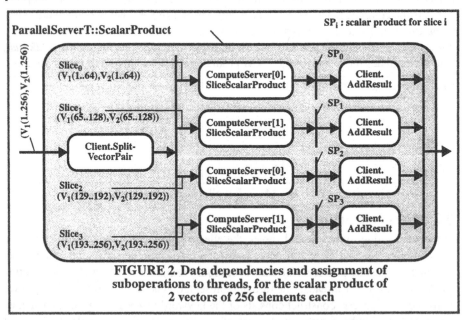

FIGURE 2. Data dependencies and assignment of suboperations to threads, for the scalar product of 2 vectors of 256 elements each

Figure 2 shows the data dependencies and an assignment of suboperations to threads for the scalar product operation, on two vectors with 256 elements each. The *Client* process splits the vector pair in slices ; the *ComputeServers* carry out the scalar product operation on the slices ; the *Client* adds the intermediate results to compute the slices' scalar product. From this assignment of suboperations to threads, it is easy to figure out the communication requirements for the proposed parallel scalar product algorithm. The slices have to be transferred from the *Client* to the *ComputeServers*, and the scalar products from the *ComputeServers* back to the *Client*. In the proposed methodology, the application programmer specifies the division of operations into suboperations, the partitioning functions, the scheduling of suboperations (the data dependencies), and the assignment of suboperations to threads. The CAP environment automatically generates the required messages and synchronizations between threads.

The basis for CAP semantics are directed acyclic graphs (DAGs), which allow to specify the data dependencies for any algorithm. The CAP extension language is rich enough to support the specification of any DAG.

The CAP formal specification of the parallel *ScalarProduct* operation (Program 3) consists of a single pipeline expression. The CAP pipeline expression consists syntactically of the keyword pipeline, 4 parameters between parentheses, and a parenthesized body. The pipeline expression divides the input token into 4 slices, using the *SplitVectorPair* partitioning function (first pipeline parameter, called the *pipeline-initialization* parameter). The *VectorPairT* tokens are sent to the first operation of the pipeline body. In this example, the pipeline body consists of a single CAP operation : *ComputeServer[Input.ComputeServerIndex].SliceScalarProduct*. The *ComputeServer* index is dynamically selected, thanks to the notation *ComputeServer[Input.ComputeServerIndex]*. *Input* refers to the input token of the current operation, in this case a *VectorPairT* token. The *SplitVectorPair* partitioning function assigns the *ComputeServerIndex* field of the input token. The allocation of extents to the various servers is application dependent. This allows to optimize dataflow for each application.

The vector pair slices are redirected to the *ComputeServers* (pipeline body : *ComputeServer[Input.ComputeServerIndex].SliceScalarProduct*). Slice scalar products are redirected to the Client (third pipeline expression parameter, called the *target-thread* parameter), to be merged using the *AddResult* merging function (second pipeline expression parameter, called the *pipeline-termination* parameter) into the Result token (fourth pipeline parameter, called the *pipeline-result* parameter).

```
operation ParallelServerT::ScalarProduct
  in VectorPairT Input
  out ResultT Output
{ pipeline ( SplitVectorPair, AddResult, Client, ResultT Result )
    ( ComputeServer[Input.ComputeServerIndex].SliceScalarProduct );
}

leaf operation ComputeServerT::SliceScalarProduct
  in VectorPairT Input
  out ResultT Output
{ // this is straight C++ code
  int i ;
  Output.ScalarProduct = 0 ;
  for (i = 0 ; i < Input.Size ; i++) {
    Output.ScalarProduct += Input.Left[i] * Input.Right[i] ;
  }
}
```

PROGRAM 3. Parallel ScalarProduct operation

The sequential *ComputeServerT::SliceScalarProduct* (bottom of Program 3) consists of a CAP leaf-operation interface, consisting of two keywords (*leaf* and *operation*), the operation thread, the operation name, and the operation input and output tokens. The body of a CAP leaf-operation consists of C++ sequential statements, used to compute the output token value based on the input token value.

In the CAP methodology, operations are distinct from threads. Threads are allocated once (usually at the server initialization, or when a new client becomes active), typically remain bound to a single processing element, and execute a large number of operations. Load balancing is achieved by careful allocation of operations to threads. The allocation of operations to threads occurs when setting the *ComputeServerIndex* of each *VectorPairT* token generated by the *SplitVectorPair* partitioning function. The distinction between operations and threads makes for a low overhead in thread management : the creation and destruction of a thread is a rare occurrence. Operations themselves entail little overhead. A typical operation descriptor is small (24 bytes in a PVM implementation), and easy to decode (little mode than a pointer-to-function dereferencing).

```
#include "cap.h"
#include "cap-scalar-product.h"
void main ()

{
  ParallelServerT Server ("ServerName") ;
  VectorPairT* VectorPairP =
    new VectorPairT (... /* initialization parameters */ ) ;
  ResultT* ResultP ;
  call Server.ScalarProduct in VectorPairP out VectorP ;
}
```

PROGRAM 4. Client program

A typical client program is described in Program 4. It opens the server called "Server-Name", and declares two tokens (*VectorPairP*, and *ResultP*). Using the call instruction, the client runs the parallel ScalarProduct operation. From the client standpoint, the parallel scalar product library looks like almost a sequential library : the only exception is that the client must use the call instruction to call a parallel library operation.

3 Generating parallel PVM programs using CAP

This section gives information on the status of the CAP project in terms of implementation and performance. It gives the current status of the CAP preprocessor implementation (section 3.1), and presents an overview of CAP's performance (overhead, and throughput, section 3.2).

3.1 CAP status

CAP is an extension language to C++. The semantics of CAP parallel expression is based on Directed Acyclic Graphs (DAGs). The CAP language extension supports the specification of any DAG, and the specification of tokens to be exchanged between operations.

A prototype CAP preprocessor has been implemented, translating mixed CAP/C++ code into straight C++ code calling PVM library routines. The CAP preprocessor generates a program matching the MPMD (Multiple Program Multiple Data) paradigm.

We have tested the preprocessor on a set of examples including an imaging library capable of storing, zooming and rotating 2-D images divided in square tiles, the Jacobi algorithm for the iterative resolution of systems of linear equations, the travelling salesman problem, and a pipelined matrix multiplication.

3.2 CAP performance

To evaluate the CAP overhead, we compare the performance of an actual CAP program with the theoretically achievable performance of the same program. For the theoretical analysis, we measure communication and computation performance, we establish timing diagrams, and derive from the timing diagram the theoretical performance of the parallel program.

Thread management introduces no overhead in CAP. OS threads are initialized when the parallel server is started up, and are statically bound to a specific processing element. The only overhead CAP introduces is due to the token header, representing 24 bytes in our prototype implementation. To evaluate the importance of the token header we compare the token header size to the number of bytes that could be transferred during a typical communication latency (section 3.2.1). We show that on a wide variety of machines, the overhead due to the CAP header is at most 15%, and in most cases below 1%. We then show that the CAP pipeline construct introduces an insignificant amount of overhead, for a wide range of message sizes and computation times.

This section analyses two aspects of CAP performance : the CAP message overhead (section 3.2.1), and the performance of CAP for a simple data fan parallel operation (section 3.2.3).

Machine/ Network	latency (μs)	throughput (MB/s)	quality (thr./lat.)	lat. byte cost (B)(thr.*lat.)
Convex SPP1200 (sm 1-n)	2.2	92	41.82	202.4*
Cray T3D (sm)	3	128	42.67	384
Cray T3D (PVM)	21	27	1.29	567
Intel Paragon SUNMOS	25	171	6.84	4275
Intel iPSC/860	65	3	0.05	195*
IBM SP-2	35	35	1.00	1225
Meiko CS2 (sm)	11	40	3.64	440
nCUBE 2	154	1.7	0.01	261.8
NEC Senju-3	40	13	0.33	520
SGI	10	64	6.40	640
Ethernet	500	0.9	0.00	450
FDDI	900	9.7	0.01	8730
ATM-100	900	3.5	0.00	3150

TABLE 1.

3.2.1 Theoretical CAP token overhead

CAP attaches a header to application-programmer-defined data. In the PVM prototype implementation of the CAP preprocessor, the header size is 24 bytes. In order to evaluate the overhead due to the CAP token header, we compare the header size with the *byte latency cost*, i.e. the number of bytes which could be transmitted during latency time on various parallel architectures and network protocols (Table 1). The numbers in the first two columns of Table 2 are all derived from Dongarra[2]. The third column is a measure of the network quality (throughput divided by latency). The lower the latency and the higher the throughput, the higher the quality. The last column list the byte latency cost of the various communication channels. In the last column, the * entries show the three lowest byte latency cost of all communication channels. The worst-case relative cost of CAP headers for zero-byte synchronization messages on channels having the smallest byte latency cost is thus around 15%. In all other cases, the CAP header overhead will be smaller. In the case where 1KBytes messages are transmitted, the token header represent only 2.4% of the total message size, and in the case of larger messages typical of storage servers (50KB), the token header size is insignificant. Notice also that the byte latency cost is completely independent of the network quality.

3.2.2 Experimental FDDI throughput with PVM protocols

We measure the throughput of the communication channel using a simple ping-pong test, sending of message of a given size back and forth between two workstations. We plot the delay for various sizes and linearize the curve to find latency and throughput figures. The program used to measure the network performance uses the PVM library for communication. The PvmRouteDirect option is turned on, to minimize the number of hops between threads. The delay represents the minimum elapsed time the first message is sent to the time the second message is received, over 20 experiments. Both messages have identical size (ranging from 0 KBytes to 48 KBytes). After linearization of the experimental results, the approximated experimental transfer time t_t is given by the formula :

$$t_t = a + \frac{S}{b} \quad \text{where} \quad \begin{aligned} a &= \text{latency} = 2.2\text{msec} \\ b &= \text{throughput} = 4.3\text{MB/s} \\ S &= \text{message size} \end{aligned}$$

If we compare with the performance numbers presented in Table 1, our performance measurements are a little under half the best FDDI performance, as measured in [2]. It represents the best performance *we* could achieve on the FDDI network through the PVM communication library.

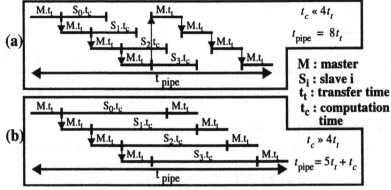

FIGURE 3. Timing diagrams for the pipeline operation

3.2.3 CAP data fan performance

In this section, we measure the performance of a 'data fan' parallel program, where (1) a master thread sends input data of size S to several slave threads running on separate processors, (2) the threads perform an active loop for duration t_c, and (3) the threads return a result of size S back to the master. The analyzed configuration consists of 4 threads receiving one message each. We perform a theoretical analysis for this program, and compare the result of the theoretical analysis to experimental results.

Theoretical analysis. Figure 3 shows timing diagrams for the pipeline operation described in the previous paragraph. Two situations are considered : the *large-message* situation, i.e. the case where the transfer time is larger than the computation time ; and the *small-message* situation, i.e. the case where the transfer time is smaller than the computation time. From the diagrams of Figure 3, we derive two total execution times $t_{datafan}$ for large messages $(t_c \ll 4t_t) \Rightarrow (t_{pipe} = 8t_t)$ (Figure 7(a)) ; and small messages : $(t_c \gg 4t_t) \Rightarrow (t_{pipe} = 5t_t + t_c)$ (Figure 7(b)).

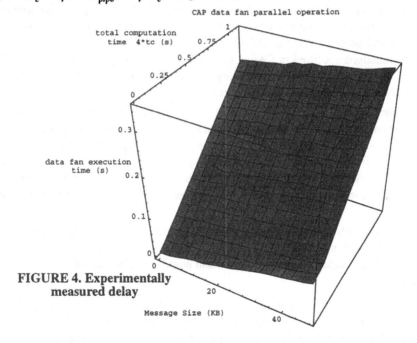

FIGURE 4. Experimentally measured delay

Experimental results confirm these formulas, as shown in Figure 4. Figure 4 plots the delay of the data fan parallel program as a function of the message size S, and the sequential computation time T. The message size S ranges from 0 to 48KB. The combined computation time of all four threads $(4*t_c)$ ranges from 0s to 1s. We discuss a few data points in Figure 4. Consider a 48KB message and 0 computation time. The transfer time is $t_t = a + S/b = 13$msec. The computation $t_c = 0$. We are clearly in the 'large message' situation, therefore the total time should be $8t_t$ or 104msec. The experimental measurement in this situation is 89.7msec, slightly better than anticipated by our simple evaluation. In the situation where the message size is 8KB and the computation time per slave is 275msec, we get a communication time $t_t = 4.2$ms. We are in the 'small message' situation. Therefore the total time should be $5t_t + t_c$, or 296msec, for a maximum achievable speed-up of $4t_c / 5t_t + t_c = 3.72$. This is the theoretically achievable speed-up, considering the computation and communication costs described above. The experimental results show a total time of 296msec. These results suggest that for problems with medium to coarse granularity, CAP does not introduce any noticeable overhead.

The data fan operation performance shown in Figure 4 is application independent. If a given parallel application fits the data fan paradigm, it is sufficient for the application programmer to know the size of the tokens exchanged between operations, as well as the computation time of each of its sequential operations to know the speed-up achievable by a given application. The CAP run-time environment will then deliver close to the theoretically achievable speed-up.

4 Conclusion

We have presented in this contribution a methodology for specifying parallel programs, based on hierarchical directed acyclic graphs. The methodology lets the designer specify a parallel problem decomposition in terms of tokens (C++ structures), data-partitioning functions specified in C++, sequential operations specified in C++, and CAP parallel operations specified as schedulings of sequential suboperations. The methodology allows to generate deadlock free parallel program.

This contribution shows that the overhead introduced by CAP is very low, and that the CAP preprocessor delivers the speed-up predicted by theoretical analyses. We have developed a simple library for parallel imaging, capable of performing zooming and/or rotation on images divided in square tiles stored on multiple disks. We have also applied the CAP methodology to the specification of parallel solutions to the travelling salesman problem and the Jacobi iterative method for solving systems of linear equations.

Future work will aim at demonstrating the performance of more complex parallel applications written in CAP, applying CAP toward distributed systems, and the developing parallel libraries for imaging, video and numerical applications.

References

[1] K. M. Chandy and C. Kesselman. CC++ : a declarative concurrent object-oriented programming notation. *Research directions in Object Oriented Programming*. MIT Press, 1993.

[2] Jack Dongarra and Tom Dunigan. *Message-Passing Performance of Various Computers*. University of Tennessee and Knoxville,Tech report 95-299, 1995. URL: http://www.netlib.org/utk/people/JackDongarra.html.

[3] Ian Foster and Carl Kesselmann. Language constructs and runtime systems for compositional parallel programming. In *Proc. COMPAR94 - VAPP VI* (B. Buchberger and J. Volkert, Eds.). LCNS 854, Springer-Verlag, p. 5-16, Sep. 1994.

[4] B. A. Gennart and R. D. Hersch. Comparing multimedia storage architectures. In *Proc. Int. Conf. on Multimedia Computing and Systems*, IEEE Press, p. 323-329, Washington 1995.

[5] Albert Y. Zomaya. Parallel Computing : paradigm and applications. International Thomson Computer Press, London 1996. URL : http://www.thomson.com/itcp.html

Load Balancing for Computation of Steady Viscous Flows Using Adaptive Unstructured Grids

E.I. Levin and A.I. Zhmakin

A.F.Ioffe Physical Technical Institute, Russian Academy of Sciences, St. Petersburg 194021, Russia

Abstract. Parallel implementation of unstructured grid solvers for steady viscous flows via domain decomposition method is considered. Various methods for grid partitioning are analysed and a new efficient realisation of a simulated annealing method is developed. The method, being still rather slow, provides grid partition structure superior to ones produced by more fast decomposition methods. This is of primary importance for stationary problems considered in the present paper: grid partition procedure is invoked only a few times (initially and after each grid refinement) during computation in contrast to dynamic problems where the execution time of domain decomposition step is a critical factor. The code has been implemented on a cluster of Sun workstations using PVM communications library. The code has been used to compute 3D internal low-Mach number viscous nonisothermal flows.

1 Introduction

Unstructured grids are extensively used now to solve fluid flow problems, with the main attention being paid to the development of solvers for Euler and compressible Navier-Stokes equations.The great complexity of fluid flow phenomena inside the modern technological devices requires the development of flexible Navier-Stokes solvers for incompressible flows.

The advantage of unstructured grids is the relative ease with which complex geometry can be treated. This approach needs the minimum input description of the domain to be discretized and is not tied closely to its topology in contrast to a block–structured grid. The required CPU time to attain the prescribed accuracy may be less than for the block–structure approach due to the much lesser total number of mesh cells as a direct sequence of the second advantage of unstructured grids — the easiness of adaptive mesh refinement, which allows one to place cells exactly where needed.

Parallel computers can be classified as Single Instruction Multiple Data (SIMD) and Multiple Instruction Multiple Data (MIMD) machines. In general the following issues are to be considered in order to develop efficient parallel codes [14]:

1. how the algorithm is to be distributed across the processors.

2. how the data is to be distributed across the processors.

3. what exchanges of data are necessary.

4. how best to perform these exchanges of data.

However, the first problem does not arise when programming for SIMD computers or when using domain decomposition methods which are of primary interest in the present study.

The efficiency of parallel algorithm can be characterized by speed up and total efficiency defined as [4] $S_n = T_s/T_p$, $E_n^{tot} = T_s/nT_p$, where n is the number of processors while T_s and T_p are the execution times for serial and parallel algorithms, respectively.

In reality total efficiency is less than 1 and can be written as a product of three factors [4],[5],[13] : parallel efficiency, numerical efficiency and load balancing efficiency. The parallel efficiency is defined as $E_n^{par} = T_{calc}/(T_{calc} + T_{comm})$, where T_{calc} is the pure calculation time and T_{comm} is the communication time. The second of these factors (increase in the number of iterations necessary to achieve convergence due to the changes in the numerical algorithm during parallelization) is not considerd in the present report. The load balance will be changing during computation when local grid refinement is used. In order to restore it one has periodically re-distribute the elements between processors and to this aim to accomplish two tasks: first, to compute the new partition of the computational domain and, second, to package all the data relevant to "foreign" elements into messages and to despatch to corresponding processors.

One of the most critical steps in implementation of parallel algorithm via domain decomposition methods is the data partitioning or mapping of computatational grid onto local memory paralell architecture. This process influences both parallel and load balancing effficiency and must provide equal load of all the processes and minimum amount and frequency of communications between processors. The problem is more complex for heterogeneous parallel arcitecture (i.e. for the case of processors of different productivity) and for the processor network topology other than a fully interconnected network. In the last case the distance of communication between processors is to be taken into account [8].

There are known different methods for partitioning of unstructered grid [2], [7], [8], [9], [14], [15], [16], [17],[18], the optimal choise being depending on the problem in question. In case of stationary flow computations using adaptive grid refinement the quality of grid partitioning is of primary importance in contrast to dynamic problems where the execution time of domain decomposition process is a critical factor.

2 Load balancing problem

A finite–volume representation of a partial differential equation can be solved most efficiently by a parallel distributed computer system if

1. the mesh is split into parts approximately proportionally for each processor productivity

2. the amount of communication between processors is minimized.

This optimization problem for mesh distribution is known as the load-balancing problem. Its solution requires the minimization of the cost function

$$H = T_{\text{calc}} + T_{\text{comm}}, \tag{1}$$

where T_{calc} is the time needed for an iteration and T_{comm} is the time then for the following inter-processor communications.

The first addend in (1) is obviously

$$T_{\text{calc}} = \max_q (N_q T_q), \tag{2}$$

where N_q is the number of the nodes assigned to the processor q, and T_q is the time in which it performs one iteration on one node (we assume that the processors may differ in productivity).

The communication time T_{comm} is determined by *communication matrix* C_{pq}. C_{pq} is the number of vertices of the mesh which belong to the processor p but are necessary to the processor q to perform an iteration.

The dependency of T_{comm} of C_{pq} is not obvious and may be strongly affected by the architecture of a multi-processor complex. We assume that for a cluster of workstations running on a Ethernet network it has the following simple form:

$$T_{\text{comm}} = \sum_{pq} (C_{pq}/b + \delta (C_{pq}) L), \tag{3}$$

where L is the *latency* of the network (time necessary to start communication between a couple of processors) and b is *effective bandwidth*:

$$b = 1/t_{\text{vertex}}, \tag{4}$$

where t_{vertex} is the time necessary to transmit through the network all information about one vertex.

The heuristic methods for domain partition fall into several categories. Nearest neighbour approaces [8] are usually implemented in two–step procedures: initial partition and subsequent boundary refinement to minimize communication. However, such methods give poor partition of genuinely unstructed grids. The key reason for this performance is that nearest neighbour methods are geometric approach whereas in reality the problem of grid partition is a topological one. In method of weighted recursive layer decomposition [14] a seed layer is selected and alternating layers for nodes and elements are grown outwards until all of the grid is allocated. In the recursive bisection approach the mesh is recursively cut into pieces, containing equal numbers of nodes. Methods for the cutting plane selection vary from a very naive *orthogonal recursive bisection* (cut along x-axis, then along y-axis, then along z, then along x again etc.) to a rather sophisticated *eigenvector recursive bisection* [18] which tries to reckon the mesh structure. The last algorithm assumes that the decomposition will produce connected sub–grids, however. this is not guaranteed. When the subdomains are connected the resultant partitition is nearest neighbour one and has the same advantages and drawbacks. The recursive clustering and extended recursive clustering algorithms [8] seem to be free from disadvantages of the above menthioned methods, however, as pointed out by Jones et.al. [8], the optimization procedure tends to get caught in local minima.

3 Simulated anneling

The simulated annealing (SA) technique [1] has been invented by S.Kirkpatrick et. al. [10]. The algorithm had come originally from the computational statistical physics. The idea is to treat the cost function eq.(1) as the full energy of some hypothetical system. Under a given temperature T the behavior of any physical system with the energy (or *Hamiltonian*) H can be simulated with the help of the following *"Metropolis algorithm"*:

1. Choose an arbitrary initial state of the system α.
2. Choose a "trial" state β by a random numbers generator.
3. If $H_\beta < H_\alpha$, then go to step 5.
4. Make a random decision: with the probability

$$p = \exp\left[-(H_\beta - H_\alpha)/T\right] \tag{5}$$

   return to the step 2, otherwise go to the step 5.
5. Move the system into the state β.

At very low temperatures, when the index of the exponent in (5) is typically large, the Metropolis procedure is exactly the "greedy" algorithm which is known to work badly for any real system. The idea of SA is to start from a high temperature T_s (when almost any change in the system is acceptable) and then to "anneal" the system by slowly lowering the temperature to a small value T_f.

Williams [18] has compared applications of recursive bisection and SA to the load balancing problem. He found that SA is the slowest method, and that the eigenvector recursive bisection, being much faster, gives results of comparable quality. Similar results have been obtained for two-dimensional problems (and essentially uniform grids) in [3]. However, Williams' tests were not quite fair, in as much as Williams used an artificial cost function H, which was ideal for eigenvector recursive bisection but very far from reality. Among the described methods only SA really use H in partitioning process, the recursive methods just relay on some general assumptions about the form of the final solution, so for real cost function partitioning quality must differ much stronger than in Williams' tests. Thus SA algorithm was chosen a the base for **balance** program. We have improved the performance of the method by using a Kirkpatrick's idea of adaptive cooling schedule [10] and our own fast implementation of the Metropolis algorithm [11].

4   Results

A wide class of non–isothermal viscous flows of practical interest is character-ized by 1) low velocities (compared to the sound speed) and 2) large tempera-ture (and, hence, density) variations. Numerical integration of the compressible Navier–Stokes equations for very low Mach numbers presents severe problems in practice while the well-known Oberbeck–Boussinesq approximation is invalid when the density (temperature) variations are greater than the mean values.

The low–Mach number Navier–Stokes equations follow from the full com-pressible Navier–Stokes equations under the following assumptions [12]: 1) Mach number is small $M^2 \ll 1$; 2) the hydrostatic compressibility parame-ter $\varepsilon = gL/R_g T_0$ is small $e \ll 1$; 3) the characteristic time τ is large compared to an acoustic time scale $\tau \gg L/a$ and for the flows in an open system can be written as

$$\frac{d\rho}{dt} + \rho \nabla \cdot \mathbf{V} = 0, \tag{6}$$

$$\rho \frac{d\mathbf{V}}{dt} = -\nabla p \hat{+} \frac{1}{Fr} \rho \mathbf{j} + \frac{1}{Re} \left[2 Div(\mu \dot{S}) -- \frac{2}{3} \nabla (\mu \nabla \cdot \mathbf{V}) \right], \tag{7}$$

$$\rho \frac{dT}{dt} = \frac{1}{RePr} \nabla \cdot (\lambda \nabla T), \tag{8}$$

$$\rho T = 1 \tag{9}$$

Approximation is based on using Ostrogradskii and related integral theorems which can be written as

$$\int\int\int_{\tau} \nabla \varphi \, d\tau = \int\int_{S} \varphi \mathbf{n} \, dS, \int\int\int_{\tau} div \, \mathbf{A} \, d\tau = \int\int_{S} (\mathbf{A} \cdot \mathbf{n}) \, dS \tag{10}$$

The detailes of the numerical method including adaptive grid refinement al-gorithm can be found in [19]. The code has been implemented on a cluster of Sun workstations using PVM communications library [6].

The flow and heat transfer in an epitaxial reactor used for growth of semicon-ductor films has been considered. A computational domain is shown in Figure. The initial grid consists of 11430 vertices and 45905 tetrahedron.

Since simulated annealing is rather slow, ortohogal recursive bisection have been used to obtain an initial decomposition of the grid. Some results of grid partitition are presented in Tables 1-3. Table 1 illustrate an influence of the ef-fective bandwidth on the optimal grid partition: its value have been increased twofold for a case B and halfed for case C compared to case A. Table 2 contains the result of grid partition for 12 and table 3 for 24 processors. Note that opti-mal grid decomposition differs greatly from one obtained by recursive bisection. One of the reasons for this is that even initial unstructured grid is highly nonuni-form: node clustering occur at the entrance, near circular substrate placed on the

bottom wall and around the outlet orifices. Two values for the number of communications points are presented. Unpacked one is just the sum of node transfers
between neighbour processors (number of edges cut [15]) while the packed takes
into account the fact that a vertex has several neighbour resident in another
processor and there is no need to duplicate information transfer.

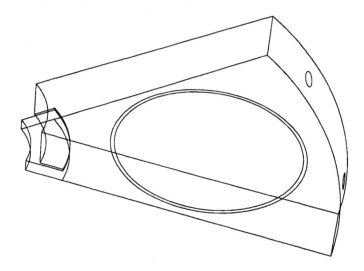

Fig. 1. Computational domain

Table 1.

Case	Host	Before annealing		After annealing		Unpacked Packed	
		N_v	N_{c0}	N_v	D	N_c	N_c
A	1	3810	11069	3320	-12.9%	2070	612
	2	3810	13549	4055	6.4%	3507	1058
	3	3810	2480	4055	6.4%	1437	443
B	1	3810	11069	3810	0.0%	2150	658
	2	3810	13549	3810	0.0%	3702	1106
	3	3810	2480	3810	0.0%	1552	495
C	1	3810	11069	3949	3.6%	1986	598
	2	3810	13549	3530	-7.3%	3528	1052
	3	3810	2480	3951	3.7%	1542	479

Table 2.

Host	Before annealing		After annealing		Unpacked Packed	
	N_v	N_{c0}	N_v	D	N_c	N_c
1	952	3745	595	-37.5 %	1047	346
2	951	3807	824	-13.4 %	1192	375
3	952	4443	147	-84.6 %	435	153
4	953	3691	374	-60.7 %	996	324
5	951	3075	812	-14.7 %	1266	436
6	952	7311	1738	82.6 %	1252	464
7	954	9205	1490	56.3 %	2809	936
8	952	8211	1740	82.6 %	3826	1299
9	953	5046	881	-7.6 %	3015	873
10	953	3856	1086	14.0 %	2861	843
11	954	3757	1109	16.4 %	2542	788
12	953	1789	634	-33.5 %	1099	304

Table 3.

Host	1	2	3	4	5	6	7	8	9	10	11	12
N_{c0}	1502	2692	2036	2217	3361	3112	2771	2500	2314	1985	1588	3496
N_c	968	792	637	658	161	262	284	124	429	710	924	1381

Host	13	14	15	16	17	18	19	20	21	22	23	24
N_{c0}	4771	5989	5142	6297	5574	3411	3531	3371	3151	3287	3028	1338
N_c	3558	2426	3234	3381	3552	694	2441	1707	510	392	2067	632

References

1. Azencott, R., ed.: Simulated annealing: parallelization techniques. John Wiley & Sons, Inc.(1992)
2. Barnard, S.T., Simon, H.D.: Fast multilevel implementation of recursive spectral bisection for partititioning unstructured problems. Concurrency: Practice and Experience 6 (1994) 101-117.
3. Bikker, S., Koshel, W.: Domain decomposition methods and adaptive flow simulation on unstructured meshes. Notes Numerical Fluid Mechanics 50 (1995) 13-25.
4. Drikakis, D., Schreck, E.: Development of Parallel Implicit Navier-Stokes Solvers on MIMD Multi-processor systems. AIAA-93-0062 (1993).
5. Durst, F., Perić, M., Schäfer, M., Schreck, E.: Parallelization of efficient numerical methods for flows in complex geometries. Notes Numer. Fluid Dyn. 38 (1992) 79-92.
6. Geist, G.A., Beguelin, A., Dongarra, J., Jiang, W., Manchek, R., Sunderam, V.S.: PVM 3 User's guide and reference manual. Oak Ridge National Laboratory TM-12187 (1994).

7. Ghosal, S., Mandel, J., Tezaur, R.: Automatic substructuring for domain decomposition using neural networks. Proc. IEEE Int.Conf.on Neural Networks (1994) 3816-3821.

8. Jones, B.W., Everett, M.G., Cross, M. Mapping unstructured mesh CFD codes onto local memory parallel architectures. Parallel Computational Fluid Dynamics '92, Eds. Pelz R.B., Ecer A., Häuser J. (1993) 311-324.

9. Kernighan, B.W., Lin. S.: An efficient Heuristic procedure for partitioning graphs. The Bell System Technical Journal, **49** (1970) 291-307.

10. Kirkpatrick, S., Gelatt, C.D., Jr., Vecchi, M.P. Optimization by simulated annealing. Science **220** (1983) 671-681.

11. Levin, E.I., Nguyen, V.L., Shklovskii, B.I., Efros, A.L. Coulomb gap hopping electric conductivity. Computer simulation. Sov. Phys. JETP. **65** (1987) 842–848

12. Makarov, Yu.N., Zhmakin, A.I.: Flow regimes in VPE reactors. J.Crystal Growth **94** (1989) 537-551.

13. Perić, M., Schäfer, M., Schreck, E.: Numerical simulation of complex fluid flows on MIMD computers. Parallel Computational Fluid Dynamics '92, Eds. Pelz R.B., Ecer A., Häuser J. (1992) 311-324.

14. Robinson, G. Parallel computational fluid dynamics on unstructured meshes using algebraic multigrid. Parallel Computational Fluid Dynamics '92, Eds. Pelz R.B., Ecer A., Häuser J. (1993) 359-370.

15. Simon, H.D.: Partitition of unstructured problems for parallel processing. Computing Systems in Engineering **2** (1991) 135-148.

16. Walshaw, C., Berzins, M.: Dynamic load-balancing for PDE solvers on adaptive unstructured meshes. Concurrency: Practice and Experience **7** (1995) 17-28.

17. Walshaw, C., Berzins, M.: Adaptive time-dependent CFD on distributed unstructured meshes. Parallel Computational Fluid Dynamics: New Trends and Advances (1995) 191-196.

18. Williams, R.D. Performance of dynamic load balancing algorithms for unstructured mesh calculations. Concurrency: practice and experience, **3(5)** (1991) 457.

19. Zhmakin, A.I. Computation of low-Mach number compressible viscous flows using 2D and 3D unstructured grids. Finite Elements in Fluids. New Trends and Applications, Eds. M. Morandi Cecchi, K.Morgan, J. Periaux, B.A. Schrefler, O.C. Zienkewicz. (1995) 483-492.

Using the ALDY Load Distribution System for PVM Applications

Claudia Gold and Thomas Schnekenburger

Institut für Informatik, Technische Universität München
Orleansstr. 34, D-81667 München, GERMANY
email: gold, schneken@informatik.tu-muenchen.de

Abstract. Load distribution is an important topic for parallel applications in distributed systems. PVM supports programming of parallel applications on distributed systems. But it does not provide load distribution among processes of a parallel program. In this paper the adaptive load distribution system ALDY is introduced. ALDY is a library of functions that provides basic mechanisms for load distribution and a collection of load distribution strategies. It can easily be used for PVM applications. This is demonstrated by an example.

1. Introduction

PVM allows the use of distributed computing power for parallel applications. The parallel virtual machine offers mechanisms for process initiation, communication and synchronization. However, PVM does not support mechanisms for assigning and migrating workload among processes of parallel programs during execution. PVM only decides before the process creation, how processes should be distributed among different hosts. One criterion for this decision is the load of different hosts. This property of PVM can be used for task-farming programs, where for each partial problem a new process is created and assigned to a suited host. In this way load is evenly distributed among the different hosts.

But task-farming is not always appropriate for efficient load distribution. For applications where each process is working with a separate part of data during a number of iterations, it is often necessary to migrate workload between the processes. The load of the different hosts of a parallel virtual machine can vary dynamically. On one hand there are probably other applications using the same hosts. On the other hand not all hosts of a heterogenous network are working with the same speed, since they have e.g. different CPU availability or memory capacity. So it may happen, that the performance of a PVM application is poor, because a PVM-task which is located on a slow host becomes a bottleneck. This can be avoided, if workload from a process running on a slow host could be transferred to a process on a faster host.

There are two different methods of load distribution distinguished: System based load distribution assigns processes to machines. This is done transparently for the application. Using application based load distribution the application programmer defines the distribution objects on his own and assigns them to application processes. This method is more flexible than the system based method, because real application objects are taken into account for distribution and the granularity of the distribution objects can be determined by the application. The disadvantage, however, is the fact, that the application has to be changed for the insertion of load distribution mechanisms.

This paper introduces the adaptive load distribution system ALDY and describes its usage for PVM applications. In Sect. 2 the concepts of ALDY are surveyed. In Sect. 3 the mechanisms for load migration in ALDY are discussed and their usage for PVM programs is described. Section 4 describes how ALDY is integrated into parallel applications. Section 5 shows an example of a PVM application demonstrating the integration of ALDY. Section 6 refers to related work. Section 7 contains a conclusion and gives an overview over future work concerning ALDY.

2 ALDY

ALDY stands for **A**daptive **L**oad **D**istribution **SY**stem. It is a hybrid between system based and application based load distribution. The programmer defines his own distribution objects, but the migration of these distribution objects to distribution units (e.g. PVM-tasks) during runtime is controlled by ALDY.

2.1 Portability

Portability is a very important factor for the design of ALDY. There are four concepts followed to reach this goal:

The first concept is to be independent from a specific implementation of the distribution objects. Therefore the application programmer maps his own objects to virtual objects of the ALDY load distribution model. Virtual objects are represented by integer identifiers. The different kinds of virtual objects are listed in Sect. 2.2.

Secondly, ALDY provides a collection of different strategies, where for individual applications the most suitable one can be chosen. It is not necessary to change the application program for testing another strategy. The new strategy can be chosen at runtime without recompilation.

To be independent from system parameters and consequently from individual resources, the ALDY strategies use only information about the states of virtual objects (see Sect. 2.3). System states like the load of hosts are not regarded.

At last ALDY is not fixed to a specific communication and synchronization platform. It only provides a kernel system for communication and synchronization. This is realized by call back functions. That means ALDY uses for its communication and synchronization functions which have to be implemented by the application programmer. By implementing these routines the programmer decides on which system the communication should be realized. ALDY uses the call back functions also for its internal communication.

2.2 Load Distribution Model

A load distribution model defines the distribution objects and distribution units. The load distribution model of ALDY is called the PAT (**Process, Agent, Task**) model (cf. Fig. 1). It consists of three classes of virtual objects:

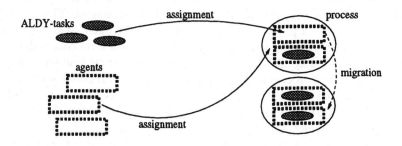

Fig. 1. PAT model

- **Virtual processes** correspond to the distribution units. The goal of ALDY is to assign workload to virtual processes. Therefore, they represent distributed resources of the system. In most cases virtual processes are real application processes.
- **Virtual agents** are objects that are assigned preemptively to virtual processes. They can be migrated dynamically to another process. Several agents may be assigned simultaneously to the same process.
- **Virtual ALDY-tasks** (not to confound with PVM-tasks) are objects that are assigned non-preemptively to agents. This means that an agent cannot be migrated, while it processes a task. Each agent processes one task after another. There is no parallelism within agents.

2.3 Library Interface

ALDY is not a complete environment for developing parallel programs. It is a library providing specific functions supporting the implementation of load distribution.

The interface between the application and ALDY consists of functions for describing virtual objects and their attributes. Some of the functions of the C-library interface are described below.

To realize efficient protocols for initialization, termination and other distributed events, ALDY requires all virtual processes to be arranged in a ring. This ring is used for the internal communication of ALDY. It is not automatically built by ALDY. Instead each virtual process has to call the library function ALDY_Init to form and initialize the ring. That means the application programmer is responsible for constructing the ring. Thus he has the possibility to organize the virtual processes in the ring corresponding to the demands of the application. For example, two virtual processes locating agents which communicate intensively should be neighbours in the ring.

A call to the function **ALDY_DefineObject** informs ALDY about the identifiers for the used ALDY-tasks and agents. ALDY allows to specify directives for virtual objects (see Sect. 2.5). After all directives are specified, a call to the function **ALDY_ReadyObject** informs ALDY that the object can be assigned.

ALDY does not use system information for its strategies. The number of active agents within a process serves as load index. An agent is active, when it executes a time consuming part of the application, and a process is defined as active, if at least one agent is active. If now e.g. the external load on a processor increases, the time a process is active for processing an ALDY-task also increases. This will cause the migration of agents from that processor to less loaded processors. Since ALDY does not know the implementation of the object which is represented by an agent, the application programmer has to inform ALDY when an agent is active. This is done with the functions **ALDY_StartAction** respectively **ALDY_EndAction**.

As long as an ALDY-task is assigned to an agent, the agent cannot be migrated (see Sect. 2.2). The call of **ALDY_NextTask** terminates the current task and requests a new task from ALDY. That means the agent to which the terminated task was assigned can now be migrated until a new task is assigned.

Other functions of the library interface are used by ALDY to inform the application about load distribution mechanisms (see Sect. 3).

2.4 Load Distribution Strategies

As mentioned in Sect. 2.1, ALDY separates the application program from the load distribution strategy. ALDY uses a generic load distribution concept that allows to integrate a collection of several parameterized specific load distribution strategies.

To realize this concept, ALDY uses an internal and an external strategy. The internal strategy is responsible for efficient global information management. It provides information about states, directives and actual locations of virtual objects. This information is used by the external strategy, which has to decide whether and where an agent should be assigned or migrated. For this decision a receiver-initiated strategy (see [2]) is currently used.

The criteria on which the external strategy makes its decisions are defined in a special parameter file that is read at the beginning of the initialization of ALDY. The individual strategy is therefore not part of the library interface.

2.5 Directives

An important facility of ALDY is the possibility to specify directives for virtual objects (library function **ALDY_AddDirective**). The task-agent-assignment directive e.g. specifies a set of agents to which an ALDY-task may be assigned. This directive can be used by applications to define the responsibility of agents for processing a specific ALDY-task.

Another directive that should be mentioned here allows the definition of a neighbourship among agents. In most cases an agent only communicates with a few other agents. With the agent-neighbourship directive tightly coupled communicating agents can be defined as neighbours. This reduces communication amount, since ALDY tries to assign neighbouring agents to the same process.

3 Mechanisms for Load Distribution in ALDY

Since ALDY works with virtual objects, it cannot migrate real application objects on its own. The mechanisms for migrating or assigning application objects are implemented by the application itself. But ALDY determines when an agent should be migrated. The call back functions ALDY_CB_AssignAgent and ALDY_CB_MigrateAgent are called by ALDY, when it decides to assign or to migrate an agent.

Further communication between the processes is needed for load distribution. On one hand ALDY has to get information about the load of a process to decide whether an agent should be migrated or not. These internal messages are coded as integer arrays, because this alleviates marshalling. On the other hand ALDY-tasks have to be sent to agents, which are migratable. Usually message-passing between migratable objects is very complex and often causes errors in parallel programs. Therefore ALDY provides an efficient protocol that allows transparent communication between migratable objects. ALDY is responsible for delivering messages (in form of ALDY-tasks) correctly to agents.

As mentioned in Sect. 2.1 ALDY communicates via call back functions. The relevant call back functions are listed below.

```
ALDY_CB_Receive          – receive message for ALDY
ALDY_CB_ObjectInfo       – receive data of object
ALDY_CB_SendInfo         – send ALDY information to another process
ALDY_CB_SendObjectInfo – send object and ALDY information to another process
```

For the implementation of these functions the same message-passing platform can be used as the one, on which the application is based. The following example (taken from [6]) illustrates an implementation of a call back function with PVM.

Example 1. Implementation of the call back function ALDY_CB_SendObjectInfo with PVM3.

ALDY_CB_SendObjectInfo has to send a message to process pid. The message contains object ob and an array of integers that keeps the internal ALDY information. The array starts at address addr and has the length n. Since ALDY does not know the real object data, these have to be defined by the application. The return value of ALDY_CB_SendObjectInfo is the number of additional bytes needed for sending object ob.

```
int ALDY_CB_SendObjectInfo(int ob, int pid, int * addr, int n) {
    pvm_initsend(PvmDataDefault); /* initialize send-buffer */
    pvm_pkint(&n,1,1);  /* put length of ALDY-message */
```

```
pvm_pkint(addr,n,1);   /* put ALDY-internals */
/* assuming object has 1000 bytes starting at address obAddr */
pvm_pkbyte(obAddr,1000,1);   /* put object into the buffer */
/* assuming that the array tids contains pvm-identifiers */
/* send ALDY message, tag 999 is used to mark ALDY-messages */
pvm_send(tids[pid-1], 999);
return 1000;   /* return number of bytes of object */
}
```

4   Programming with ALDY

Figure 2 shows the integration of ALDY into parallel programs. Load distribution with ALDY is performed on application level. That means, from the operating system point of view, ALDY and the application form together a single parallel program. The ALDY system is linked to each process of the parallel program. The application interacts only locally with the corresponding ALDY instance. The library interface (described in Sect. 2.3) connects the application and ALDY.

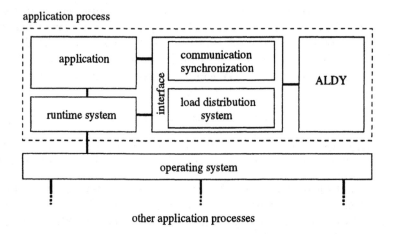

Fig. 2. Integration of ALDY

Which are the steps for programming a parallel algorithm with ALDY?

The first step is the mapping of application objects to virtual ALDY objects (see Sect. 2.2). The problem here is to find suitable dynamic data structures for the original static data structures. Migratable parts must be separated from the original static data structures. Also the granularity of the migratable parts (agents) must be considered, because the granularity determines the grade of load balance that can be reached.

In the next step the virtual objects must be defined and supplied with attributes. The task of the application programmer is to find unique integer identifiers for all virtual objects. Also the corresponding calls to the ALDY library functions must be inserted into the application program.

At last the communication system which ALDY should use is chosen by implementing the call back functions.

5 Example

PVM Relax is a parallel implementation of a Relaxation algorithm [5] using PVM. Given is a grid of metal with a constant temperature distribution around the grid. The problem is to determine the temperature distribution that will be reached inside the grid after a while.

To solve this problem for each point inside the grid the average temperature value of all surrounding points is computed for a number of iterations. The number of iterations can be reduced by using a numerical method called Relaxation, that also led to the name PVM Relax.

For the numerical solution of this problem a matrix is used as basic data structure, where each component of the matrix contains the temperature of a point of the metal grid.

5.1 Parallizing with PVM

The Relaxation algorithm can be parallized by partitioning the matrix into several strips (for stripwise decomposition see [3]), where each strip keeps some adjacent lines of the original matrix. For the computation of each strip one PVM-task is responsible.

Before a new iteration can be computed, each strip needs the lower and upper border line of its neighbouring strips containing the data of the previous iteration (see Fig. 3). That means, before each iteration a PVM-task has to wait for the border temperature values of its strip with **pvm_recv**. After each iteration the new computed values of the border lines are sent to the neighbouring tasks with **pvm_send**. The borders are only exchanged for information. They are not computed by the neighbouring PVM-tasks.

5.2 The Integration of ALDY

The pure PVM solution does not offer the possibility to migrate workload from one PVM-task to another. The execution of PVM Relax can be slowed down, if only one PVM-task is located on a slow host. This is avoided by using ALDY.

For the integration of ALDY PVM Relax has to be transformed according to the PAT model (see Fig. 3). To achieve more accurate load balance the matrix strips are partitioned again in smaller ones. The idea for load distribution is to assign several strips to one process and to migrate strips from overloaded to underloaded processes. Therefore the strips are represented by agents. Several agents are located on a virtual process. Virtual processes represent PVM-tasks. In the pure PVM solution the border exchange of each layer is realized

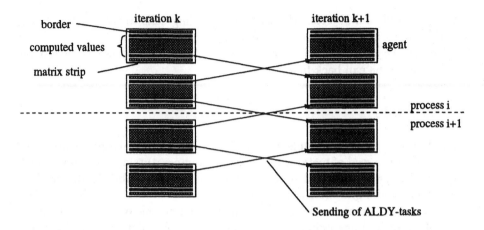

Fig. 3. Border exchange of PVM Relax and corresponding virtual ALDY-objects

with PVM-messages. Since now these data have to be sent between migratable agents, ALDY-tasks must be used for sending borders. Another advantage for using ALDY-tasks is the fact, that during processing of ALDY-tasks no agent migration is possible. This reduces overhead caused by stopping an agent anywhere during the computation of an iteration. Furthermore, the implementation of an agent is eased, because less information is needed, if an agent can only be transferred at a few exact defined execution points.

After the mapping of the real application objects to the virtual objects has been done, the calls to the ALDY library functions are inserted and the call back functions have to be implemented (see Sect. 3). The communication is again done by PVM.

Since agents which are processing neighbouring strips communicate intensively, these are defined as neighbours (see Sect. 2.5).

5.3 Results

With the integration of ALDY, the workload of the different PVM-tasks can dynamically be adapted to the varying workload of the application and to the actual speed of individual nodes. All difficult parts concerning the migration of objects are automatically provided by the ALDY system. Only the call back routines for sending and receiving objects have to be implemented by the application programmer.

Another advantage of ALDY is the fact, that different load distribution strategies for PVM Relax can easily be exchanged and tested without any modification of the application code for finding e.g. the most appropriate strategy.

Experiments [4] concerning the efficiency of the integration of ALDY into PVM Relax were conducted on an Ethernet network of nine SUN workstations. On each workstation one process was located to which eight agents were assigned at the beginning of program execution. Since the workload of PVM Relax is distributed equally among the processes, the workload was artificially unbalanced for the experiments. Therefore the values of the different lines of the matrix were computed a different number of times for each iteration. The used matrix consisted of 400 lines and rows. The execution of the pure PVM solution lasted between six and nine minutes. The experiments led to the following results:

- The overhead caused by ALDY amounts to about 5%. This means that the pure PVM solution is about 5% faster than the solution with ALDY, if no distribution of agents is allowed.
- Measurements with a receiver-initiated strategy resulted in an improvement of about 30%. This means that the solution using the load distribution of ALDY was about 30% faster than the solution without ALDY. For the given parameter settings the maximal improvement that could be reached theoretically is 50%, of which about 60% were obtained.

6 Related Work

Casas et.al. [1] describe three load migration systems for PVM – ADM, MPVM and UPVM. In contrast to ALDY all these systems are fixed to PVM. That requires reimplementation whenever the available communication and synchronization system changes, though the concepts still remain the same.

The load distribution system ADM (Adaptive Data Movement) is a pure application based system. Since migration events should be answered as soon as possible, migration checks have to be inserted in the inner compute loops of the program by the application programmer. Each process of the application has to check on its own, if a migration event has occurred. These checks induce a considerable overhead (about 23%, see [1]). On the contrary using ALDY the whole redistribution including all migration checks is done by ALDY and does not lie in the responsibility of the application programmer.

The systems MPVM (Migratable PVM) and UPVM follow pure system based concepts. The distribution objects of MPVM are processes and the distribution objects of UPVM are User Level Processes (ULP). ULPs are smaller distribution objects than processes and have some of the characteristics of a process and some of a thread. The migration of processes is expensive, since the process context has to be saved and a new process has to be started on the other host, which resumes the work of the migrated process. Therefore the migration of agents as provided by ALDY is more efficient. Furthermore, the application itself has the possibility to specify the granularity of the migration objects. A disadvantage of ALDY is the fact, that ALDY has to be integrated by hand for each new application. However, this can be done almost straightforward, if the structure of the application program allows to separate agents from the context as shown for PVM Relax.

7 Conclusion and Future Work

This paper describes the ALDY load distribution system. The example PVM Relax shows that ALDY is well suited for PVM applications.

One basic and central concept, where ALDY differs from earlier load distribution systems, is the use of virtual objects according to the PAT model. This allows the use of many different kinds of distribution objects. If later versions of PVM would e.g. provide threads, these could also be mapped to agents. Other new concepts of ALDY are the implementation of data exchange on top of the application's communication mechanism via call back functions and the collection of load distribution strategies.

At the moment ALDY is implemented as a prototype. There are still some functions missing. Also this prototype includes only one strategy out of a variety of strategies. The current strategy is a receiver-initiated one. This strategy is very efficient, but it would be interesting to test a corresponding sender-initiated strategy. Strategies with e.g. microeconomic or physical analogies should also be considered.

Further we are looking for more applications to test ALDY. We need more information about when and where it is useful to apply ALDY. Also some more measurements about the efficiency of ALDY are interesting.

References

1. Casas, J., Konuru, R., Otto, S. W., Prouty, R., Walpole, J.: Adaptive Load Migration systems for PVM. In *Supercomputing'94 Proceedings*, pages 390-399, 1994.
2. Eager, D. L., Lazowska, E. D., Zahorjan, J.: A Comparison of Receiver-Initiated and Sender-Initiated Adaptive Load Sharing, In *Performance Evaluation 6*, pages 53-68, 1986.
3. Hanxleden, R. V., Scott, L. R.: Load Balancing on Message Passing Architectures. In: *Journal of Parallel and Distributed Computing*, Vol. 13, pages 312-324, 1991.
4. Korndörfer, M.: Anwendung einer Lastverteilungsbibliothek für ein paralleles numerisches Programm. Diplomarbeit, Technische Universität München, Institut für Informatik, November 1995.
5. Pleier, C., Stellner, G.: PVM in der Praxis: Temperaturverteilung parallel berechnet. In *iX – Multiuser Multitasking Magazin*, pages 186-191, Heinz Heise Verlag, December 1994.
6. Schnekenburger, T.: The ALDY Load Distribution System. Technical Report SFB 342/11/95 A, Technische Universität München, May 1995.
7. Schnekenburger, T.: Supporting Application Integrated Load Distribution for Message Passing Programs. In *ParCo (Parallel Computing)*, Gent, Belgium, September 1995.
8. Sunderam, V. S., Geist, G. A., Dongarra, J., Manchek, R.: The PVM concurrent computing system: Evolution, experiences, and trends. In: *Parallel Computing*, Vol. 20, No. 4, pages 531-546, 1994.

Parallel Virtual Machine for Windows95

Marcos J. Santana[*]   Paulo S. Souza[**]   Regina C. Santana[*]   Simone S. Souza[**]
e-mails: mjs / pssouza / rcs / rocio@icmsc.sc.usp.br

[*] Instituto de Ciências Matemáticas de São Carlos (ICMSC/USP)
Departamento de Computação e Estatística
São Carlos - SP - Brazil

[**] Universidade Estadual de Ponta Grossa (UEPG)
Departamento de Informática
Ponta Grossa - PR - Brazil

Abstract: This paper describes the implementation of PVM-W95 (Parallel Virtual Machine for Windows95), that comprises a message passing environment (similar to the PVM), allowing the creation of a parallel virtual machine by using personal computers (working as workstations in a distributed computing environment), interconnected through a communication network and running the Windows95 operating system. Preliminary studies aiming the validation of PVM-W95 were performed. The results obtained showed that the PVM-W95 behaves stable and the parallel applications developed reached speedups, according to the hardware adopted.

1 Introduction

Distributed computing systems applied to parallel computing bring a better cost/benefit relation for actual implementation of parallel software [15]. They offer adequate computing power to those applications that although not requiring the power of massively parallel architectures, need more processing power than it is feasible with sequential machines [4]. PVM (Parallel Virtual Machine) and MPI (Message Passing Interface) comprise two successful examples of message passing libraries, widely discussed in the related literature, that allow the building of virtual parallel machines upon a cluster of workstations (normally RISC machines executing the UNIX operating system) [5][8].

Although personal computers (PC) and Windows have been utilized by a large number of companies, research and teaching institutions and public departments, until recently there was no discussion available in the literature about parallel computing on this platform.

Therefore, considering the main objective of having a better cost/benefit relation for parallel computing, this paper addresses the suitability of building parallel software on a cluster of personal computers running the Windows95 operating system. This clearly increases the number of potential users for parallel computing and improves the cost/benefit relation [10].

The creation of a virtual parallel machine in Windows95 environments is possible by means of the PVM-W95 (Parallel Virtual Machine for Windows95), a message passing environment built according to the original PVM specification[4] [5] [12].

The choice of PVM for this implementation is mainly due to its widely utilization, being considered a pratical standard for distributed parallel computing. Furthermore, PVM is sufficiently robust, simple and efficient to be also a success in Windows95 environments.

Although high performance is always something related to parallel computing, the aim of PVM-W95 is to offer a tool-kit for parallel computing in Windows95, even if the performance reached is lower than the one observed in UNIX environments. Even, with this aim, preliminary results using the PVM-W95 presents a good overall performance.

The WPVM is another PVM for Windows95 version, under development in the University of Coimbra (Portugal) [1]. The WPVM and the PVM-W95 have the same objectives (allow the implementation of parallel applications in Windows environment, using PVM) and were developed independently and simultaneously, having some basic differences in their implementation approach.

2 Parallel Virtual Machine for Windows95

The PVM-W95 offers the required resources to develop concurrent programs in C (or C++) language and to execute them in personal computers and running the Windows95 operating system (connected by a communication network), similarly to the PVM [10].

The development and implementation of PVM-W95 used the C language and the Borland C++ 4.5 programming environment. The C language was an obvious choice due to the implementation of the original PVM; the Borland C++ 4.5 environment was used because it has several helpful resources and allows the generation of object-oriented applications, concurrent process, 32 bits applications, etc.

The PVM-W95 was completely developed based on the PVM source code. Every PVM procedure was carefully studied and afterwards most of them was adapted to execute on the new platform. An overall amount of 33.258 lines of code, organized through 58 files (.c and .h) were analyzed and translated to the new system. This strategy allows to reduce the development and implementation times required to reach high portability and heterogeneity [10].

The final code produced for the PVM-W95 was organized by using object-oriented concepts: abstraction, encapsulation, polymorphism and inheritance [9][13][14]. The adoption of object orientation aims to give to the PVM-W95 code the required organization to facilitate future reusability.

The PVM-W95 (as the original PVM) is composed of two major parts: a daemon (Pvmd) and a procedure library with the PVM interface (Libpvm) [10][11][12]. The Pvmd is executed on each host that composes the overall virtual machine working as a machine manager and a message router. The Libpvm has a set of primitives that acts as a link between a task and the virtual machine (the Pvmd and other tasks) [4][5].

Analyzing the PVM code one can observe that it is organized according to the Pvmd and the Libpvm. Thus, considering to this organization and taking into accounting the concepts of object orientation, the PVM-W95 code is organized with four main classes: PvmBase, TaskCom, PvmLCtrl e PvmDCtrl.

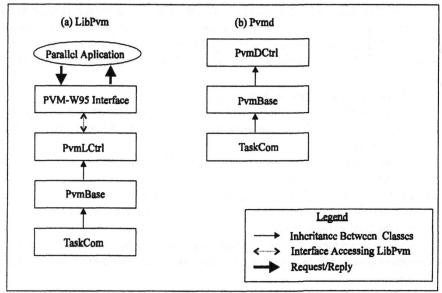

Figure 1 - Class Structure. (a) Libpvm (b) Pvmd

The PvmBase class is responsible for the code portion utilized both by Pvmd and Libpvm. This class implements:
- the methods and data structures common to Pvmd and Libpvm, which handle the messages and packets;
- the methods which insert and remove the message and packet headers;
- the location of directories and files required for the PVM-W95;

According to Figure 1, besides encapsulating the procedures and data described above, the PvmBase class inherits the TaskCom class, providing all methods required to the PVM communication. The PvmBase class will be inherited by PvmLCtrl and the PvmDCtrl classes, respectively responsible for the Libpvm and Pvmd.

2.1 Communication in the PVM-W95

The TaskCom class is responsible for the interprocess communication mechanisms (IPC). This class was built to make the PVM-W95 code independent of the communication protocol adopted (the TCP/IP is used in this implementation). Aiming to facilitate future maintenance, 28 methods were created, all inherited by the PvmBase class, and responsible for the interface between PVM-W95 and the communication protocol [10] [11].

The PVM-W95 uses the Windows Sockets (Winsock 1.1) [2][3][6], to have access to the TCP/IP stack. The choice of the WinSock 1.1 and TCP/IP is related to the original PVM implementation.

2.2 PVM Daemon (Pvmd)

The Pvmd is composed of the PvmDCtrl class (Figure 1-b), which inherits the methods and data structures of PvmBase required to handle the message and packets.

The first Pvmd (started by the user) is called the master and the others (started by the master) are the slaves. This difference only appears when management operations, such as the creation of new slave Pvmds and the inclusion of those into the virtual machines are required, once only the master Pvmd can perform these operations.

The PvmDCtrl class is responsible for:

- to configure every Pvmd (master and slave) starting to execute (e.g. to issue the IP address, to create the communication mechanisms between the required processes, etc.);
- to control every message routing performed by the Pvmd both between its tasks and between other Pvmds.
- to keep every message and packet queue received or sent by the Pvmd.
- to create and update the host and task table, making the virtual parallel machine configuration correct.
- to establish and execute the Pvmd⟺Pvmd and Pvmd⟺Task communication protocols, ensuring that requests and responses are transmitted and received correctly.
- to create and keep the wait contexts to ensure the Pvmd multitasking execution;
- to create new Pvmds (if it is the master), allowing the dynamical modification of the virtual parallel machine.

2.3 Communication Library (Libpvm)

The Libpvm is a dynamical link library (.dll) composed of the PvmLCtrl class and the functions responsible for the interface with the parallel application (Figure 1-a). The PvmLCtrl class inherits the methods and data structures required to handle the message and packets from PvmBase class. The functions responsible for the interface just access the methods available in the PvmLCtrl.

The PvmLCtrl class is responsible for:

- to implement the encoders and decoders used for heterogeneous platforms.
- to connect and disconnect a task with PVM-W95
- to create the required IPC mechanisms;
- to keep the Pvmd⟺Task and Task⟺Task communication protocols;
- to implement the procedures for task group management;
- to avoid the parallel application access to the methods and/or structures not allowed.

2.4 UNIX x Windows95: Relevant Differences from the PVM-W95 View Point

Although the UNIX and Windows95 operating systems have several differences, most of them do not affect the PVM-W95 operation [7]. This occurs because some of the UNIX resources not present in the Windows95 are not used by PVM software.

On the other hand, some of the UNIX resources, not present in the Windows95 and used by PVM, can be supplied by similar commands in the Windows95.

The differences emphasized here are the lack of rsh( ), fork( ) and kill( ) in the Windows95.

The lack of rsh( ): When inserting a new host in the virtual machine, the PVM needs to execute remotely the Pvmd and this is performed by means of the UNIX syscall rsh( ). There is no similar Windows95 syscall. The solution was to develop an application called Remote, similar (although simpler than) to the rsh( ). The objective is to include in Remote the required features in order to allow the PVM-W95 to execute remotely a new Pvmd. Remote is composed of two modules: r_daem and r_exe. The r_daem is executed in all machines that can be included as a PVM-W95 host. The r_daem receives messages from the r_exe requiring the concurrent execution of a new process. The r_exe module is executed by the user either from the command line, or from the application (such as Pvmd'). Using the r_exe, the Pvmd' asks a slave Pvmd to be executed.

The lack of fork( ): The PVM utilizes the UNIX syscall fork( ) to create concurrent processes. This syscall duplicates completely the current process (father process) by copying the executable program, its data, program counter and all context information. The new process is the child process.

When Pvmd' is started by the master Pvmd, the Pvmd' initialization protocol uses the fork( ) syscall to configure the Pvmd' with the information received from the father process.

This is not possible in the Windows95, using the Borland C++4.5, once there is no available syscall performing this job. There is the command spawn( ) that can create 32-bit concurrent processes in the Windows95 but this syscall always executes a new process which starts from the beginning, as if it had been started from the operating system prompt.

This constraint could be overcome by modifying the new host insertion protocol, with the Pvmd' being executed not only when new hosts are required. The Pvmd' is created in the beginning of the master Pvmd execution and it stops only when the virtual machine is "dismounted". The Pvmd' could also be executed only when new hosts are required, but this would certainly have an undesirable performance impact, due to the overhead to configure the process from its beginning.

As Pvmd' is always executing, some modifications in the original protocol were necessary. Now, it is not possible to ask its termination after a host inclusion. The procedure responsible for handling crashes of the virtual machine also had to be modified allowing that master and slave Pvmds and Pvmd' were canceled.

The lack of kill( ): The original PVM uses the syscall kill( ) and signal( ) every time a failure in the virtual machine occurs and a process has to be informed about. For instance, before Pvmd terminates, it sends a signal for all tasks under its supervision ensuring that they will also terminate. In Windows95, using the Borland C++ 4.5, it is only possible to receive a signal from the same process by means of the command raise( ).

The solution adopted was to have different alternatives for each case found in the PVM. Therefore, for instance, in the case discussed above, the task will kill themselves when the Pvmd is not present. This, perhaps, is not an ideal solution but overcomes the problem!

2.5 What is missing in the PVM-W95

Very few PVM modules were not included in the first version of the PVM-W95, which do not affect the objectives of this work. They are the PvmConsole, XPVM and XDR.

3 Validation

The PVM-W95 was validated according to the following steps:

1. *Partial validation of the primitives*: during the development of the PVM-W95, each primitive was tested following the same approach used for the program "testall" available in the original PVM. The results show that each individual PVM-W95 primitive worked properly;

2. *PVM user's guide examples*[4][5]: three examples presented in the PVM user's guide were run in the PVM-W95. The program *hello* exchanges a string between the hosts. The program *spmd* builds a virtual token ring with PVM tasks. The program *matrices multiplication* executes a parallel multiplication version of two matrices. The results obtained for these examples in the PVM-W95 were exactly the same described in the PVM user's guide.

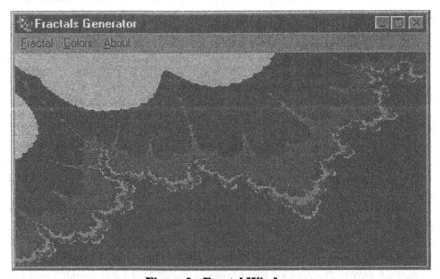

Figure 2 - Fractal Window

3. *A graphical example*: the validation of the PVM-W95 with the Windows95 graphical interface, was performed through the implementation of a fractal system (Figure 2). The generation of this fractal is obtained through the calculation of each pixel. The example developed considers a window with 640x480 pixels. In the sequential version of this example, these 307,200 pixels are evaluated in a sequential order. In the parallel version, four hosts were considered: a master and three slaves.

The master defines the fractal dimensions, spawns the three parallel tasks and shows the fractal. Each slave is responsible for the generation of 640x120 pixels. In this way, the pixels are evaluated in parallel by three slaves.

The sequential version was executed in a 486DX2-66, taking 30.1 seconds. The parallel version was executed in a 486DX2-66 as the master and a 486DX2-66, a 486DX4-100 and a PENTIUM-100 as the slaves, taking 20.9 seconds. Thus, the speed-up and efficiency reached are respectively 1.44 and 36%.

4 Conclusions

This paper presented the main features and implementation issues of the PVM-W95, emphasizing the modifications performed on the original PVM code in order to reach the objectives established in this project.

The adoption of object-oriented programming techniques is one of the main advantages of the PVM-W95, because this allowed to build a very modular and reusable code in many aspects. The TaskCom class is the major example since it introduces clear advantages in terms of future maintenance.

The existing differences between UNIX and Windows95 are responsible for most of the difficulties found during the porting. The lack of fork( ) and rsh( ) is an example of this.

The results obtained are considered completely successful, once the system built allows the perfect virtual parallel machine construction, with stable behaviour and reasonable performance.

5 Acknowledgements

We would like to thank Luciano J. Senger, Márcio A. Souza and Robson Picinato for their help with the PVM-W95 validation. We also thank the Brazilian Foundations FAPESP, CAPES and CNPq for the financial support given to the Distributed Systems and Concurrent Programming Research Group at ICMSC-USP.

References

1. Alves A., Silva, L., Carreira, J., Silva, J., G.: WPVM: Parallel Computing for the People. Procedings of HPCN'95, High Performance Computing and Networking Conference, in Springer Verlag Lecture Notes in Computer Science, Milan, Itália, May (1995) pp. 582-587.
2. Bonner, P.: Network Programming with Windows Sockets, Prentice-Hall, (1996).
3. Dumas, A.: Programming WinSock, Sams Publishing, (1995).
4. Geist, A., Beguelin, A., Dongarra J., Jiang, W. Manchek, R., Sunderam V.: PVM: Parallel Virtual Machine. A User's Guide and Tutorial for Networked Parallel Computing, The MIT Press, (1994).
5. Geist, A., Beguelin, A., Dongarra J., Jiang, W., Manchek, R., Sunderam V.: PVM3 User's Guide and Reference Manual, Oak National Laboratory, September, (1994).
6. Hall, M., Towfiq, M., Geoff, A., Treadwell, D., Henry, S.: Windows Sockets. An Open Interface for Networtk Programming under Microsoft Windows, version 1.1, (1993).

7. King, A.: Desvendando Windows95, Rio de Janeiro, Ed. Campus, (1995).

8. McBryan, O. A.: An overview of message passing environments, Parallel Computing, v. 20, (1994) 417-444.

9. Pinson, L. J.,Wiener, R. S.: Programação Orientada para Objeto e C++, São Paulo, McGraw-Hill, (1991).

10. Souza, P. S. L.: Máquina Paralela Virtual em Ambiente Windows, MSc Dissertation, Instituto de Ciências Matemáticas de São Carlos (ICMSC), Universidade de São Paulo (USP), May, (1996).

11. Stevens, W. R.: UNIX Network Programming, Prentice Hall International Inc.,(1990).

12. Sunderam, V. S., Geist, A., Dongarra, J., Manchek, R.: The PVM concurrent computing system: evolution, experiences and trends, Parallel Computing, v. 20, (1994) 531-545.

13. Takahashi, T., Liesenberg H. K. E., Xavier, D. T.: Programação orientada a objetos: uma visão integrada do paradigma de objetos, São Paulo, IME-USP, (1990).

14. Wiener, R. S., Pinson, L. J.: C++ Programação Orientada para Objeto. Manual Prático e Profissional, São Paulo, McGraw-Hill, (1991).

15. Zaluska E. J.: Research lines in distributed computing systems and concurrent computation, Anais do Workshop em Programação Concorrente, Sistemas Distribuídos e Engenharia de Software, (1991) 132-155.

Using PVM in Wireless Network Environments

Christian Bonnet

Institut Eurécom, 2229 route des Crêtes,
BP193, F-06560 Sophia Antipolis Cedex, France
E-mail : bonnet@eurecom.fr

Abstract. The PVM (Parallel Virtual Machine) system has been used during the past years in various networked environments as well as parallel and vector computers. PVM relies on a message passing model to federate an heterogeneous computing platform in a very powerfull virtual machine. The development of wireless local area network (WLAN) technologies allows us to build flexible networks which can be configured to operate independently or in connection with a fixed network. In this paper, we describe the study of PVM behaviour over the WLAN technology.

1 Introduction

The emergence of the WLAN technology allows a more flexible way to build a Local Area Network by federating computers that can move freely. One can consider a set of wireless computers communicating via radio links as a parallel machine. The radio link which provides this mobility is subject to disturbancies due to varying indoor propagation conditions [2]. The distance between antennas, the multiple reflexions of the radio signal, and the different natures of the construction materials can lead to large attenuations of the transmitted signals. Under these conditions are PVM based applications still viable? This paper presents experiments done in a wireless environment which address performance, robustness and dynamicity issues. The first experiments on PVM based applications compare the performance between wired and wireless environments. The second step was to study the benefits of using PVM to make robust applications: can PVM overcome radio link failure situations? Finally, we discuss the suitability of PVM for mobile applications.

2 Testbed Hardware and Software Configuration

The testbed hardware is composed of PCs equipped with 2Mbps ATT WaveLAN [1] radio cards. WaveLan is an indoor WLAN which uses spread spectrum technology in the 2.4GHz ISM band. PCs can communicate directly with each other provided they are reachable via radio. In order to provide a communication link with fixed machines a special node named 'the access point' is provided. This access point acts as a bridge between the radio segment and the wired segment

(10 Mbps Ethernet in our case) of the communication. Each machine runs PVM on top of Linux [3] operating system.

In such context we choose to run different typical applications in order to test the performances of the overall system. The first type of application has a Master-Slave structure : the Mandelbrot curves drawing. In that kind of application the master is in charge of work allocation to the slave tasks: the master splits the total image in small pieces and sends the orders of computation to the slave tasks. When the slave tasks have achieved their computations, they send the results back to the master task which displays the resulting image. A version of this application compares the performance between local and distributed computations.

The second type of application has a distributed structure : the travelling salesman problem. The program finds the shortest cycle in a graph using brute force : all cycles are tested. The program relies on a recursive tree structure, all tasks are not created by a single master but by the individual nodes of the subtrees. Each single task is in charge of the length computation for one cycle. Other test programs have been used to monitor the performance: Matrix Multiplication which gives the speedup compared to local computations ,'Test Timing'and'test 43' [5] which evaluate the latency and throughput of PVM messages passing for various message lengths.

3   Performance of wireless PVM

The different experiments have been done under the following configurations :

- 1 PC
- 2 PCs on Ethernet
- 3 PCs on Ethernet
- 2 PCs on WLAN
- 3 PCs on WLAN
- 1 Ethernet PC, 1 access point, 1 radio PC
- 2 radio PCs, 1 access point

Tests results are given for Matrix multiplication and'Test Timing'.

Table 1 gives the relative speed compared to a local computation. For a configuration with 3 PCs the speedup is greater than 2 for a matrix dimension of 500 in the case of WLAN and 100 in the case of Ethernet. We can notice that for the 50x50 matrix, the hybrid configuration (1 Ethernet PC, 1 access point, 1 radio PC) behaves as in the single PC configuration. This is due to the extra work performed by the access point bridging the messages from the WLAN to the Ethernet segment. When the dimension of the matrices is increased this configuration leads to the same level of performance compared to that of 2 radio PCs. Table 2 gives the performance results in terms of latency time and throughput. In this test a master task sends messages to a slave task. Messages are fragmented into packets if necessary. For each received packet, PVM sends an acknowledgement message. The master retransmits the message if it does not

Table 1. results of matrix multiplications tests

Dimension	20		50		100		500	
configuration	speed	Mflops	speed	Mflops	speed	Mflops	speed	Mflops
1 PC	1.0	0.37	1.0	0.36	1.0	0.36	1.0	0.33
2 PC Ethernet	0.3	0.12	1.2	0.42	1.5	0.55	1.8	0.61
2 PC Radio	0.3	0.12	1.1	0.41	1.5	0.54	1.8	0.61
3 PC Ethernet	0.3	0.10	1.6	0.58	2.2	0.77	2.8	0.91
3 PC Radio	0.3	0.10	1.5	0.54	1.9	0.68	2.7	0.89
3PC access point	0.3	0.11	1.0	0.37	1.5	0.52	1.8	0.61

Table 2. results of 'Test Timing'

Nb user bytes	0	100	1000	10000	100000	1000000
configuration	latency (μ sec)	throughput	(Mbps)			
1 PC	3553	0.23	1.2	6.2	6.5	6.6
2 PC Ethernet	3659	0.22	2.0	6.1	6.5	6.6
2 PC Radio	4673	0.16	0.82	1.4	1.4	1.4
2 PC Radio	4757	0.16	0.82	1.2	1.3	1.3
3 PC, access point	5561	0.15	0.69	0.79	0.72	0.72

receive this acknowledgement before a timeout. The results show that the latency time increases dramatically in the WLAN configuration for long messages. Unsurprisingly the WLAN rate of 2Mbps limits the throughput measurements compared to the Ethernet configurations.

The different tests showed that the WLAN configuration is still appropriate for PVM applications provided that they do not require considerable message to be transferred between processors.These tests were undertaken under good radio propagation conditions. Measurements in our laboratory showed that an attenuation up to 60 dB does not affect the throughput of the WLAN. However an attenuation of 66dB results in very low throughput. Attenuation of up to 60 dB was measured when the distance between the antennas was more than 18 meters.This means that WLAN does not suffer from attenuation in a room of reasonable size. Fading is an other problem : indoor propagation conditions may vary quickly due to door openings, movement of people etc. This may lead to radio link failure if the duration of the fade is too long.

4 Wireless PVM robustness

PVM applications must overcome different kinds of failures including radio link failure, slave failure and master failure. We do not consider the latter in this paper. In a typical Master/Slave application, the master task waits for the results of the slave tasks by invoking a blocking primitive "pvm_recv". If the radio link is not operational, the master task waits until recovery. The crucial issue is how to distinguish between a transient radio link failure that is correctly recovered by PVM, and a slave host failure. As we wanted to work at the host level instead of the task level, a notification service ("pvw_notify") is provided by PVM allowing a task to know if other tasks are alive or not. A possible solution is to test periodically the actual configuration supporting the application. Making the master task poll the configuration while waiting for the slave tasks via a non blocking receive primitive, results into waste of efficiency. It is easier to insert an additional task (a "manager task") into the application while keeping the blocking reception. The manager task knows the initial machine configuration at the beginning. The manager task is in charge of periodically polling the configuration and notifying the master task when a change occurs. In case of a slave failure, the master task has the choice of restarting the corresponding computations on an other host machine.

5 Dynamicity

One key benefit of the wireless technology is its ability to provide mobile user services. PVM provides features for dynamic group management : one element task may dynamically join and leave a given group of tasks. Communications between tasks occur without knowledge of the exact group composition or topology. This concept of group communication is well adapted to mobility needs : a host machine may perform some computations in a given group then move to an other place for other computations and eventually come back. This kind of behaviour can be used to provide computation power on demand with an additional flexibility : an independent network of machines can be built quickly using the radio technology. If the WLAN is not connected to a fixed infrastructure then the radiolink bandwidth is dedicated to a given application. Dynamicity could be enhanced by providing a means to integrate new hosts which had not been initially declared to the PVM host configuration file. This can be done via an authentication server. When a new host wants to dynamically join the set of PVM machines, it should register with the authentification server to ensure its visibility to potential PVM applications. Applications may use the resources of this new host if they receive the authorization from the authentification center. Thus the authentication center should maintain two kinds of information : a list of credential information that allows the authentication of incoming hosts, and an access control list that allows applications to benefit from the resources of the new hosts.

6 Conclusion and perspective

The aim of this study was to analyse the performance of PVM applications in a wireless environment. The results of the different tests showed that if the application requires extensive computations compared to message transmissions, the resulting performance of the overall system is still acceptable on the WLAN. This may be acceptable if the WLAN technology is used only as a means to replace the Ethernet-like network. If the WLAN technology is used for mobility purposes then battery consumption problems of the mobile hosts must be taken into consideration. Therefore large computations should be avoided in such an environment. However PVM presents some interesting features for mobile computing in terms of robustness and dynamicity. Radio links are subject to fading, it was tested that failures due to bad radio propagation conditions are treated in a smooth way by PVM after radio link recovery. Dynamicity through group communications is provided by PVM. Host migration and dynamicity in setting up the system have to be investigated by introducing new services in addition to the PVM services.

7 Acknowledgements

I am grateful to Luis Aldaz, Alexandre Iseppi and Pascale Mondoloni for their valuable support during the test campaigns.

References

1. Bruce Tuch : An engineers story of WaveLAN,NCR Corporation, Utrecht the Netherlands
2. Homayoun Hashemi : The Indoor Radio Propagation Channel. Proceedings of the IEEE Vol.81, No.7 (July 1993) 943–967
3. Shahid H. Bokhari : The Linux Operating System, IEEE Computer, (August 1995) 74–79
4. Al Geist, Adam Beguelin, Jack Dongarra, Weicheng Jiang, Robert Manchek, Vaidy Sunderam : PVM: Parallel Virtual Machine, A User's Guide and Tutorial for Networked Parallel Computing. The MIT Press, Cambridge, Massachussetts, (1994)
5. Martin Do, Emmanuel Jeannot : PVM 3.2.x Testing Routines, (1994)

Parallel Image Compression on a Workstationcluster Using PVM [*]

A. Uhl[1] and J. Hämmerle[2]

E-mail: {uhl,jhaemm}@cosy.sbg.ac.at

[1] Research Institute for Softwaretechnology and Department of Computer Science and System Analysis, Salzburg University, Austria
[2] Research Institute for Softwaretechnology, Salzburg University, Austria

Abstract. Parallel image compression on a workstation-cluster is discussed. We introduce a parallel meta-algorithm for image compression which is well suited for every block based image compression scheme. This algorithm is explained in more detail in the case of fractal image compression. We present experimental results of a host/node implementation on a FDDI interconnected workstationcluster using PVM.

1 Introduction

One of the ironies to come out of image compression research is that as the data rates come down the computational complexity of the algorithms increases. This leads to the problem of long execution times to compress an image or image sequence. This shows the "need for speed" in image and video compression [9].

Unfortunately many compression techniques demand execution times that are not possible using a single serial microprocessor. The use of general purpose high performance computers for such tasks is therefore suggested (beside the use of DSP chips or application specific VLSI designs). In this paper we focus on a workstation-cluster and the parallel programming environment PVM ([10]) for the lossy compression of still images. Such a configuration can be seen as a moderate parallel distributed memory MIMD architecture with high communication cost.

It has already been shown that general convolution algorithms in image processing tasks (e.g. edge detection, morphological image processing, feature extraction, template matching, etc.) can be performed more or less efficiently on a cluster of workstations [6]. In this paper we focus on lossy compression algorithms, which include of course more complex computations than just convolutions.

Lossy compression algorithms can be classified in various ways – one possibility is to distinguish between algorithms that operate on image subblocks (e.g. JPEG [12], Vector Quantization [7], Block Truncation Coding [2], Fractal Compression [3]) and algorithms that operate on the entire image (e.g. Wavelets

---

[*] This work was partially supported by the Austrian Science Foundation FWF, project no. P11045-ÖMA.

[1], Subbandcoding [13], DPCM [8]). Here we introduce a meta-algorithm for image compression algorithms that operate on image subblocks which is especially suited for MIMD parallelization.

2 A meta algorithm for image-block based compression on workstationclusters

For all the compression techniques considered, we use the host/node programming paradigm: A host-process is responsible for the partition of the image into subblocks and the distribution of these blocks among the node-processes which perform the actual compression. Load balancing is essential for efficient cluster-computing because of either unequal computational demand of the subtasks (e.g. adaptive techniques) or additional processes running on the machines. Therefore the distribution is organized in a dynamical way, e.g. using the image subblocks as a pool of tasks when using an asynchronous single pool of tasks load balancing technique [5]. When a node-process has finished its calculations it sends the encoded subblock back to the host-process which assigns a new subblock to that node-process. See figure 1.

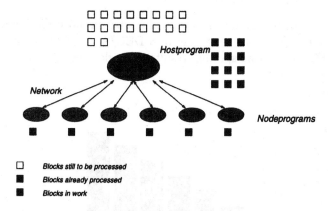

Fig. 1. Host/node meta-algorithm for image-block coding

Before we go into details of the parallelization of the different techniques we want to point out that the parallelization is especially designed for a workstation-cluster implementation. If different MIMD architectures are used (especially with a large number of processors or small memory capacity or low processor speed on a processor element) different parallelization strategies have to be applied.

1. *JPEG*: The DCT, quantization, and entropy coding of the blocks can be carried out on the nodes independently.
2. *Vector Quantization*: Each node-process has the whole codebook in its memory and is able to evaluate the best match for its imageblocks indepen-

dently. Only the index of the best codebook-vector is transmitted to the host-process.

3. *Block Truncation Coding*: Mean, variance, and bitplane can be evaluated independently of the node-processes. The bitplane is also encoded there.

4. *Wavelet Block Coding*: In contrast to a common wavelet coder it is possible to use an adaptive block-based wavelet coding scheme [11]. The host-process calculates the adaptive variance based quadtree and sends the resulting leaves (imageblocks with different size) to the node-processes where the wavelet transform, quantization, and coding is performed independently. The encoded imageblock is sent back to the host-process.

5. *Fractal Compression*: According to the classification in [4] a simple parallelization strategy may be applied since workstations are usually equipped with a large memory. The node-processes have the whole image in their main memory out of which the predecimated images are produced. They calculate the best transformation for the actual rangeblock which is an image subblock that is found via indexing in the original image. In this case the host-process only organizes the distribution of the indices. The calculated transformation and position of range and domain are sent to the host-process.

3 Experimental results

We present results of an implementation of fractal compression on a workstation-cluster at our department consisting of 8 DEC/AXP 3000/400 interconnected by FDDI. Figure 2 shows speedup-results in single user mode for an increasing number of processors.

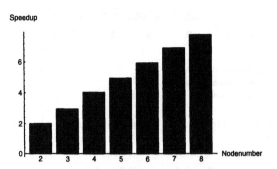

Fig. 2. Speedup of fractal coding

The real strength of the parallelization is exhibited if the application of dynamic load balancing is necessary. We produce an artificial load distribution on the cluster with one machine running load 3.0, one machine running load 2.0, three machines running load 1.0, and three idle machines. We achieve speedup 4.837 in this artificial load-environment which is a very good result if compared to the maximal speedup 2 using optimal static load distribution.

4 Conclusion

Concluding we emphasize that various block-based image compression techniques may be implemented quite efficiently on workstation-clusters. The parallelization techniques introduced here may not be suitable for other parallel MIMD architectures. The centralized approach – for example – creates a bottleneck for massively parallel systems. Architectures with little memory on the processor elements reqire more sophisticated parallelization for vector quantization and fractal coding.

References

1. M. Antonini, M. Barlaud, P. Mathieu, and I. Daubechies. Image coding using wavelet transform. *IEEE Trans. on Image Process.*, 1(2):205–220, 1992.
2. E.J. Delp and O.R. Mitchell. Image compression using block truncation coding. *IEEE Transactions on Communications*, 27(9):1335–1342, 1979.
3. Y. Fisher, editor. *Fractal Image Compression: Theory and Application to Digital Images.* Springer Verlag, New York, 1995.
4. J. Hämmerle and A. Uhl. Parallel algorithms for fractal image coding on MIMD architectures. In *Proceedings of the International Conference on Visual Information Systems 96*, pages 182–191, 1996.
5. A.R. Krommer and C.W. Überhuber. Dynamic load balancing – an overview. Technical Report ACPC/TR92-18, Austrian Center for Parallel Computation, 1992.
6. C. Lee and M. Hamdi. Parallel image processing applications on a network of workstations. *Parallel Computing*, (21):137–160, 1995.
7. N.M. Nasrabadi and R.A. King. Image coding using vector quantisation: a review. *IEEE Transactions on Communications*, 36(8):957–971, 1988.
8. D.J. Sharma and A.N. Netravali. Design of quantizers for DPCM coding of picture signals. *IEEE Transactions on Communications*, 25(11):1267–1274, 1977.
9. K. Shen, G.W. Cook, L.H. Jamieson, and E.J. Delp. An overview of parallel processing approaches to image and video compression. In M. Rabbani, editor, *Image and Video Compression, Proc. SPIE 2186*, pages 197–208, 1994.
10. V.S. Sunderam, G.A. Geist, J. Dongarra, and R. Manchek. The PVM concurrent computing system: evolution, experiences, and trends. *Parallel Computing*, 20:531–545, 1994.
11. A. Uhl. Adaptive wavelet image block coding. In H.H. Szu, editor, *Wavelet Applications III, Proc. SPIE 2762*, pages 127–135, 1996.
12. G.K. Wallace. The JPEG still picture compression standard. *Communications of the ACM*, 34(4):30–44, 1991.
13. J. Woods and S.D. O'Neil. Subband coding of images. *IEEE Trans. on Acoust. Signal Speech Process.*, 34(5):1278–1288, 1986.

Genetic Selection and Generation of Textural Features with PVM

Thomas Wagner, Christian Kueblbeck, Christoph Schittko

Fraunhofer-Institut fuer Integrierte Schaltungen,
Am Weichselgarten 3, 91058 Erlangen, Germany
Email: {wag,kue,csc}@iis.fhg.de

Abstract. Automatic classification of textured images is crucial for both surface inspection problems in quality control systems and medical imaging applications. It requires sophisticated textural image features in order to distinguish between different defects or image classes.

In the following paper, we report on experiments in which we automatically select adequate subsets of textural features from a large set of potential candidates. For the underlying problem of tumor cell identification, conventional selection techniques as well as genetic algorithms have been investigated on. The Gallops PVM package has been running on up to 48 machines in order to find the best feature subset.

1 Introduction and Background

Automatic image processing systems suffer from high engineering costs necessary to adapt the configuration of the system to the problem under consideration.

As a consequence, this paper presents an approach to reduce these costs by applying optimization methods to the setup and the configuration of an image processing system (cf. Fig. 1).

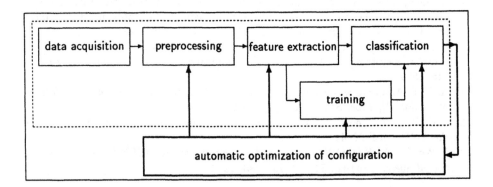

Fig. 1. Automatic optimization of system configuration: conventional image processing systems [10] (inside the dashed box) can be submitted to a global optimization procedure in order to determine the optimal configuration of the system.

Since the search for a solution of such optimization problems can take weeks on a single workstation, a cluster of 48 computers has been employed. In general, automatic optimization of image processing systems can cover any of the following areas:

- The selection of image features or classifiers under use,
- the selection of adequate training features (critical for successful classification results), and
- the construction of new image features[1].

The most important aspect of any successful optimization lies in an adequate choice of the underlying fitness function, which measures the individual quality of each configuration. When optimizing image processing systems, the fitness function can be the recognition rate of the total system measured on a test data set. An additional training data set is used to adapt the classifier to a given configuration [10]. In order to allow an unbiased judgement of the selected subset, the final performance of the feature subset is calculated on a separate verification sample.

Size of samples for tumor cell classification

class	training	test	verification	sum
no tumor	72	72	71	215
tumor	101	101	100	302
sum	173	173	171	517

Table 1. Size of training, testing, and verification sample for the classification of potential tumor cells.

Since both training and testing of the classifier have to be performed for each evaluation of a configuration, automatic optimization gets very time consuming: The typical execution time for one evaluation of a fitness function is of the order of 1 to 100 seconds[2]. Fortunately, parallelization of this sort of problem is an easy task.

Due to the high computational load connected with such configuration problems, PVM seems to be a good platform for performing such complex optimization tasks.

---

[1] Each feature is basically a mathematical equation leading to some sort of real number. Considering the equation as consisting of a set of construction elements, the elements can be selected and ordered randomly by applying means of genetic programming.

[2] For the tumor sample an average calculation time of one fitness function with 20 features was 2.3 seconds on a PC 133 MHz. All textural features had been calculated in advance. The total number of images in training and testing sample was 346.

2 Selection of Features

In our paper, we present the results for the selection of textural features for a medical imaging problem, the detection of tumor cells. The sample is described in Tab. 1.

Classification results for different textural feature sets:

feature set	recognition rate in in % (test set)	recognition rate in % (verification set)
Haralick	72,2	69,5
Unser	67,6	60,8
Galloway	63,5	66,8
Laine	56,1	63,7
Local features	61,8	61,8
Chen	73,4	69,6
All features	71,2	77,2
Best with all strategy (17 features)	82,6	79,5
Genetic selection (47 features)	86,1	73,1

Table 2. Recognition rates for feature sets and different intelligent subsets on test and verification samples: All selection strategies can further improve the recognition rate.

About 100 textural features from seven different feature sets (Haralick [6], Galloway [3], Unser [12], Chen [2], Laine [9], fractal features [11], local features [1]) have been calculated from each of these images. A 1-nearest neighbour algorithm has turned out to be a simple but good choice to tackle the classification problem.

Our main goal was to find an optimal subset of features for both improving recognition ratio and reducing computation time. Since it is not possible to test all feature subsets (for 100 features a total of $2^{100} \approx 10^{30}$ possibilities exist), several methods have been suggested to reduce the number of guesses. Both conventional algorithms (variance analysis [8], the Bhatthacharya distance [10]), intelligent search strategies ("best with all" [11], sequential backwards [7]) and genetic algorithms (GAs [4]) have been studied. Previous experiments in our group have shown that only intelligent search strategies and genetic methods yield satisfying recognition results.

In general, GAs use bitstrings to encode the parameters describing a problem. In the case of feature selection with tumor cells, a string of 100 bits can be used to encode which textural feature is selected (bit value 1) and which is omitted (bit value 0). An initial pool of such bitstrings, which is called a *population*, is varied by genetic operations called *mutation* and *crossover*. *Selection* determines which individuals are considered for reproduction into the next generation. The choice is dependant on the fitness of the individual which in our case is given by the recognition rate on the testing sample. Only the textural features with actual bit-code 1 contribute.

While conventional strategies such as "best with all" selection run on single machines, a cluster of about 50 workstations has been used to run the experiments for genetic selection. PVM version 3.3.0 and the parallel genetic simulation package Galopps PVM 3.0 β [5] serve as simulation tools.

3 Experiments, Performance, and Results

Tab. 2 shows the recognition results for both conventional features sets such as e. g. Haralick features and for automatically selected subsets. The different rates for testing and verification sample can be explained by the fact that the selection is only optimized for the testing sample and not for the verification images. This is a known problem for any selection strategy.

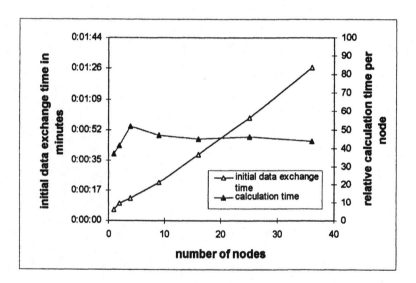

Fig. 2. Scaling features for both initialization phase and calculation of one generation (50 individuums) on each platform. Since there is almost no data interaction, the problem paralellizes well during the calculation phase.

Scaling features of the algorithms under use are shown in Fig. 2. As one would expect, during the initialization phase the perfomance is limited by the bandwith of the ethernet. During this phase, the samples (about 140 kB in this case) have to be transferred to each node.

During the calculation period, there is almost no interaction between the nodes; the migration of the best individuums between neighbouring nodes is usually not critical.

The algorithm is also robust with respect to partial process breakdown. If some processes die during execution, the computing power available is reduced and the convergence time raises. Nevertheless, the stability of the calculation as a whole is not affected.

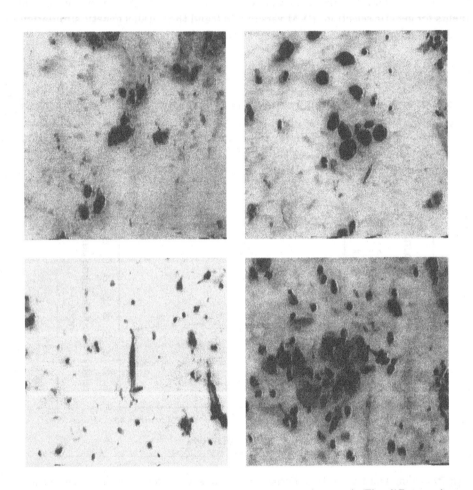

Fig. 3. Examples of normal cells (top) and tumor cells (bottom). The difference is not obvious for untrained human inspectors.

4 Discussion

With respect to the question of finding the most adequate feature subset, genetic selection strategies are outmatched by the intelligent search strategy "best with all". While recognition results of both algorithms are comparable (a higher recognition rate of the genetically selected features on the test sample is compensated by a lower rate on the verification sample; this is an obvious overfitting phenomenum), the comparison of computational effort (about 5000 fitness function evaluations (ffes) for "best with all" and a maximum of 480.000 ffes for genetic selection) clearly shows that "best with all" is the more relevant strategy in terms of computer ressources and applicability.

Nevertheless, when going to examine genetic programming for the construction of new genetic feature sets, there is no conventional pendant which could be used instead of the computationally challenging genetic strategy. We hope that the experiences from genetic selection will be of some help for this. At least for the simulation of genetic algorithms, powerful tools are available which allow the unexperienced user to profit by the calculation power of workstation clusters without writing parallel programs himself.

In order to provide dedicated low-cost equipment for parallel computing, a network of PC motherboards connected with ethernet links is currently under construction at our site which will hopefully provide an adequate testing platform for further experiments.

References

[1] BLANZ, W. E.: Design and Implementation of a Low-Level Image Segmentation Architecture - LISA. In: *Machine Vision Applications* (1993), Nr. 6, S. 181–190

[2] CHEN, Y. ; NIXON, M. ; THOMAS, D.: Statistical Geometrical Features for Texture Classification. In: *Pattern Recognition* 28 (1995), Nr. 4, S. 537–552

[3] GALLOWAY, M. M.: Texture analysis using gray level run lengths. In: *Computer Graphics and Image Processing* 4 (1975), S. 172–179

[4] GOLDBERG, David E.: *Genetic algorithms in search, optimization and machine learning.* Reading, MA : Addison-Wesley, 1989

[5] GOODMAN, Erik D.: An Introduction to Galopps / Michigan State University. 1995 (95-06-01). – Tech. Report

[6] HARALICK, M. ; SHANMUGAM, K. ; DINSTEIN, I.: Textural Features for Image Classification. In: *IEEE Transactions on Systems, Man and Cybernetics* SMC-3 (1973), Nr. 6, S. 610–621

[7] KITTLER, J. Feature Selection and Extraction. 1986

[8] KREYSZIG, Erwin: *Statistische Methoden und ihre Anwendungen.* Goettingen : Vandenhoeck und Ruprecht, 1975

[9] LAINE, A. ; FUN, J.: Texture Classification by Wavelet Packet Signatures. In: *IEEE Ttransactions on Pattern Analysis and Machine Intelligence* 15 (1993), Nr. 11, S. 1186–1191

[10] NIEMANN, H.: *Klassifikation von Mustern.* Berlin/Heidelberg : Springer, 1983

[11] RO, Th. ; HANDELS, H. ; BUSCHE, H. ; KREUSCH, J. ; WOLFF, H. H. ; PPPL, S. J.: Automatische Klassifikation hochaufgelster Oberflchenprofile von Hauttumoren mit neuronalen Netzen. In: SAGERER, G. (Hrsg.) ; POSCH, S. (Hrsg.) ; KUMMERT, F. (Hrsg.): *Mustererkennung 1995.* Bielefeld : Spinger Verlag, 1995

[12] UNSER, Michael: Sum and Difference Histograms for Texture Analysis. In: *IEEE Transactions on Pattern Analysis ans Machine Intelligence* 8 (1986), S. 118–125

{0, 1}-Solutions of Integer Linear Equation Systems[*]

ANTON BETTEN, ALFRED WASSERMANN

Department of Mathematics
University of Bayreuth
Germany

Abstract. A parallel version of an algorithm for solving systems of integer linear equations with {0, 1}-variables is presented. The algorithm is based on lattice basis reduction in combination with explicit enumeration.

1 The algorithm

A parallel version of an algorithm proposed by KAIB and RITTER [4] has been implemented with PVM to find all {0, 1}-solutions of integer linear equation systems. For example such systems are of interest in the construction of block designs, see [1, 2, 3, 8]: It is possible to find block designs if one finds {0, 1}-vectors x and $\lambda > 0$ with

$$A \cdot x = \lambda(1, 1, \ldots, 1)^\top, \tag{1}$$

where A is a matrix consisting of nonnegative integers. Our problem is also related to cryptography [6] and theory of numbers [7].

The algorithm – using lattice basis reduction [5] – constructs a basis for the equation kernel that consists of short integer vectors. Then the integer linear combinations of these basis vectors are enumerated and tested if they yield {0, 1}-solutions of (1). For an explicit description of the algorithm see [8].

2 Parallelization

The backtracking algorithm is now implemented using PVM. The algorithm runs along a search tree with an unpredictable number of childs at each vertex. Actually, we search all solutions of a discrete optimization problem, the optimal value of the objective function being well known in advance.

[*] This research was supported by PROCOPE.

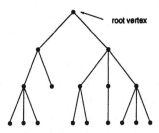

The parallel version of the backtracking algorithm is easily given: We fix the maximal number of tasks allowed to be processed by the virtual machine. This number depends for instance on the number of machines or on the amount of main memory available.

There is a principal task controlling every other task, that we will call "master". After some preprocessing, i.e. LLL reduction, the master creates subordinate tasks, called "slaves", that possess their own search loop. Each slave enumerates a certain part of the search tree. After enumerating all branches of its subtree the slave is allowed to die.

If the fixed maximal number of tasks is not reached during some stage of the algorithm, a new task will be created by the master. Therefore, the slave who seems to have gained the least progress is told to *split*, i.e. to create a new slave, who does part of the work of the old task. The search tree of the old task is shrunken.

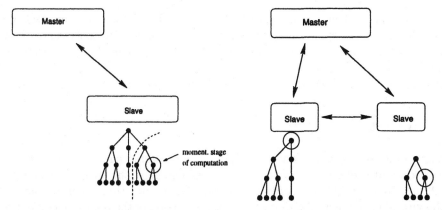

The slave who got the split message still finishes the branch of the root tree in which he is currently computing (rightmost branch in the figure). But just before doing so he creates a new slave and initiates a task containg the remaining branches (of the root vertex). So, the original tree of the slave is divided into two parts, the momentary subtree is cut off and will be finished by the slave himself, the remaining branches are carried over to the new slave. Note that splitting is always done at the level of the root vertex. Otherwise, the shape of the subtrees would become too difficult to handle. Here, the root vertex and its first edge to a deeper node is all one needs to know for defining a subtree: the tree is defined to be all those subtrees of the root vertex which start at the special edge and continue with the following branches – remember that we enumerate a basis of a kernel so there is a notion of left and right in each level. The new slave informs the master of his existence and both slaves start to work on their two smaller subproblems.

Implicitely, this strategy implies dynamical load balancing: Each machine receives a new task as soon as it has finished the previous one. This reduces the size of other trees. Therefore slower machines get help by faster machines.

Note that perhaps a split request cannot be carried out. Namely, if the search (sub-) tree does not possess any more branches leading down from the root vertex except that one containing the momentary position of computation. The next figure shows the messages needed for a task split.

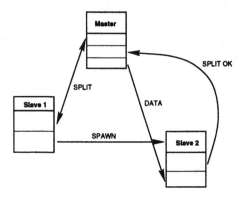

In a first version of our algorithm, we made PVM to choose by itself the machine where to spawn the new task. But we noticed bad behaviour of the algorithm, because some machines got too much load. Therefore we decided to put machines and tasks under control of the master. So each split message of the master names a certain machine where the new task should be spawned.

3   Results

Since we do not have an homogenous pool of computers we indicate the speed of each computer of our virtual machine by percentage of the speed of a Pentium 90 running under Linux. In order to measure the running time, each computer of the virtual machine was tested with the serial version of the program, computing just a small example.

machine type	P90 speed
HP 9000 755 / 99 MHz	303 %
HP 9000 712 / 80 MHz	240 %
Silicon Graphics SGI5	181 %
Intel Pentium 90 MHz	100 %

Several tests with PVM were done with the input matrix KM_PGL23plus_t6_k8 and $\lambda = 36$ from [2]. This is a 28×119 matrix with dimension of the search space equal to 93. Each solution is a 6-(25,8,36) design. For a detailed explanation of the automorphism group see also [2]. The following table lists the results of our tests on four different configurations of virtual machines. The second column lists the number of computers of each type we used. The third column contains the types of these computers. The fourth column contains the collected percentage of Pentium 90 speed of the computers in the virtual machine. In the column "No. Proc." the maximal number of slave processes is noted which were allowed to run simultaneously on the virtual machine. The last column gives the time after which the result was printed on the screen.

	No.	Computer	P 90 speed	No. Proc.	Time
1.	3	Silicon Graphics SGI5	181 %		
	3		543 %	12	323 min
2.	2	HPPA 9000/755 99 MHz	303 %		
	1	HPPA 9000/712 80 MHz	240 %		
	2	Silicon Graphics SGI5	181 %		
	5		1208 %	24	160 min
3.	3	HPPA 9000/755 99 MHz	303 %		
	1	HPPA 9000/712 80 MHz	240 %		
	6	Silicon Graphics SGI5	181 %		
	10		2235 %	45	89 min
4.	6	HPPA 9000/755 99 MHz	303 %		
	1	HPPA 9000/712 80 MHz	240 %		
	6	Silicon Graphics SGI5	181 %		
	13		3144 %	58	56 min

Moreover, with the fourth configuration we were able to find 10008 solutions for the same matrix KM_PGL23plus_t6_k8 and $\lambda = 45$ which were previously not known to exist. The computing time we needed was 7:34 hours.

References

1. A. BETTEN, A. KERBER, A. KOHNERT, R. LAUE, A. WASSERMANN: The Discovery of Simple 7-Designs with Automorphism Group $P\Gamma L(2,32)$. *AAECC 11* in *Lecture Notes in Computer Science* **547** (1995), 281–293.

2. A. BETTEN, R. LAUE, A. WASSERMANN: Simple 7-Designs With Small Parameters, Spetses, 1996.

3. D. L. KREHER, S. P. RADZISZOWSKI: Finding Simple t-Designs by Using Basis Reduction. *Congressus Numerantium* **55** (1986), 235–244.

4. M. KAIB, H. RITTER: Block Reduction for Arbitrary Norms. Preprint 1995.

5. A. K. LENSTRA, H. W. LENSTRA JR., L. LOVÁSZ: Factoring Polynomials with Rational Coefficients, *Math. Ann.* **261** (1982), 515–534.

6. J. C. LAGARIAS, A. M. ODLYZKO: Solving low-density subset sum problems. *J. Assoc. Comp. Mach.* **32** (1985), 229–246.

7. C. P. SCHNORR: Factoring Integers and Computing Discrete Logarithms via Diophantine Approximation. *Advances in Cryptology – Eurocrypt '91* in *Lecture Notes in Computer Science* **547** (1991), 281–293.

8. A. WASSERMANN: Finding Simple t-Designs with Enumeration Techniques, submitted.

Extending Synchronization PVM Mechanisms

P. Theodoropoulos, G. Manis, P. Tsanakas and G. Papakonstantinou

National Technical University of Athens
Dept. of Electrical and Computer Engineering
Zografou Campus, Zografou 15773, Athens Greece

1. Introduction

Several synchronization techniques have been proposed and many of them have been adopted by parallel and distributed operating systems or parallel programming platforms. The most widely used synchronization techniques are the barriers and the message passing model. Semaphores represent another synchronization technique that is mainly used by traditional stand-alone operating systems.

Barriers represent a rather limited synchronization model, which requires that all processes reach at a specific meeting point in execution. Intel Paragon operating system[1] requires at least one process per node to participate in the synchronization. In other words, it is not possible for every subset of the running processes to be synchronized. Another system providing synchronization is the Clouds[2] system which implements semaphores. However, semaphores in Clouds are implemented as a part of Clouds DSM model. One of the systems supporting synchronization is the Orchid[3] platform, which supports synchronization using semaphores and barriers. In Orchid, light-weight processes are the basic unit of parallelism; synchronization takes place amongst threads of control.

PVM supports process synchronization using barriers and message passing. On the other hand, PVM does not support global semaphores. This work aims at that target, using the relevant experience from the Orchid platform. We are next presenting the design and implementation of global semaphores on top of PVM, along with the interface of global semaphore operations.

It may be thought that in a message passing programming model, the implementation of global semaphores is redundant. This can be further justified by the existence of the barrier synchronization model. The semaphore model is more generic, even compared with the message passing model, as it allows an arbitrary number of processes to be synchronized, not only by meeting a certain point of execution, but also by allowing the execution of their code in a controlled way, without the need to know the exact ids of the other participating processes. Moreover, the main benefit of the semaphore mechanism is the ability to synchronize processes for sharing resources. In a parallel virtual machine such a resource can be a distributed shared memory. Thus, the implementation of the global semaphore mechanism on top of PVM can be thought of as an extension in future versions of PVM, supporting a shared memory model. This is a future target of our research work.

2. Design and Implementation

The key concept in the implementation of the global semaphores mechanism is that the access of a global semaphore is reduced to the access of a local one. At every processor participating in our topology, a process is created and kept responsible for the management of the global semaphore operations on that processor. This process will be further referenced as the semaphore server. Every semaphore server maintains a list of all the semaphores that were created on the hosting processor. Each element of the list consists of the semaphore's name, the semaphore's id, the semaphore's current reference count, the semaphore's current value, a pointer to its waiting list and, finally, a pointer to the next element in the list.

The global semaphore library uses a general semaphore descriptor, shown in fig. 1, to address a semaphore within the parallel virtual machine. The descriptor is similar, in structure, with the task identifiers (TIDs, [4]). For purposes of efficiency, it is made to fit into the largest data type (32 bits) available on a wide range of machines. The H field is the same with the corresponding field in the TID, namely it contains a host number relative to the virtual machine. The ID field contains an 18-bit integer number identifying the different semaphores that can be opened by a specific semaphore server. Bits 30 and 31, denoted by the letter N, are presently not used in the current implementation.

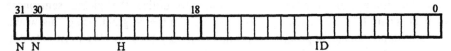

Fig. 1. General semaphore descriptor

All semaphore servers are members of a global server group identified by the name "GSSERVER_GROUP". Furthermore, each semaphore server on startup creates a new group with the name "GSSERVER_XXX", where XXX stands for the H field of its TID. In the current implementation, there is no reason to create a semaphore server on each processor of the virtual machine. When a particular processor does not own any semaphores, there exists no semaphore server. The creation of servers is a dynamic process. Initially no server is running nowhere. With the first open request for a global semaphore on a specified host, a server is started on the local host to handle the request and if the request must be forwarded to another host, a server is started on that host also. With this arrangement, if he have a virtual machine of a vast number of processors, but we want to use the semaphore facilities on a few processors only, we achieve savings in resources and a more efficient implementation, as we use collective operations between all servers in some occasions.

In order to achieve the maximum performance, we have tried to minimize the communication overhead of message transmission between processes residing on different processors. This is because in a loosely coupled parallel machine, the most significant overheads are due to network latencies. For all semaphore operations, where we already know the semaphore descriptor, we send an appropriate message to the corresponding server immediately. We obtain its TID by queering for the TID of the process with instance 0 in group "GSSERVER _XXX" where the XXX field is extracted from the H field of the descriptor. We could have used a more direct approach by coding the whole semaphore server TID in the descriptor. But this would have resulted in a longer data structure. With the method used, we could have achieved this benefit with the incorporation of the semaphore mechanisms in PVM, so that a specific id is reserved for the semaphore server on every processor.

3. Interface

The interface to the presented semaphore mechanism consists of eight functions. These functions constitute the API for the semaphore mechanism that is available to the parallel application programmer, as a library. The manipulation of a valid semaphore is done via a semaphore descriptor, as outlined above. This descriptor is a long integer (32-bit value) and is positive for all valid semaphore ids. In the following we explain the syntax and the function of the available interface calls:

- *int gs_init(void);*
  This function is used to initialize the environment of the global semaphore library so that successive calls to the other library functions will work as appropriate. This function should be the first to call, ahead of all other functions of the library, in a program that wants to utilize the global semaphore mechanism. If this function is called repeatedly in the program's code this would result in no damage in program execution or library behavior. We must note that the current implementation is not tread-safe, but we can provide a multithread-safe implementation with minor changes to all other library functions and by totally removing this call from the interface. The effect in the library execution speed will be negligible.

- *int gs_open(char *gsemname, char *hostname, int initvalue);*
  This function creates a global semaphore with name *gsemname*, at the processor with name *hostname* and initializes its value to *initvalue*. The call, if successful, returns a positive integer number, which is the descriptor for the new semaphore. This number must de used in successive calls to the library when we want to refer to this semaphore.
  If on processor *hostname* a semaphore with name *gsemname* has already been created then we don't create a new semaphore but we return the descriptor for the existing one. If the parameter *hostname* has the value "localhost", then the semaphore will be opened locally on the processor on which the calling program is running, while if it has the value "anyhost" we will try to find a corresponding semaphore with the same name located on any processor of our parallel virtual machine. If we succeed then we return the descriptor of the corresponding semaphore, otherwise we create a new semaphore locally on the processor that the program is running on.
  The call returns an error code, a negative integer number, if a new semaphore cannot be created, due to insufficient memory or too many already opened semaphores, or if one of the function parameters has a bad value.

318

- *int gs_close(gsem_t gsem);*
  This function is used to close an active semaphore with descriptor *gsem*. This descriptor has been returned by a previous successful call to gs_open().

- *int gs_reset(gsem_t gsem, int value);*
  This function empties the waiting queue of the semaphore with descriptor *gsem*, unblocking all processes that have been blocked on this semaphore by a previous call to gs_wait(). The value of the semaphore is then reset to *value*.

- *int gs_count(gsem_t gsem, int *value);*
  This function returns in *value* the current value c of the semaphore with descriptor *gsem*.

- *int gs_wait(gsem_t gsem);*
  The P operation on a semaphore. This function decrements the semaphore *gsem* by one, potentially blocking the calling process.

- *int gs_signal(gsem_t gsem);*
- *int gs_signaln(gsem_t gsem, int n);*
  The V operation on a semaphore. The function gs_signal() increments the semaphore *gsem* by one, potentially unblocking a waiting process. A call to function gs_signaln() is equivalent to calling function gs_signal() n times successively.

Except for function gs_open(), for which the returned value has already been explained above, all other functions return the value PvmOk (0) for successful completion or a negative integer indicating the corresponding error if the call has failed for some reason. Effort has been made the corresponding error value to be consistent with PVM's error codes so that it can represent errors such as network failures or process breakdowns detected by a corresponding pvmd.

4. Conclusions

The main advantage of our implementation is its direct handling of all semaphore operation requests. More over, it does not require the existence of semaphore primitives in the underlying operating system. Thus, it can as easily be implemented in environments that do not already have semaphores.

References

[1] Intel Corporation, Paragon User's Guide, June 1994.
[2] Partha Dasgupta, Richard J. LeBlanc, Jr. Mustaque Ahamad and Umakishore Ramachandran, "The Clouds Distributed Operating System", November 1991, IEEE.
[3] C. Voliotis, G. Manis, Ch. Lekatsas, P. Tsanakas and G. Papakonstantinou, "ORCHID: A Portable Platform for Parallel Programming", to appear in Euromicro Journal of System Architecture.
[4] Al Geist, Adam Beguelin, Jack Dongarra, Weicheng Jiang, Robert Manchek, Vaidy Sunderam, "PVM: Parallel Virtual Machine", The MIT Press, 1994.

PLS: A Parallel Linear Solvers Library for Domain Decomposition Methods

José M. Cela[1], José M. Alfonso[1] and Jesús Labarta[1]

Centro Europeo de Paralelismo de Barcelona (CEPBA),
c/ Gran Capitán s/n, 08034 Barcelona, Spain.

Abstract. In this paper we describe a parallel library (PLS) to solve linear systems arising from non overlapped Domain Decomposition methods. Preconditioned Krylov subspace iterative methods are considered as linear solvers. PLS had been implemented on the top of PVM using FORTRAN 77, additional library which allows the use of dynamic memory allocation.

1   Introduction

The library covers a relative extensive set of numerical methods. Table 1 shows the complete set of methods implemented at the present. Two main options are considered, use the whole matrix or use the Schur complement matrix. In both cases symmetric and non symmetric matrices are considered. The set of preconditioners are:

- A simple Diagonal scaling in the whole matrix solvers. In the Schur matrix solvers the diagonal of the A_{BB} submatrix is used for this purpose.
- Relaxed Incomplete Factorizations (LU and Cholesky) for the whole matrix solvers. The ILUt preconditioner defined in [3] was used as basic algorithm to define these incomplete factorizations.
- A preconditioner based in the diagonal blocks of the A_{BB} submatrix is used in the Schur matrix solvers. This preconditioner had been derived from a preconditioner defined in [1], and it has been proved successfully in some industrial problems [2].

If a domain decomposition has been performed then ordering firstly internal unknowns and secondly boundary ones, a block arrow structured matrix like the following is obtained:

$$\begin{pmatrix} A_{ii}^{(1)} & & & A_{ib}^{(1)} \\ & \ddots & & \vdots \\ & & A_{ii}^{(P)} & A_{ib}^{(P)} \\ A_{bi}^{(1)} & \cdots & A_{bi}^{(P)} & A_{BB} \end{pmatrix} \begin{pmatrix} x_i^{(1)} \\ \vdots \\ x_i^{(P)} \\ x_B \end{pmatrix} = \begin{pmatrix} f_i^{(1)} \\ \vdots \\ f_i^{(P)} \\ f_B \end{pmatrix} \tag{1}$$

If a parallel process is associated to a single domain, the k-th process stores the $A_{ii}^{(k)}$, $A_{ib}^{(k)}$ and $A_{bi}^{(k)}$ sub-matrices. The sub-matrix A_{BB} is distributed by

Table 1. Available linear solvers in PLS

	Whole Matrix Solvers		Schur Matrix Solvers			
	Iter. Meth.	Precond.	Iter. Meth.	Precond.	Inner solver	
					Method	Precond.
No	GMRES(m)	None	GMRES(m)		LU	
Symmetric	BiCGstab	Diag. Scal.	GGS	None		None
Matrix	CGS	RILUt	BiCGstab	Diag. A_{BB}	GMRES(m)	Diag. Scal.
				Block Diag.	BiCGstab	RILUt
Symmetric					Cholesky	
Matrix	CG	RICHt	CG		CG	RICHt

blocks of rows among the processes. The group of rows stored by one process are those associated with the boundary nodes assigned to this process/domain. We call $A_{bb}^{(k)}$ the sub-matrix of the same size that A_{BB} formed by the set of rows of A_{BB} assigned to the k-th process and zero somewhere else. All the sub-matrices $(A_{ii}^{(k)}, A_{ib}^{(k)}, A_{bi}^{(k)}, A_{bb}^{(k)})$ are stored in a sparse format to exploit their internal sparsity.

With respect to the storage of the vectors, each process stores the part associated with its internal unknowns. The part of the vectors associated with boundary unknowns is replicated among all the processes. Note that one process does not need all the boundary vector, it only needs those parts of the boundary vector associated with its neighbor domains. We use the following notation for the vectors: $y_i^{(k)}$ is the vector associated with the internal unknowns of the k-th process, $y_b^{(k)}$ the vector associated with the boundary unknowns of the k-th process, and $y_B^{(k)}$ is the set of y_b vectors stored by the k-th process, i.e., it is the union of $y_b^{(k)}$ with the vectors $y_b^{(i)}$ where $i \in Neigh(k)$. We denote by $Neigh(k)$ the set of neighbors of the k-th domain.

An iterative method to solve (1) can be applied directly to the matrix A using the Domain Decomposition data distribution to parallelize the algorithm. We call this option Whole Matrix Solvers (WMS). A second option to solve (1) is to apply an iterative method to the Schur complement system,

$$\left(A_{BB} - \sum_{k=1}^{P} A_{bi}^{(k)} A_{ii}^{(k)^{-1}} A_{ib}^{(k)}\right) x_B = f_B - \sum_{k=1}^{P} A_{bi}^{(k)} A_{ii}^{(k)^{-1}} f_i^{(k)} \qquad (2)$$

At the end of the iterative method only the boundary part of the solution is obtained. The internal unknowns can be computed in a parallel backward substitution. We call this option Schur Matrix Solvers (SMS).

2 Whole matrix solvers

The two main operations in all the iterative methods are matrix times vector product, and apply the preconditioner. The matrix by vector product can be done with a parallel algorithm which requires only communications between pairs of processes. A scheduling of these communications can be computed in the pre-processing phase of the application, in such a way that they are done in parallel as much as possible without deadlocks. Clearly, the performance of this scheduling depends on the decomposition, but for problems where a quasi-optimal decomposition is possible, this communications are not a bottleneck and a good scalability of the matrix-vector product is possible.

We have implemented a parallel incomplete factorization called PILUt as preconditioner. This preconditioner is based in the original idea of the ILUt preconditioner . The main difference between ILUt and PILUt is how the fill-in is controlled in the factorization process. By construction $A_{ib}^{(k)}$, $A_{bi}^{(k)}$ and $A_{bb}^{(k)}$ matrices have no null entries only in those rows/columns associated with the boundaries of the neighbors of the k-th domain. When the factorization algorithm is applied the block level fill-in is concentrate in the A_{BB} matrix, i.e., although there is fill-in also in the matrices A_{II}, A_{IB} and A_{BI}, the internal block structure of those matrices remains the same. The block fill-in in the A_{BB} matrix can be see as the addition of two consecutive block fill-ins, the first one originated by factorization of the A_{BI} matrix, and the second one due to the factorization of the A_{BB} matrix itself. The first class of block fill-in is restricted by construction to the blocks of $A_{bb}^{(k)}$ associated with the boundaries of the neighbors of the neighbors of the k-th process. Note that a fill-in in a block which originally was a null block in the A_{BB} matrix means a communication between no neighbor processes when the triangular system must to be solve. We have introduce a new parameter bfil in the PILUt algorithm to select the allowed block fill-in. If bfil = 0 the block structure of the L_{BB} and U_{BB} matrices is the same that the block structure of the A_{BB} matrix. Then, the communications of the k-th process are restricted to its neighbors. If bfil = 1: The first block fill-in is allowed. Then, the k-th process has communications with its direct neighbors and with the neighbors of its neighbors. If bfil > 1: All the block fill-ins are allowed.

3 Schur matrix solvers

Working with the Schur complement instead of the whole matrix has the main advantage that for many problems the Schur matrix has a better condition number than the whole matrix. Because an iterative method requires only the product of the matrix by a vector, this can be computed using the implicit definition of S. As in the WMS the Schur matrix times vector algorithm has communications which can be scheduled in the pre-processor, overlapping a considerable number of them. The product of the Schur matrix by a vector algorithm requires the solution of a local linear system within each domain for each iteration of the solver. The first approach is to perform a LU decomposition of the $A_{ii}^{(k)}$ matrices in the initial phase of the solver, and solve two triangular systems each time. This method has the drawback of requiring a significant amount of time at the start up of the solver. It also requires a significant amount of memory to store the $L_{ii}^{(k)}$ and $U_{ii}^{(k)}$ matrices. Although the use of local reordering techniques, like Minimum degree, may reduce these drawbacks it will not eliminate them. Another attempt may be solving the local linear systems through an iterative method. Different local iterative methods combined with different local preconditioners were tested. To ensure the convergence of the Schur method, the local iterative method must have more accuracy than the one required for the external iterative solver.

For SMS a preconditioner based on the diagonal blocks of the matrix A_{BB} was used. We call this preconditioner Block Diagonal (BD). The preconditioner is defined as a diagonal block matrix where each block is a block in the diagonal of the A_{BB} matrix, i.e, each block of the preconditioner is associated with the connections between the boundary nodes of a single domain. The entries in the A_{BB} matrix associated with connections between the boundaries of different domains are neglected. The justification of this preconditioner is based on the experimental observation that for many problems the reduced interface operator has a strong spatial local coupling and a weak global coupling. This observation was described by T. Chan and it was the base of the MSC(k) preconditioner. Other related preconditioner was proposed by O. Axelsson [1]. He proposed to use the complete A_{BB} matrix as preconditioner, neglecting the contribution of the term $A_{BI}A_{II}^{-1}A_{IB}$. Because, in general the A_{BB} matrix will be not block diagonal, this preconditioner requires again to solve triangular systems.

Our preconditioner use the main information available in the A_{BB} matrix, neglecting the connections between different boundaries. Note that in 3-D meshes there are only two possible boundaries between domains. *Boundaries between only two domains* (surfaces), and *Boundaries between several domains*, (curves). Nodes on surfaces can be numbered in such a way that half of the surface is assigned to one domain a the other half to the other domain, then the connection between the boundary of the two domains is reduced to a line of nodes. The nodes in the curves can be number again in a strip decomposition which minimize the connections between the boundaries assigned to each domain touching the curve. We have observed that the improvement in the convergence when all the A_{BB} matrix is used as preconditioner does not compensate the communication cost needed to take into account all connections.

References

1. Axelsson, O.: Iterative Solution Methods. Cambridge University Press (1994).
2. Cela, J. M., Dürsteler, J. C., Labarta, J.: Manufacturing Progressive Addition Lenses using Distributed Parallel Processing. IRREGULAR96, to be published (1996).
3. Saad, Y.: ILUT: a dual Threshold Incomplete LU factorization. Numerical Linear Algebra with applications, Vol. 1, No. 4, (1994) pp. 387–402.

Parallel Integration Across Time of Initial Value Problems Using PVM

René Alt and Jean Luc Lamotte

MASI and Institut Blaise Pascal
4 Place Jussieu 75252 Paris Cedex05
France

Abstract. This paper shows the efficiency of PVM in the solution of initial value problems on distributed architectures. The method used here is a collocation method showing a large possibility of parallelism across time. The solution of the linear or non linear system, which is a part of the method is obtained with Picard's iterations using divided differences. This solution is also obtained in parallel as well as the values of the approximated solution at different times. Numerical examples are considered. The tests have been performed on a network of workstations and on the Connection Machine CM5.

Keywords : P.V.M., parallel methods, initial value problems, ordinary differential equations, collocation methods.

1 Introduction

Let us consider the classical m-dimensional initial value problem:

$$y'(t) = f(t, y(t)) \qquad f : [t_0, b] \times \mathbb{R}^m \longrightarrow \mathbb{R}^m$$
$$y(t_0) = y_0 \tag{1}$$

f is supposed to verify a Lipschitz condition on the variable y in the domain of integration so that the problem has a unique solution. Let us also define a mesh of n values t_0, t_1, t_{n-1} in a sub-interval of $[t_0, b]$. An approximate m-dimensional solution of problem (1) on the given sub-interval is a set of m polynomials $P_1(t), P_2(t), ..., P_m(t)$ with degree n satisfying the following collocation conditions:

$$P'_j(t_i) = f(t_i, P(t_i)) \qquad i = 0, 1, ..., n-1.$$
$$P_j(t_0) = y_{j,0} \qquad\qquad j = 1, ..., m. \tag{2}$$

Hence, a collocation method consists in finding a set of polynomials satisfying the initial value problem at each point of the mesh. such a method is inherently implicit and requires the solution of a linear or non linear system according to the nature of problem (1). More details about collocation methods can be found in [1], [2], [5], [6] or [4].

2   Solving the non linear system

2.1   The collocation polynomials

The approximating polynomials are very often obtained thanks to the Lagrange formulae (see [2]). In the present algorithm they are computed in a straightforward manner and expressed with their coefficients in respect to their powers of t. More precisely, let an approximating polynomial be:

$$P_j(t) = \sum_{j=0}^{n} a_{ij} t^{n-j} \tag{3}$$

The coefficients a_{ij} are directly computed as being the solution of system (2).

2.2   The algorithm for solving the system

The algorithm which is proposed here to solve the implicit system is a fixed point iterative method and is the following :

For the sake of simplicity it is described for a 1-dimensional differential initial value problem.

(a) Initialize an approximate solution at points t_i using Euler's method.

(b) At each point t_i compute the derivatives $f(t_i, y(t_i))$

(c) Compute the values $P'(t_i), i = 0, ..., n - 1$ of polynomial $P'(t)$ using divided differences.

(d) Compute the primitive polynomial $P(t)$ and its values points t_i

(e) Loop at b until $\max_{i=0,n-1} |f(t_i, P(t_i)) - P'(t_i)| \leq \epsilon$, for a given ϵ

3   Parallelism in the algorithm

3.1   The parallel tasks

As a matter of fact, at each step b), c), d) of the above iterative process, the values of $P'(t_i)$, $P(t_i)$ and $f(t_i, P(t_i))$ at each point t_i can be computed in parallel. The degree of parallelism across space is thus the equal to the degree of the collocating polynomials. This clearly to n parallel identical tasks, each of them having to perform steps b),c),d) and to compute the difference $|f(t_i, P(t_i)) - P'(t_i)|$ at a single point t_i.

3.2   Architecture and communications

The logical architecture is composed of a master process a n computing slave processes. Each slave is linked to the masster and is independant of the other slaves. The iterations are synchronised by the master and the communications are between the master and the slaves. The load of communication increase linearly with the degree of the collocation polynomials.

In the general case of a large linear or non linear differential system, all the components of the vectors $P_j(t_i))$, $P'_j(t_i))$ and $y_j(t_i))$ are computed in the same task T_i. The algorithm and logical architecture are thus very well adapted to the fat tree physical architecture with vector processing nodes of the Connection machine CM5. A SPMD programming using PVM is also very easy.

3.3 Complexity in communications and function calls

Let l be the number of sub-intervals on which the code is performed. On the ith sub-interval the number of points is called n and the number of necessary iterations to reach the solution is called k_i. The number of communications from and to the master and from and to each slave process can be easily calculated and are given in the following table :

master	slave
$(n-1)(1+2*\sum_{i=1}^{l} k_i)$	$2*\sum_{i=1}^{l} k_i$

Table 1

Concerning the number of function calls, the initialisations are performed on the master whereas the iterations are performed in parallel on the slaves. Consequently the number of function calls are given by the formulas in the following table.

sequential mode	parallel mode
$n(l+\sum_{i=1}^{l} k_i)$	$n.l+\sum_{i=1}^{l} k_i$

Table 2

From tables 1 and 2 it can be seen that the parallel code is efficient as soon as :

$$n(l+\sum k_i)\theta_f > (n.l+\sum k_i)\theta_f + (n-1)(1+2\sum k_i)\theta_c$$

θ_f and θ_c are respectively the computing function time and the time of a communication. If $\sum k_i$ is much greater than 1, this formula leads to :

$$\theta_f > 2\theta_c$$

4 Numerical experiments

Various differential systems quoted from classical test problems have ben solved. The method appears particularly efficient in the cases for which the load of computation required by the second member $f(x,y)$ is important. In that case, the communication time is negligible in respect to the computing time.

As an example let us report the numerical tests relative to the well known Van der Pol oscillator defined by:

$$y''(t) = 3\ (1 - y^2(t))\ y' - y(t) \qquad t \in [0, T]$$
$$y(0) = 2$$
$$y'(0) = 0 \tag{4}$$

The numerical solution has been found identical to the one obtained with LSODE [3]. In order to be able to compare the communication time and the computing time, a delay has been artificially introduced in the computation of the function. Then, the solution has been computed sequentially and in parallel with exactly the same conditions and with increasing delays. The number of parallel processes was seven ($n = 7$).

The sequential t_s and parallel t_p computing times in seconds are reported in table 3.

t_s	0.3	5	9	17	23	34	68
t_p	3	5	6	8	9	13	23

Table 3

5 Conclusion

In this paper it has been shown that the collocation methd for solving ODEs can be easily performed in parallel on a distributed memory architecture. The parallelism is straightforward and the number of communications as well as the number of parallel function calls are linear in respect to the number of points in the interval. As a function call of the sequential code is replaced by two communications (a call and a receive), it appears that the parallel code is efficient as soon as the computing time of the function is greater than twice the time of a communication. This is confirmed by experience.

References

1. R.Alt, J.Vignes, Validation of collocation methods with the CADNA library, *Math. Comp. in Sim.*, 22, 1996, (to appear).
2. E Hairer, S.P. Norsett and G. Wanner, Solving Ordinary Differential Equations I: Nonstiff problems, *Springer Series in Computational Mathematics 8* (Springer, Berlin 1987).
3. A.C. Hindmarsh, ODEPACK, a systematized collection of ODE solvers,*scientific computing*, R. S. stepleman et al. (eds.), north-holland, amsterdam, 1983, pp. 55-64.
4. H.B. Keller, Numerical Methods for two-point boundary-value problems. Dover Publications, Iue. New-York (1992), 225-227.
5. J. Villadsen and M.L. Michelsen, Solution of differential equation models by polynomial approximation. Prentice Hall, Englewood Cliffs, R.N.Y., 1978.
6. R. Weiss, The application of implicit Runge Kutta and collocation methods to boundary value problems.*Math. Comp.*, 28, (1974), 449-464.

Distributed Computation of Ribbon Tableaux and Spin Polynomials

Sébastien Veigneau*

Institut Gaspard Monge, Université de Marne-la-Vallée, 2, rue de la Butte Verte,
93166 Noisy-le-Grand Cedex, France

Abstract. Recent works in algebraic combinatorics have brought up to date the importance of certain planar structures, called ribbon tableaux, which are generalizations of Young tableaux. This paper gives an algorithm to efficiently distribute, using PVM, the computation of the set of all ribbon tableaux of given shape and weight. It also provides a way to compute the spin polynomials associated to those sets of ribbon tableaux, these polynomials leading to generalizations of Hall-Littlewood functions.

1 Introduction

Ribbon tableaux, introduced by Stanton and White [9] can be used to construct a basis of highest weight vectors in the Fock space representation of the quantum affine algebra $U_q(\widehat{sl}_n)$ [7]. This basis plays a crucial rôle in the computation of the canonical basis of the q-deformed Fock space. It is conjectured in [8] that this basis gives the decomposition matrices of q-Schur algebras at roots of unity. Also, generating functions of ribbon tableaux lead to generalizations of Hall-Littlewood functions, which are q-analogues of products of Schur functions [7].

We present a distributed algorithm for generating sets of ribbon tableaux or computing the generating functions according to a statistic called spin. We describe a PVM implementation of this algorithm and discuss experimental results.

Basic definitions relative to partitions, Young tableaux and symmetric functions can be found in [3] and [6].

2 Ribbon Tableaux

A ribbon tableau is obtained by filling a Ferrers diagram with *ribbons*. A *k-ribbon* is a connected skew diagram of boxes containing no 2×2 square of boxes. Not all Ferrers diagrams are tilable with k-ribbons. An obvious necessary condition is that the number of boxes should be divisible by k but this is not sufficient. For instance, the shape $\lambda = (3, 2, 1)$ is not tilable with dominoes.

Having filled a shape λ with k-ribbons, we label each ribbon by a positive integer in such a way that labels are weakly increasing across rows from left

---

* Supported by PROCOPE.

to right and strictly increasing across columns bottom-up. More precisely, let the *root* of a ribbon be its rightmost and lowest box. The exact condition on columns is that the root of a ribbon labeled by i must not be above any box of a ribbon labeled by $j \geq i$. The *weight* of a ribbon tableau is defined as in the case of Young tableaux. Let R be a k-ribbon, $h(R)$ its *height* and $w(R)$ its width. Observe that $h(R) + w(R) = k + 1$. The *spin* of R is $s(R) := \frac{h(R)-1}{2} \in \frac{1}{2}\mathbb{N}$. The *spin* of a tableau T is $s(T) := \sum_R s(R)$ where the sum is over all ribbons of the tableau. For two partitions λ and μ such that $|\lambda| = k|\mu|$, we denote by $\text{Tab}_k(\lambda, \mu)$ the set of all k-ribbon tableaux of shape λ and weight μ. The most important information is the *spin polynomial* $G_{\lambda,\mu}^{(k)}(q) := \sum_T q^{s(T)}$ in which the sum is over all $T \in \text{Tab}_k(\lambda, \mu)$. More details about ribbon tableaux can be found in [3] and [7].

We encode a ribbon tableau $T \in \text{Tab}_k(\lambda, \mu)$ by a matrix $M(T)$ with r rows and s columns, r being the length of the partition μ and s the length of the conjugate partition λ' of λ. This matrix is filled with entries 0 or k, the i-th row containing the value k exactly μ_i times. To obtain the i-th row, we look for the indices of columns containing the roots of ribbons labeled i, write k in these columns and other entries are set to 0. To rebuild T from $M(T)$, we first insert the ribbons labeled by 1. The first row of the matrix $M(T)$ indicates the positions of the roots of the ribbons but we have to "fold up" these ribbons. Assume the matrix has s columns and consider the vector $\rho_s := [s - 1, s - 2, \ldots, 2, 1, 0]$, then add this vector to the first row of the matrix. Now, sort the resulting vector in decreasing order, subtract ρ_s and obtain the conjugate of the shape occupied by ribbons labeled by 1. A given shape has at most one tiling with ribbons carrying the same label. To build the shape occupied by ribbons labeled by 2, we start with the sorted vector obtained at the previous step and add the second row of the matrix. Then, we proceed exactly as for the first row to obtain the shape occupied by ribbons labeled by 1 and 2. As we already know the shape occupied by ribbons labeled by 1, we recover the shape occupied by ribbons labeled by 2. We repeat this construction for ribbons labeled by 3, 4 and so on.

This decoding allows to compute the spin of a ribbon tableau. Indeed, having added ρ_s to the first row of $M(T)$, we obtain a new vector and we have to save the number of inversions I of the permutation that sorts this vector in decreasing order. This number is equal to $I = \sum_R (w(R) - 1)$ so that $I = \sum_R (w(R) - 1) = \sum_R (k - h(R))$, which leads to $s(T) = (n(k - 1) - I)/2$ where n is the number of k-ribbons in the partial ribbon tableau T.

3   Recursive Computation of Spin Polynomials

The matrix coding leads to a recursive formula for the polynomials:

$$F_{\alpha,\mu}^{(k)}(q) := \sum_{T \in \text{Tab}_k(\lambda,\mu)} \left(\sum_{R \in T} q^{(w(R)-1)} \right), \tag{1}$$

where $\alpha = \lambda' + \rho_s$, with $s \geq \lambda_1$. Let $\mu = (\mu_1, \ldots, \mu_r)$. We obtain the possible positions of the μ_r ribbons labeled by r in a ribbon tableau of shape λ and weight

μ by reversing the previous decoding algorithm: the global shape containing all ribbons labeled by $1, \ldots, r-1$ are obtained by subtracting from α all distinct permutations of the vector $v_r := [0, \ldots, 0, k, \ldots, k]$ (k repeated μ_r times), sorting these vectors in decreasing order and storing inversion numbers. We thus build a tree, rooted by α, with edges indexed by all distinct permutations v_r^σ of v_r and with vertices labeled by the sorted vectors $\alpha - v_r^\sigma$, stored with the number of inversions of the sorting permutation. If a vertex contains a vector with repetitions, it cannot belong to a matrix describing a ribbon tableau, so that all edges leading to such vertices are eliminated. Let $\alpha_1, \ldots, \alpha_m$ denote the remaining vertex labels, and $I_1, \ldots, I_m$ the corresponding inversion numbers. Let also $\mu^* = (\mu_1, \ldots, \mu_{r-1})$. Then, the recurrence formula is:

$$F_{\alpha,\mu}^{(k)}(q) := \sum_{p=1}^{m} q^{I_p} F_{\alpha_p,\mu^*}^{(k)}(q) \ , \tag{2}$$

with the condition $F_{\rho_s,\emptyset}^{(k)} := 1$. Now, the spin polynomial is given by the formula:

$$G_{\lambda,\mu}^{(k)}(q) = q^{\frac{|\mu|(k-1)}{2}} F_{\alpha,\mu}^{(k)}(q^{-\frac{1}{2}}) \ . \tag{3}$$

This is actually a polynomial (no fractional exponent) if all parts of λ are divisible by k (the most interesting case).

4 Parallel Implementation with PVM

We now present a parallel implementation of the previous algorithm on a network of computers clustered into a parallel virtual machine described via PVM.

We divide the computation into two parts, one being managed by the so-called *master* and the other one by the *slaves*, the master and the slaves being tasks in the virtual machine.

On the one hand, the master has to initiate, administrate and terminate the computation of the spin polynomial. Initiate the computation means that it has to read the input data λ, μ and the maximal number of slaves. It also has to administrate the computation of the spin polynomial, *i.e.* to agree or disagree with the requests of creation of new tasks. Then finally, it has to terminate the computation, that is, to give the spin polynomial $G_{\lambda,\mu}^{(k)}(q)$.

On the other hand, slaves investigate all possibilities. They correspond to the nodes of the previous tree, including the root. Each task has to enumerate all distinct permutations v_r^σ of v_r (labeling the edges) and to eliminate the non admissible ones. Then, it asks the master whether it should create new tasks corresponding to the remaining edges, creates the accepted new tasks with their parameters, informs the master that it is dying and finally dies.

The master reads the input data and creates the root of the tree, *i.e.* spawns the first task. Then, it enters a while loop that terminates when there is no more alive task in the virtual machine. The master is thus waiting for any message from any task. It can be a message informing that a task is dying which corresponds either to a successful result at the lowest level in the tree, or simply to a node that has terminated its work. The master also receives messages carrying requests for new task creations. The answer depends upon several parameters. The master has to check whether the maximal number of spawned tasks is already reached or not. If this number is already reached then the master informs the task not to spawn new tasks but to recursively compute its subtree. Furthermore, at a given level of the tree, several sorted vectors may be equal, and the master has to manage the tasks so that a given subtree is computed only once, even if two – or more – tasks have sent to the master the same request. This remains true even if a task is sequentially computing its subtree, requests being then sent before recursive calls. In other words, it is transparent for the master, which only has to count the number of spawned tasks among the total number of tasks.

When all tasks are terminated, the master builds the spin polynomial from the partial results obtained from tasks of the lowest level in the execution graph.

We can for example obtain the spin polynomial for 6-ribbon tableaux of shape $(18, 18, 18, 18, 18, 18, 18, 18)$ and weight $(4, 4, 4, 4, 4, 4)$ (1079706100 elements) with about 7 minutes on a cluster of 10 HP9000 workstations. This computation takes 4 minutes and 30 seconds with 15 computers.

A detailed version of this paper can be found on the Internet at the URL: *http://www-igm.univ-mlv.fr/~veigneau/private/HTML/publications.html.*

References

1. Betten, A., Veigneau, S., Wassermann, A.: Parallel Computation in Combinatorics using PVM, preprint, Université de Marne-la-Vallée, 1996.
2. Carré, C., Leclerc, B.: Splitting the square of a Schur function into its symmetric and antisymmetric parts. J. Alg. Comb. 4 (1995), 201–231.
3. Désarménien, J., Leclerc, B., Thibon J.-Y.: Hall-Littlewood functions and Kostka-Foulkes polynomials in representation theory. Séminaire Lotharingien de Combinatoire, Université de Strasbourg, 1993 (URL: *http://cartan.u-strasbg.fr:80/~slc/opapers/s32leclerc.html*).
4. Geist, A., Beguelin, A., Dongarra, J., Jiang, W., Manchek, R., Sunderam, V.: PVM: Parallel Virtual Machine, A users guide and tutorial for networked parallel computing. MIT Press, 1994.
5. Kirillov, A. N., Lascoux, A., Leclerc, B., Thibon, J.-Y.: Séries génératrices pour les tableaux de dominos. C.R. Acad. Sci. Paris **318** (1994), 395–400.
6. Knuth, D.: The art of computer programming. Vol. **3**, Addison-Wesley, 1981.
7. Lascoux, A., Leclerc, B., Thibon J.-Y.: Ribbon tableaux, Hall-Littlewood functions, Quantum affine algebras and unipotent varieties. preprint IGM **96.2**, q-alg/9512031.
8. Leclerc, B., Thibon J.-Y.: Canonical Bases of q-Deformed Fock Spaces. International Math. Research Notices, to appear.
9. Stanton, D., White, D.: A Schensted algorithm for rim-hook tableaux. J. Comb. Theory A **40** (1985), 211–247.

Parallel Model-Based Diagnosis
Using PVM

Cosimo Anglano, Luigi Portinale

Dipartimento di Informatica - Universita' di Torino
C.so Svizzera 185 - 10149 Torino (Italy)
ph: +39 11 7429111 fax: +39 11 751603
e-mail: {mino,portinal}@di.unito.it

Abstract. The present paper outlines the PVM implementation of a particular approach to model-based diagnosis which uses a Petri net model of the system to be diagnosed. Parallel backward reachability analysis on the state space of the net is used to explain the misbehavior of the modeled system. The analysis algorithm is based on the automatic identification of the parallelism present in the structure of the Petri net model. Starting from the above information, an MIMD message passing program is automatically constructed. PVM has proved to be a useful tool for implementing the above parallel programs; in particular, we tested the approach by implementing a parallel program for a car fault diagnosis domain.

1 Introduction

Solving a diagnostic problem usually involves finding the primary causes that produce a particular malfunction, whose effects are observed in the behavior of a given system. In the model-based approach to diagnosis, a model of the structure and/or of the behavior of the system to be diagnosed is supposed to be available; the solution to a diagnostic problem is obtained by tracking back the observed misbehavior on the available model, looking for a possible explanation of such a misbehavior (see [3] for a complete survey). A Petri net [4] representation of the causal behavior of the system to be modeled has been proposed in [5, 6] with the introduction of the model of *Behavioral Petri Nets* (BPN). Unlike standard Petri nets, BPNs comprise two different types of transitions: *regular transitions*, behaving like ordinary transitions of Petri nets, and *OR transitions*, having a slightly different behavior with respect to the classical firing rule of Petri nets. In particular, while regular transitions may fire when all their input places are marked, an OR transition may fire when at least one of its input places is marked.

Thanks to some particular features of the BPN model, reachability graph analysis can be suitably adopted in the search for the solutions to a diagnostic problem (see [5]). In particular, backward reachability analysis on the state space of the net is used, since diagnosis is a problem of post-diction rather than prediction. The observed information about the state of the system under examination (i.e. the symptoms of a particular diagnostic problem) is modeled by means of a marking from which backward reachability analysis starts. This analysis is performed by defining a backward firing rule for each kind of transitions of the BPN formalism, and by applying these rules, starting from the marking corresponding to the observed symptoms, in order to generate in a backward fashion the reachability graph of a given BPN. The markings of the reachability graph corresponding to the presence of primary causes of the model will represent the possible diagnoses.

2 Parallel Backward Reachability Analysis

The generation of the backward reachability graph is a computationally expensive problem. In order to alleviate it, in [2] we have proposed an approach in which the backward reachability graph, corresponding to a particular initial marking for a given BPN, is generated by means of an MIMD, message-passing parallel program which is automatically constructed starting from the structure of the above BPN. The derivation of this parallel program is accomplished by first individuating the available parallelism in a given BPN, by means of an algorithm which partitions the BPN into subnets (see [2] for more details on the partitioning technique), and then by generating a process for each subnet. Each process is in charge of performing the backward firing for the transitions in the subnet it corresponds to, so its behavior depends on the type of transitions it contains. A special process, called *ENV*, is in charge of assembling the final marking from the partial marking produced by the processes. Communications occur among processes whose corresponding subnets share a place. In particular, if a given place is an input place for a transition of one of the processes and an output place for a transition of the other one, the former process sends messages to the latter one (details can be found in [2]). The computation evolves according to the *macro-dataflow* paradigm, that is each process starts its computation as soon as enough inputs are available. Each time a process is "activated" by the presence of some inputs, it produces some output data which are sent to other processes and so on, until no more inputs remain to be processed.

The application of the technique discussed above yields to a parallel program whose granularity is minimal. In order to achieve adequate performance, granularity needs to be adjusted according to the features of the target parallel system. This is accomplished by means of a clustering step, in which a set of coarser-granularity processes is obtained by coalescing several processes into a single one. The clustering step is based on the estimation of the computation and communication costs of each initial process. Since the behavior of the processes is highly data-dependent, a probabilistic analysis is carried out in order to obtain average values for the above quantities. The clustering operation is performed by using the SCOTCH [1] package, which is also used to compute the mapping of the resulting processes on the processors of the target architecture.

3 PVM Implementation of the Backward Reachability Analysis

Over the recent years, clusters of workstations have been increasingly used as cost-effective alternative to supercomputers, because of the favorable cost/performance ratio they offer. In this work we have used PVM to implement, on a cluster of workstations, the parallel programs performing the backward reachability analysis for arbitrary BPN models. PVM version 3.3 has been used to implement these programs; this choice has been quite natural, since PVM features fit well the program requirements. Moreover, once the program corresponding to a given BPN has been constructed, it can be ported with little or no efforts on other platforms, because PVM is available on a broad range of architectures. At the moment of this writing, the PVM program performing the analysis of a given BPN is constructed manually starting from the partitioning of the net, but work is in progress to construct an automatic generator. The basic structure of the program encompasses a master process (ENV), whose purpose is to read the inputs provided by the user, to start the computation of processes, to collect the results, and to detect the termination of the computation. Initially, ENV spawns all the processes of the parallel programs on the hosts of the system, according to a particular (static) mapping, and then it enters a loop in which it waits for the user requests (i.e. for an initial marking from which the backward reachability analysis may start). After this marking has been specified, ENV computes the set of processes which will participate to the solution of the corresponding diagnostic problem, and will disable the other ones in order to avoid the arise of deadlocks (the set of active processes depends on the marking from which the backward analysis starts [2]). After the computation of a particular solution is completed, the cycle is repeated again, until the user decides to terminate the entire program.

As a proof of concept, and to gain insights in the automatic construction process, we have implemented the parallel program corresponding to a particular testbed domain (car engine fault diagnosis) for which we have a knowledge base available. Such a knowledge base corresponds to a BPN model with more than 100 places and transitions, resulting in a set of about 50 processes at the finer level of granularity. We have used a cluster of Sun SPARC workstations (model 5, 10, and 20) connected by a thin wire Ethernet. Experiments have been carried out to assess the efficiency of the parallel program mentioned above with respect to a sequential implementation of the backward reachability analysis. Preliminary results indicate that:

a) keeping the granularity at the finer level yields to unacceptable overheads which highly affect the efficiency of the program, so the effectiveness of the clustering strategy is crucial;

b) the parallel backward analysis is effective when the computation of a given solution requires the "exploration" of large portions of the net.

4 Conclusions

In the present paper we have briefly described the PVM-based environment for the support of a parallel approach to model-based diagnosis. This approach relies on the use of a Petri net model from which an MIMD, message passing program is automatically derived. The possibility of exploiting a cluster of Sun workstation made PVM a natural candidate for implementing the approach on a testbed domain (car engine fault diagnosis). Preliminary experiments have confirmed the suitability of PVM for the implementation of the approach. However, more experiments are still needed; in particular, since a sequential version of backward reachability analysis is available for our testbed domain, we plan to compare more deeply the PVM implementation with it, in order to have a more detailed evaluation of the parallel approach.

As a final consideration, note that since the parallel program is automatically constructed starting from the net description, we can build a flexible environment for diagnostic problem solving in which the user represents the causal behavior of the system by means of a BPN model and the environment automatically generates the parallel program performing the diagnosis. Its portability on a wide range of target systems, made possible by PVM, is in our belief one of the points of strength of this environment.

References

1. SCOTCH Home Page: http://www.labri.u-bordeaux.fr/~ pelegrin/scotch.
2. C. Anglano and L. Portinale. B-W analysis: a backward reachability analysis for diagnostic problem solving suitable to parallel implementation. In *Proc. 15th Int. Conf. on Application and Theory of Petri Nets, LNCS 815*, pages 39–58, Zaragoza, 1994. Springer Verlag.
3. W. Hamscher, L. Console, and J. de Kleer. *Readings in Model-Based Diagnosis*. Morgan Kaufmann, 1992.
4. T. Murata. Petri nets: Properties, analysis and applications. *Proceedings of the IEEE*, 77(4):541–580, 1989.
5. L. Portinale. *Petri net models for diagnostic knowledge representation and reasoning*. PhD Thesis, Dip. Informatica, Universita' di Torino, 1993. Available as anonymous ftp at ftp.di.unito.it in /pub/portinal.
6. L. Portinale. Petri net reachability analysis meets model-based diagnostic problem solving. In *Proc. IEEE Intern. Conference on Systems, Man and Cybernetics*, pages 2712–2717, Vancouver, BC, 1995. Extended version to appear on IEEE Trans. on SMC.

Running PVM-GRACE on Workstation Clusters

Fukuko Yuasa, Setsuya Kawabata, Tadashi Ishikawa,[1] Denis Perret-Gallix,[2] and Toshiaki Kaneko[3]

[1] KEK, National Laboratory for High Energy Physics,
1-1 OHO Tsukuba Ibaraki, 305 Japan
[2] LAPP, Laboratoire d'Annecy-le-Vieux de Physique des Particules,
74941 Annecy-le-Vieux CEDEX, France
[3] Meiji-Gakuin University,
Kamikurata-cho 1518, Totsuka, Yokohama 244, Japan

Abstract. We have implemented PVM in GRACE: an automatic Feynman diagram computation package to benefit from the available cpu power distributed over a cluster of workstation. It is clearly shown that PVM-GRACE gives excellent results in term of reduction of elapsed time when the network overhead is kept small as compared to the processing time.

1 Introduction

We present a PVM [1] based automatic Feynman diagrams calculation system, PVM-GRACE. GRACE [2] is a powerful software package for the calculation of high energy physics process cross-sections and for event generation.

Using the Feynman diagrams technique, all observables can be computed for any process at any order of the perturbative theory. However, the calculation becomes rapidly tedious and involved as a number of final state particle increases for example. Automatic computation systems have been developed to cover the increasing need for such calculations. The Minami-Tateya group has developed the GRACE package which generates automatically the FORTRAN code corresponding to the process matrix element. Integration over the multi-dimensional phase space leads to cross-section and event generation. The larger the center of mass energy and the number of final state particles, the larger the cpu time involved.

The basic idea behind the PVM-GRACE project is to distribute the amplitude calculation over several PVM children spawned by the PVM parent. This simultaneous calculation of the total amplitude reduces the total elapsed time of GRACE drastically as well as the required memory allocation on each workstation.

2 PVM-GRACE

GRACE consists of four subsystems, the graph generation, the source generation, the numerical integration, and the event generation subsystems. When the initial and final states of the process, the physics model, and the kinematics are set by input datacards, GRACE generates the complete set of Feynman diagrams and the FORTRAN code with the appropriate Makefile. After running make, the total cross section is computed. A multi-dimensional grid mapping the integration phase space is produced so as to be suited for the behavior of the differential cross section and contains the event generation probabilities. Events are then generated by randomly selecting one of these hypercell. GRACE uses the program package BASES/SPRING [3] for multi-dimensional numerical integration and event generation. Of all steps, the numerical integration subsystem is the most cpu time consuming part.

Generated FORTRAN subroutines are grouped into three categories, the parent part, the utility part, and the amplitude calculation part. The amplitude calculation contains as many subroutines as the number of Feynman diagrams plus utility subroutines. In the parent part, the amplitude calculation is spawned over the available cpu's, after the calculation is finished, the results of numerical integration are sent back to the parent.

The sample points are represented by real*8 variables and the results are stored in an array of complex*16. Since the amount of data to be transferred between tasks affects deeply the performance of PVM-GRACE, we have tried the following two approaches for the data transfer:

- Each sample point selected by the parent task is broadcast to the children process in charge of the amplitude computation, then each result is returned to the parent:
- A full set of phase space points are sent to the various children process and an array of results is returned.

The merit of the first approach is that the basic version of BASES can be used. But the problem is the large number of transfer operations needed. On the other hand, the merit of the second is that the number of transfer setup is much reduced. But the problem, here, is that BASES must be modified so as to generate sets of appropriate phase space points. For a network such as Ethernet or FDDI, the second approach shows better performance than the first one. Therefore, in this paper we will concentrate on the second approach.

3 Performance Measurement

Performance of PVM-GRACE is measured for the two physics processes:

- Process 1 : $e^+e^- \rightarrow WW\gamma$
- Process 2 : $e^+e^- \rightarrow e^-\bar{\nu}_e u\bar{d}$

In the first process, 28 Feynman diagrams have to be calculated. In the second one, there are 88 Feynman diagrams requiring much more processing time. To carry out the numerical integration with a good accuracy ($\leq 1\%$), the amount of the data per each transfer is roughly 1.5MB (*Process 1*) and 3MB (*Process*

2). We configured PVM (version 3.3.10) on the workstation cluster for nine HP 9000/735s connected by FDDI.

N_{child} is the number of the PVM child process, each running on a different cpu. $N_{child} = 0$ is the non-PVM program running on 1 cpu. The measured elapsed times are shown in Fig. 1 and Fig. 2 for *Process 1* and *Process 2*, respectively. In Fig. 1, the elapsed time is the smallest for $N_{child} = 4, 5$ being reduced by more than a factor 2 when compared to $N_{child} = 0$. In Fig. 2, the elapsed time is the shortest for N_{child} over 9, less than 1/5 of the $N_{child} = 0$ case.

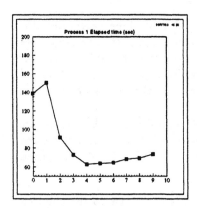 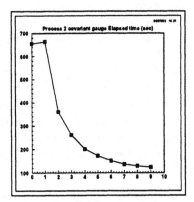

Fig. 1. Elapsed time (sec) versus number of child (N_{child}) for *Process 1*. **Fig. 2.** Elapsed time (sec) versus number of child (N_{child}) for *Process 2*.

The increase of the elapsed time for $N_{child} = 1$ is due to the overhead of PVM (packing/unpacking data, sending/receiving between tasks and network waiting time). The performance curves saturate when the overhead due to the data transfer compensates the time reduction obtained by the addition of new cpu. For *Process 1*, the saturation appears faster than *Process 2* as the processing time in each cpu is shorter.

4 Discussion

This behavior can be understood as follows:

$$\frac{T_{process}}{N_{child}} + N_{child} \cdot T_{network} + \alpha = T_{PVM-GRACE} \tag{1}$$

where $T_{process}$ is total processing time and $T_{network}$, the transfer time for returning results to the parent process. α is the overhead constant. Equation (1) shows $T_{PVM-GRACE}$ has minimum at $N_{child} = \sqrt{\frac{T_{process}}{T_{network}}}$. Since $T_{process}$ is fixed for each physics process, N_{child}^{opt} is depending on the characteristics of the physics process. Therefore, we have to choose the adequate number of N_{child} for each physics process so as to make PVM-GRACE most effective.

The data was fitted by using Equation (1). We obtained $T_{process} = 134$, $T_{network} = 5.2$, and $\alpha = 11.5$. Using these values, $N_{child}^{opt} = 5$ which is consistent with the expected value.

5 Conclusions

We have implemented PVM in GRACE successfully by modifying the integration program BASES. The performance of PVM-GRACE for the two physics processes have measured and we have got the following results:

- PVM-GRACE provides an efficient way to cut down the elapsed time by using ideal cpu cycle on a workstation cluster.
- the time reduction is strongly depending on the ratio, $R = \frac{T_{process}}{T_{network}}$, and therefore on the physics process.
- the best reduction time corresponds to an optimum $N_{child}^{opt} = \sqrt{R}$.
- however, to avoid network saturation, one should select a configuration with $N_{child} \leq N_{child}^{opt}$. For example, for Process 2 $N_{child} \sim 6$.

Since the results obtained with PVM-GRACE is encouraging, we have added a special option in the code generation to generate the FORTRAN code for PVM automatically.

References

1. Al Geist et al.: PVM3 user's guide and reference manual. ORNL/TM-12187.
2. MINAMI-TATEYA group: GRACE manual. KEK Report 92-19.
   T.Kaneko: A Feynman-graph generator for any order of coupling constants. Computer Physics Communication **92** (1995) 127-152.
3. S.Kawabata: A new Monte Carlo event generator for High Energy Physics. Computer Physics Communication **41** (1986) 127-153.
   S.Kawabata: A New Version of the Multi-dimensional Integration and Event Generation Package BASES/SPRING. Computer Physics Communication **88** (1995) 309-326.

Search of Molecular Ground State via Genetic Algorithm; Implementation on a Hybrid SIMD-MIMD Platform

N.Pucello[1], M.Rosati[3], M.Celino[2], G.D'Agostino[1], F.Pisacane[1], V.Rosato[1]

[1] ENEA, Divisione Nuovi Materiali, C.P. 2400 - 00100 Roma A.D. (Italy)
[2] ENEA, HPCN Project, C.P. 2400 - 00100 Roma A.D. (Italy)
[3] CASPUR, Università "La Sapienza" di Roma, P.le A.Moro, 00100 Roma (Italy)

Abstract. A genetic algorithm for ground-state structure optimization of a Palladium atomic cluster has been developed and ported on a SIMD-MIMD parallel platform. The SIMD part of the parallel platform is represented by a Quadrics/APE100 consisting of 512 floating point units while the MIMD part is formed by a cluster of workstations. The proposed algorithm contains a part where the genetic operators are applied to the elements of the population and a part which performs a further local relaxation and the fitness calculation via Molecular Dynamics. These parts have been implemented on the MIMD part and on the SIMD one, respectively. Results have been compared to those generated by using a Simulated Annealing technique.

1 Introduction

Genetic algorithms are optimization techniques, inspired to the evolution of living beings [1]. A population of guess solutions undergoes to successive transformations which allow its "evolution" to the optimal solution. The applied transformations are called "genetic operators": selection and mating. In the selection phase, each individual of the current population is evaluated on the basis of a given property ("fitness"); individuals with higher values of the "fitness" have a larger probability to be selected to produce the new individuals of the further generation. Mating is performed using suitable crossover and mutation operators which guarantee the formation of a new generation and a rapid evolution to the minimum.

A genetic algorithm has been recently applied to the determination of the lowest energy configuration of a Carbon cluster of 60 atoms; the proposed algorithm was composed by the combination of crossover transformation (two selected individuals were cut by a random plane and the formed halves glued together to form two parent clusters of the next generation) and a local relaxation procedure performed to relax the cluster structure after the crossover procedure and to evaluate the fitness (i.e. the cohesive energy) via a Tight-Binding hamiltonian [2]. In the present case, this algorithm has been partially modified, applied to a metallic cluster of $N = 147$ Palladium atoms and designed for a hybrid SIMD-MIMD parallel platform [3] composed by a SIMD Quadrics APE100 and by a cluster of workstations SUN Sparc20.

2 Description of the genetic algorithm

The initial population is formed by a number n of individuals (clusters) whose atomic coordinates are randomly generated. The potential energy per atom E_i of each cluster is the "fitness" which determines the selection phase. It is evaluated by using a n-body potential derived by a Second Moment approximation of a Tight Binding Hamiltonian (SMTB) [4, 5].

The algorithm can be schematically presented as follows. Let us assume $\{P_i\}_k$ a set of coordinates of the individuals of the k-th generation. $P_i \in \Re^3$, N being the (fixed) number of atoms of each cluster; $i = 1, n$ spans the different individuals of the population. Using this notation, the algorithm can be schematically represented as consisting of several steps in a loop:

1. Relaxation $R : \{P_i\}_k \rightarrow \{P_i'\}_k$
2. Fitness evaluation $H : \{P_i'\}_k \rightarrow \{E_i\}_k$
3. Selection $S : \{E_i\}_k \rightarrow \{p_i\}_k$
4. Crossover $C : p_i^{(k)} \otimes p_j^{(k)} \rightarrow p_i^{(k+1)} \otimes p_j^{(k+1)}$
5. Mutation $M : \{P_i\}_{k+1} \rightarrow \{P_i'\}_{k+1}$.

The procedure starts by generating the coordinates of the n individuals constituing the first generation. This is done by generating random sequences with the constraint of avoiding unphysical interparticles distances. On this configurations, the fitness evaluation H is performed. The Molecular Dynamics (MD) code evaluates the potential energy of the elements of the population.

The relaxation task (1) is accomplished in several steps. In the first one, the excess energy, eventually stored into the clusters, is removed by using a suitable modification of the classical MD technique (where particles velocities are suppressed as soon as their scalar product with the corresponding forces is negative). In the second step, the clusters are allowed to relax at a constant temperature T_{MD} for some thousand of time steps (each of them $\tau = 10^{-15}$ sec). The relaxation temperature T_{MD} represents a first parameter which has been optimized to provide the best efficiency of the algorithm.

Task (3) provides the mating probability p_i of each cluster of the current population

$$p_i = \frac{\exp\left[-(< E > -E_i)/K_B T_{CO}\right]}{\sum_{i=1}^{n} \exp\left[-(< E > -E_i)/K_B T_{CO}\right]} \qquad (1)$$

where $< E >= \frac{1}{n}\sum_{i=1}^{n} E_i$. The selection rule introduces the selection temperature T_{CO} which is a further parameter for the algorithm.

Task (4), introduced by Deaven and Ho [2], allows to generate two new individuals starting from of the current generation selected according to the mating probability. Each of the two selected clusters is bisected by a common plane of random orientation which separates the clusters in equal halves. Halves of different clusters are then glued together to form the new individuals. The process is iterated as many times as to generate an equal number n of new individuals (which will form the new generation). Each individual of the new generation undergoes to a further genetic operator (mutation, task (5)), which introduces, according to given probability μ (which is a further parameter of the algorithm) a random fault in the cluster coordinates by displacing a given atom of a random amount. The new generation is thus formed and ready to start a new optimization cycle. A further parameter of the entire process is n, the (fixed) number of individuals of the generation. At the end, the genetic algorithm is thus defined by a set of four parameters which will be adapted to provide the fastest possible convergence and a minimum energy configuration: n, T_{CO}, T_{MD} and μ.

We have tested the algorithm on a cluster of 147 Palladium atoms whose ground-state configuration [6] has an icosahedral structure with a potential energy $E_p = -3.6398$ eV/atom.

3 Description of the parallel platform and of the algorithm implementation

The implementation of the algorithm has been performed by using a hybrid (SIMD-MIMD) platform whose MIMD part is formed by two Sun Sparc20 workstations (connected by a FDDI network) exchanging information by means of PVM libraries [7]. One of the MIMD nodes is connected to a massively parallel SIMD machine (Quadrics/APE100 with a selectable number of floating point units (FPU hereafter) namely 128 and 512 [8]) via a transputer link which ensures a sustained throughput of 600 kbytes/sec. C+TAO libraries [8] allow the host-SIMD connection. This particular arrangement of massively parallel platform has been named GENESI (GEneralized NEtworked SImd) [3].

As most of the computation time is required for the relaxation and the fitness evaluation tasks, these have been demanded to the SIMD platform. MD is, in fact, efficiently parallelizable on the SIMD machine as each node contains the coordinates of a given individual of the current population and it is able to perform its relaxation without data exchanges with other FPUs. Following the relaxation and fitness evaluation, the SIMD machine returns clusters coordinates and fitness to the host workstation which provides the broadcasting of these informations to the other nodes of the MIMD platform. Each node performs the cross over and the mutation operators up to produce a subset of the whole new generation. The as-prepared individuals for the new generation are then sent to the host workstation which loads the SIMD machines with the new coordinates and starts the new relaxation and fitness evaluation phase.

4 Discussion

The main results achived by the present work can be summarized in the following points:

- A suitable revised version of the genetic algorithm proposed in [2] for the cluster geometry optimization has been devised. Modifications are essentially related to the implementation of local relaxation algorithm which allows, through the use of both gradient descent and thermal relaxation, to evacuate large excess energy introduced by the cross-over and mutation operations. It has been demostrated that the presence of the thermal relaxation phase highly increases the rate of convergence of the whole algorithm.

- Parallel efficiencies higher than 40% have been reached, thus leading to a sustained power > 10 Gflops (on 512 nodes).
- The proposed algorithm has shown a faster convergence to the region of the optimum solution altough simulated annealing provided, in this particular case, a more accurate solution (in a given wall clock time).
- A prototype version of an hybrid (SIMD-MIMD) platform has been realized and used, for the first time, in a scientific context. The platform has shown a great deal of potentiality, as well as a high flexibility to deal with different algorithm complexities. A further version of the platform is going to be installed in which the MIMD part will be replaced by an integrated Meiko CS2 machine. Each MIMD node will then dispose of a SIMD platform with a different number of FPU's in a way to allow a further flexibility in partitioning the whole platform in order to contain the most suitable number of FPU's for the given application.

References

1. G.J.E. Rawlins, "Foundations of Genetic Algorithms", Morgan Kaufmann Publ., San Mateo, California, 1991.
2. D.M.Deaven and K.M.Ho, Phys.Rev.Lett. 75 (1995) 288.
3. In ENEA (Italian National Agency for New Technologies, Energy and Environment), a project, named GENESI (GEneralized NEtworked SImd) is in progress as basis for the PQE2000 project that will produce the hardware and the software of the tera-flop platform based on hybrid hyper-nodes MIMD-SIMD (for more information: http://www.enea.it/hpcn/hpcn01e.html).
4. V.Rosato, M.Guillope and B.Legrand, Phil.Mag. A 59 (1989) 321-336.
5. F.Cleri and V.Rosato, Phys.Rev.B 48 (1993) 22-33.
6. G.D'Agostino, Mat.Sc.Forum 195 (1995) 149.
7. A.Geist, A.Beguelin, J.Dongarra, W.Jiang, R.Manchek, V.Sunderam, "PVM 3 user's guide and reference manual", Oak Ridge National Laboratory, Oak Ridge, Tennessee (1994).
8. A.Bartoloni et al., Int. J. of Mod. Phys. C, vol. 4 N. 5 (1993) 955-967.

Using PVM in the Simulation of a Hybrid Dataflow Architecture

M.A.Cavenaghi[1]   R.Spolon[1]   J.E.M.Perea-Martins[1]   S.G.Domingues[1]
A.Garcia Neto[2]

[1] State University of São Paulo - Department of Computer Science
C.P. 467 - 17033-360 - Bauru - SP - Brazil
e-mail: marcos@ifqsc.sc.usp.br
[2] University of São Paulo - Department of Physics and Informatics
São Carlos - SP - Brazil

Abstract. The aim of this work is to propose the migration of a hybrid dataflow architecture simulator developed on an uniprocessor system (sequential execution) to a multiprocessor system (parallel execution), using a message passing environment. To evaluate the communication between two process, we have been simulating an optical interconnection network based on WDM technics.

1   Introduction

The hybrid dataflow architecture Wolf is inspired in the Manchester Dataflow Machine (MDFM) [1] proposed by J.Gurd and I.Watson in the 70's. One of the most important feature of the Wolf architecture is the capability of detecting and executing code with a close data dependency between two subsequent instructions (sequential code) without using a special compiler (we compile code written in the functional language Sisal using the MDFM compilers).

Saw [6] is a time driven simulator developed in C++ that implements the Wolf characteristics. As a logic continuation of our research on the Wolf architecture, we intend to implement a parallel version of the Saw simulator (SawP). We will use the Parallel Virtual Machine (PVM) [3, 4] to implement the new simulator.

One of the most degradable problem in dataflow architectures is the communication between to nodes. To improve this communication, we will propose a theoretic model based on the Wavelength Division Multiplexing (WDM) [5] optical interconnection system. We intend to apply this interconnection model to the SawP simulator and we expect to obtain an improvement in inter-nodes communication.

2   The Wolf Architecture

The architecture is made of the units shown in Figure 1. The tokens are inserted in the Wolf through the input. When a token is read, it is sent to the Collecting Network (CN). The CN receives the tokens and sends them to the Data Memory

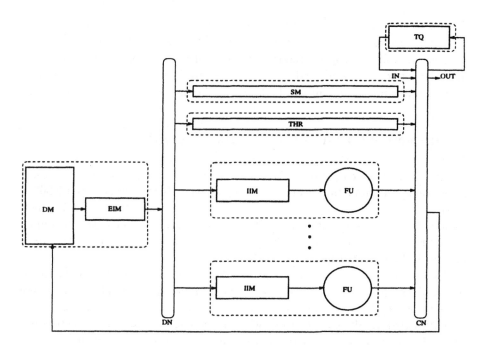

Fig. 1. Wolf Architecture

(DM) whether it is free. Otherwise the tokens are sent to the Token Queue (TQ) where they are stored. The TQ is a memory that stores tokens avoiding deadlocks.

The DM is responsible for the token's matching. When a token arrives in the DM, a package consisting of the incoming token and the partner data is sent to the External Instruction Memory (EIM). The EIM verifies if the package is directed to a data structure instruction. If so, the EIM generates an executable package by fetching the opcode and sent the new package to the Distribution Network (DN). Otherwise if the package is not directed to a data structure instruction, the EIM sends the incoming package without any modifications to the DN.

The DN distributes tokens to the Structure Memory (SM) if the package is executable or to the first idle Internal Instruction Memory (IIM) if it is a common package. The IIM stores information about the node to whom data is directed. With these informations and the incoming package, the IIM makes an executable token and sends it to its associate Functional Unit (FU) where the token is executed and up to two result tokens can be produced. Eventually, these tokens are sent to the CN and the path repeats again.

The SM is responsible for the execution of instructions referring to vector manipulations. This SM is identical to the Structure Store [1]. The SM stores vectors and executes instructions such as reads, writes, allocation and deallocation of space. The reads and writes can occur in any order due to the implementation of the I-Structure mechanism. In its final form, this memory will be based on a vector pipeline processor capable of processing vectors with fetch to the FU's.

When an executable token arrives in the FU, up to two tokens can be produced. These tokens may be of two kinds: free or chained. Free tokens are sent to the CN recirculating to the pipeline. Chained tokens remain in the FU internal register set to be used as operands of subsequent instructions. The THR unit is not relevant for this work and its functional description is not presented here. More details about this unit can be found in Garcia Neto [6].

3   Implementation of the Simulator SawP

SawP is a time driven parallel simulator developed to implement the Wolf characteristics. This simulator has been developed under IBM PowerPC and RS6000 platforms using the programming language C. Dashed rectangles in Figure 1 show how each part of the original Wolf architecture was rearranged to implement the architecture simulated by SawP. Each dashed rectangle represents a process to be executed in parallel.

Due the hybrid characteristics of the Wolf architecture,i.e. the ability to detect and execute sequential code without the need of a special compiler, the units named IIM and FU were put together in a single process. The other units were arranged in different processes whereas there are no close relations between them. The communication between two processes is implemented using PVM primitives. Global synchronization mechanisms can be implemented using the characteristics of the dataflow model.

3.1   Simulation of an Optical Interconnection Network

To improve communication between to processes we are simulating an optical interconnection network due its bandwidth (around 30 THz), noise immunity and high security. These features has the potential to solve some problems of electrical networks such as signal processing speed, latency and the amount of high-speed electronic processing required for transporting and routing.

One of the most promising technology of optical devices, applied in interconnection networks is the WDM (Wavelength Division Multiplexing) coupler, whose function is to multiplex or to demultiplex optical signals into one optical fiber. In fact, WDM combines or separates many channels in parallel, dividing the optical fiber bandwidth into many non-interfering channels, each one operating using an unique wavelength. Each node in the system has a number of optical transmitters (lasers) and optical receivers (filters), thus it can send or receive signals to or from the WDM coupler.

WDM devices has attractive characteristics such as passivity, i.e. don't need electrical power, the same device can perform both multiplexing and demultiplexing, the optical channels are independent of one another and it is transparent to data format. It enables the implementation of an optical interconnection network with large bandwidth, scalability, modularity, reliability, easy to implement virtual topologies, reconfigurability, protocol transparency and facilitated broadcasting.

On a single-hop architecture a node may communicate to each other in one direct hop and the network don't need high-speed transit routing, but use tunable transivers. Our optical interconnection network is based on group division of WDM channels [7] and our purpose is to develop a system with a simple software control using single-hop architecture, but avoiding the time dependence of tunable transivers. The tunable transivers are replaced by a set of fixed transivers, it increase the node degree but eliminate the tuning time. Considering a WDM coupler with C communication channels and N fan-out, the architecture has C groups with N/C nodes per channel.

To avoid dependence of transiver tuning time, each node has one fixed transmitter and C fixed receivers with one buffer memory per receiver. Each transmitter is assigned to one unique channel and each one of the C receivers of a node is assigned to a unique specific channel, thus each node can receive informations simultaneously from all channels.

To avoid collision is used a common clock and the channel access is regulated by time division. In the same slot time, the system can have C nodes transmitting simultaneously, each one on its specific channel. If two or more nodes using different channels send informations to the same receiver node, all information are received by the assigned receiver and stored in its buffer memory. Each data packet has a fixed size and is initialized with the address of the destination node, thus the receiver node processor recognizes the time slot and verifies the local buffers, searching informations addressed for it.

References

1. Gurd, J. R., Watson, I., Glauert, J.R.W. A Multi-Layered Dataflow Computer Architecture. Internal Report, DCS, Univ. Manchester, march, 1980.
2. Ferscha, A., Tripathi, S. Parallel and Distributed Simulation of Discret Event Systems. Technical Report 3336, Dept. of Computer Science, University of Maryland, College Park, August, 1994.
3. Geist, A. et al. PVM 3 User's Guide and Reference Manual. Oak Ridge National Laboratory, Tennesse, September, 1994.
4. Geist, A. et al. PVM: Parallel Virtual Machine - A User's Guide and Tutorial for Networked Parallel Computing. The MIT Press, 1994
5. Mukherjee, B. WDM-Based Local Networks. Part I: Single-Hop Systems. IEEE Network, May,1992.
6. Garcia Neto, A., Ruggiero, Carlos A. The Proto-Architecture of Wolf, a Dataflow Supercomputer. Actas da XVIII Conferencia Latinoamericana de Informatica, Centro Latinoamericano de Estudios en Informatica, Las Palmas de Gran Canarias, Espanha, August, 1992.
7. Ghose, K. Performance Potentials of an Optical Fiber Bus Using Wavelength Division Multiplexing. Proceedings of the SPIE, Vol. 1849, pages 172-182, 1993.

Managing Nondeterminism in PVM Programs

Michael Oberhuber

TU München
Institut für Informatik
Lehrstuhl für Rechnertechnik und Rechnerorganisation (LRR–TUM)
email: oberhube@informatik.tu-muenchen.de

1 Introduction

Clusters of workstations gain more and more success and acceptance. One reason are emerging standard libraries for interprocess communication, like PVM and MPI. But despite all success it is still hard to exploit the available computing power of these systems. Beside the fact that design and coding of parallel programs is more complex than in the sequential case, testing and debugging is more complicated since nondeterminism caused by interprocess communication plays a major role.

Nondeterminism implies two unpleasant properties: successive executions of the same program with the same input values often do not exhibit identical behavior and watching the program influences the execution. The former effect is called *unpredictable behavior*, the latter one is called *probe effect*. Consequences of these properties are poor test coverage, only partially useful regression tests, and the impossibility of applying methods based on cyclic debugging.

Of course, there exist several approaches for debugging parallel programs. But the problems which are entailed by nondeterminism are only partially addressed. The only tool focusing on nondeterminism is mdb [1]. However, none offers an appropriate methodology for debugging nondeterministic programs in a systematic way.

Facing this fact we investigated a new idea addressing the problems of nondeterminism in a different way. In this paper we present our first approach to this problem, which followws the concept of controlled execution driven by patterns. The first implementation of this approach is currently being realized for PVM within the scope of the TOOL–SET environment [2].

2 Nondeterminism in Parallel Programs

Parallel programs consist of a set of sequential processes that have to exchange information in order to work together. The information a process receives is another kind of input which is not provided by the user, but by other processes. In contrast to sequential programs, whose behavior is solely determined by its input data, there is another parameter, the interprocess communication, that influences the execution of a parallel program. In general, the communication behavior is not totally predefined, i.e. there is no total synchronization, which is the reason why nondeterminism is an inherent property of parallel programs. Due to unpredictable influences, there may be different communication behaviors related to one input.

As a result of this fact, the cycle of development is incomplete or even interrupted. The phases we want to address here are *testing* and *debugging*. For testing it is necessary to consider different structures of communication in addition to a set of input data. In order to avoid random testing an environment is needed to enforce as much different communication structures as possible. If an error is discovered while testing, you have to switch to the debugging phase. But often it is hard to reproduce the erroneous execution because of unpredictable behavior and the probe effect.

The goal of our investigations is to provide the user with the ability to specify not only input data for tests, but also the sequence of interprocess communications. This also guarantees reproducibility for debugging and regression testing.

3 Common Objects and PVM

The approach we sketch in this paper is not focused on PVM, but tries to deal with non-determinism in general. Therefore, the approach is based on an abstract communication model called *common objects*. Common objects represent a primitive but sufficient model of interprocess communication that can be mapped to all known implementations like message passing or shared memory.

When sending a message with PVM it has to be tagged for identification. Together with the target process, which is a mandatory parameter for a send operation, it forms a sort of an unidirectional channel. A channel is an incarnation of a common object. It is identified by a tuple consisting of a message tag and a target process.

The objects can be seen from the point of object oriented design. In that way processes invoke methods of these objects to get their states or to manipulate it. The basic methods of common objects are a *modify* operation and an operation to get the current state of an object, which we call *read* operation. While a send operation is equivalent to a *modify* operation, a receive call is composed of a *read* and *modify* operation. The reason is that messages are consumed by a receive call.

In PVM the receiving process is always determined. Either a task identifier is given explicitly, or a group identifier maps the operation to a set of tasks. This results in a communication model, where free access to common objects is restricted to send operations. This is shown in Figure 1.

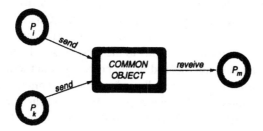

Fig. 1. Competition between senders

4 Managing Nondeterminism with Communication Patterns

The management of nondeterminism is based on events which are related to common objects. Each access to a common object triggers an event. According to PVM we are talking about communication events, i.e. about send and receive events.

If there is no total synchronization, several degrees of freedom exist in relation to the order of communication events. Regarding the situation in figure 1, you can see that it depends on the sequence of the send events which message P_m eventually receives. Generally, two sequences exist: $P_i \succ P_k$ and $P_k \succ P_i$.

During the phases of testing and debugging it would be of considerable profit to be able to enforce one or the other sequence of events. Therefore, the intention of our approach is to order specified events at runtime. To avoid the effort of specifying a complete communication sequence for an execution we decided to use patterns. The pattern serves as a rule about the order of events. The sequence, or parts of the sequence, are being created at runtime in a generic way based on the given pattern. The patterns may be given by the user or by additional tools. During runtime an instrumented library is responsible for the control of the PVM communication.

To minimize the control mechanism the number of relevant events is reduced. There are two properties of communication events that are suitable to be exploited.

First, we will focus on common objects. That means, only those events are regarded the relating operations of which affect the same communication object. In the sense of PVM the approach filters out messages that are dedicated to the same process with identical message tags. Other messages are of no interest.

Second, we want to concentrate on events that are independent [3]. Independent events may occur in any order. This property can, at least partially, be ascertained at runtime. But for this first approach support will be provided by post mortem race detection or previous control runs. A drawback of this approach is, that the analysis of trace data may provide insufficient information, since it only describes one single execution.

The remaining questions are how to define communication patterns and how to control an execution according to patterns. A pattern's task is to describe an order of accesses to a common object. It is not necessary to specify each single access. Unimportant accesses may be omitted in the specification of a pattern. There exist two basic templates for communication patterns: one cyclic pattern and one priority driven pattern. The templates are filled with a total ordered set of process identifiers.

Let us show the use of patterns in a short example. To realize a repeated alternate sending of process P_1 and P_2 to the common object EX_OBJECT choose the cyclic template and specify $P_1 \succ P_2$. The effect is shown in figure 2 on the left. The priority driven template tries to give access rights to the process with highest priority as long as the whole application is not blocked or the process has not terminated. Otherwise the process with next lower priority is allowed to access. Coming back to the example above it could be meaningful to arrange a communication behavior where process P_1 sends all its messages before P_2. This is an appropriate task for the second template. Figure 2 shows the specification and the result on the right.

Additionally it is possible to define more complex patterns by combining different templates or identical templates with different specifications for one common object.

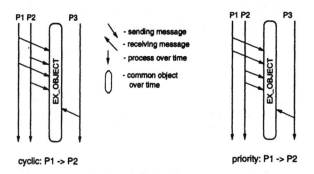

Fig. 2. Examples for the use of templates

The transition from one pattern to another is triggered by a given arbitrary event. By combining patterns it is possible to specify all possible behaviors of a parallel program.

Deterministic finite automata (DFA) are responsible for the control of the communication. There are two levels of automata. On the first level the templates are monitored. The second level consists merely of one automaton that cares for the transitions between templates.

5 Future Work

The parts of the approach that have to be active at runtime will be realized as a distributed tool extension of OCM [4], the OMIS (On-line Monitoring Interface specification) compliant monitoring system of LRR-TUM. Since all requirements like an event/action mechanism, operations to start and stop tasks, are integral parts of OMIS it is straightforward to use the extension facilities of OCM to build the control mechanism on top of OMIS services. Another topic which is not yet addressed are tools to support the user with information about possible spots of nondeterminism.

References

1. Suresh K. Damodaran-Kamal and Joan M. Francioni. mdb: A Semantic Race Detection Tool for PVM. In SHPCC94 [5], pages 702–709.
2. T. Ludwig, R. Wismüller, R. Borgeest, S. Lamberts, C. Röder, G. Stellner, and A. Bode. THE TOOL-SET – An Integrated Tool Env ironment for PVM. In J. Dongarra, M. Gengler, B. Touracheau, and X. Vigo uroux, editors, *Proceedings of EuorPVM'95 Short Papers*, Lyon, France, September 1995. Ecole Normale Supérieure de Lyon. Technical Report 95-02.
3. E. Fromentin, N. Plouzeau, and Michel Raynal. An Introduction to the Analysis and Debug of Distributed Computations. In ICAP95 [6], pages 545–553.
4. T. Ludwig, R. Wismller, and M. Oberhuber. OCM - An OMIS Compliant Monitoring System. In *To appear in Proceedings of EUROPVM96*, Munich, oct.
5. IEEE. *Proceedings of SHPCC94*, Knoxville, May 1994.
6. IEEE. *Proceedings of First Int. Conf. on Algorithms and Architectures for Parallel Processing*, Brisbane, March 1995.

Parallelization of an Evolutionary Neural Network Optimizer Based on PVM

Thomas Ragg

Institut für Logik, Komplexität und Deduktionssysteme
Universität Karlsruhe, 76128 Karslruhe, Germany
email: ragg@ira.uka.de

Abstract. In this paper the parallelization of a evolutionary neural network optimizer, ENZO, is presented, that runs efficiently on a workstation-cluster as a batch program with low priority, as usually required for long running processes. Depending on the network size an evolutionary optimization can take up to several days or weeks, where the overall time required depends heavily on the machine load. To overcome this problem and to speed up the evolution process we parallelized ENZO based on PVM to run efficiently on a workstation-cluster using a variant of dynamic load balancing to make efficient use of the resources. The parallel version surpasses other algorithms, e.g., Pruning, already for small to medium benchmarks, with regard to performance and overall running time.

1 Introduction

In the last years we developed ENZO [3, 2], an evolutionary neural network optimizer that surpasses other algorithms, e.g., Pruning or Optimal Brain Surgeon [1] with regard to performance and scalability [6]. ENZO is freely available software that runs under various platforms. It is typically used from researchers on workstations in a local area network. Since an evolutionary optimization can take up to several days or weeks depending on the network size, it is usually required to run the program in the background with the lowest priority. For this purpose we parallelized ENZO based on PVM [4] to run efficiently on a workstation-cluster but as a batch program with low priority.

2 Evolution as Framework for Neural Network Optimization

Neural networks are most commonly used as non-linear function approximaters, which learn the underlying functional relationship from data. This is usually done by gradient descent methods like Backpropagation or Rprop [8, 7]. Optimizing the topology of neural networks is an important task, when one aims to get smaller and faster networks, as well as a better generalization performance. Moreover, automatical optimization avoids the time consuming search for a suitable topology, i.e., the underlying model is optimized through the evolution. Derived from the biological example the basic principles of the evolutionary optimization may be summarized as follows (Fig.1):
The population is initialized with candidate solutions, e.g., neural networks, which are interpreted as individuals. Each individual is evaluated according to a fitness function which in turn serves as optimization criterion. Additional constraints can be embedded in the fitness function like penalty terms for size or performance.

New candidate solutions, called offsprings, are created using current members of the population, called parents, through mutation or re-combination. The mutation operator creates a copy of the parent with small variations, whereas the re-combination mixes the offspring's properties from two parents. In terms of neural network optimization the mutation operator is more important, and may remove or add weights resp. neurons to the offspring network. The selection of parents is randomly but biased, preferring the fitter ones, i.e., fitter individuals produce more offsprings. For each new inserted offspring another population member has to be removed in order to maintain a constant population size.

ENZO's Main Loop

```
pre_evolution();
repeat
    selection();
    mutation();
    optimization();
    evaluation();
    survival();
until stop_evolution();
```

Figure 1: The basic algorithm of ENZO. The pre-evolution and optimization phase are marked as time consuming parts.

This selection is usually done according to the fitness of each member (Survival of the fittest). Before evaluating the fitness of each member it is locally optimized (trained). Thus ENZO can be regarded as a global meta heuristic acting on locally optimal solutions. To find a good global solution about d generations are necessary with a population of 30 networks producing 10 offsprings each generation, where d is the dimension of the input vector for the neural network. The time complexity for an evolutionary optimization can be determined to $O(\lambda epn^{1.5})$, and if we regard Pruning as a special case of evolution with one parent and one offspring each generation, we get $O(epn^2)$, where e is the number of training epochs, p equals the number of patterns, n the number of weights and λ the number of offsprings (cf. [6]). Since offsprings can be trained independently from each other this most time consuming part of the evolution is worth being parallelized (Fig.1), thus reducing the time cpmlexity of evolution to $O(epn^{1.5})$.

3   Parallel Implementation

The design of the system leads straightforward to a master/slave model, where the standard ENZO program acts as master, while the training of networks is done in parallel on the available hosts (See also Fig.2). Two problems concerning load balancing arise in this context. Firstly, the offsprings are of different quality and do not require the same training duration, i.e., the number of epochs necessary to learn the patterns can vary from a few to several hundreds. Secondly, since the training time of the networks ranges from minutes to days, the overall computing time depends heavily on the load of the machine during that time. Several strategies for dynamic load balancing have been proposed recently [9], where random polling mechanisms has been proved to be simple and efficient [5].

To minimize the effects described above dynamic host assignment is used as illustrated in Fig.3, which has similarities to random polling. At the beginning of each generation as many offsprings are trained by slave tasks as hosts are available. If a task terminates, the next untrained offspring is selected. If all offsprings are either fully trained or a slave task for them is still running, the "slowest" process is determined and the corresponding offspring is trained additionally on the free host. This is indicated in Fig. 3 by the vertical dashed lines (offsprings 4, 5 and 10 restarted on hosts 1,

3 and again 1). The "slowest" process is determined in the following way: 1. Select the offsprings with at most k corresponding slaves, where k is a user-definable constant. 2. Determine among their corresponding slaves the one with the lowest ratio of trained epochs to the overall time since the task started.

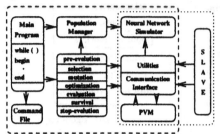

Figure 2: Design of ENZO. The parts surrounded by the dashed lines form the master program, while the parts surrounded by the dotted lines form the slave program. The PVM-library is hidden behind a communication interface.

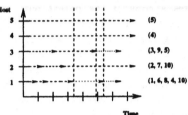

Figure 3: Assignment of slave process (1-10) to hosts (1-5, y-axis) over time. The arrows indicate the time needed for computation. The vertical dashed lines mark the time, where the slowest process is determined and forked on a second host.

4 Results and Discussion

In previous work [6] we showed that using evolutionary algorithms as a meta heuristic as done in ENZO is concurrently the best available optimization method with regard to network size and performance. The sequential program is about one order of magnitude slower than Pruning. We compare the time performance on two benchmarks of different sizes with respect to the number of weights in the network: Breast cancer diagnosis, a classification problem from the UCI repository of machine learning databases, and learning a winning strategy for the endgame of Nine Men's Morris (sometimes called "mill"). For both benchmarks we chose a population size of 30 networks generating 10 offsprings each generation. The time measurements were made on our workstation cluster containing 18 machines ranging from SUN Sparc 2 to SUN Sparc 20. The average computing power is 0.74, if the computing power of a Sparc 20 is set to 1.0 as reference value. All following results are averaged over 11 runs.

The starting network for breast cancer diagnosis had 108 weights and the topology 9-8-4-1, where the first number gives the size of the input vector, the last number is the size of the output vector, and the other numbers correspond to the size of the hidden layers. Both the training set and test set consists of about 350 patterns. The average classification performance on the independent test set was 98.1% for ENZO and 96.7% for Pruning, while ENZO decreased the network size to 10 weights compared to 29 weights achieved by Pruning.

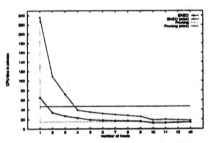

Figure 4: Absolute time requirement for ENZO compared to Pruning for the breast cancer benchmark. Measurements were made for processes with low as well as with high priority.

Figure 4 shows the average computing time for Pruning compared to ENZO dependent on the number of hosts. If more than 4 machines are available, both algorithms need about the same computing time. Furthermore, if all machines were used, ENZO running with low priority was only slightly slower than the process with high priority (18 min. resp. 12 min.). That the gain of performance using ten machines is rather small compared to 5 machines is due to the fact that we have exactly 5 Sparc 20 machines in our pool. Figure 5 shows the results for the strategy learning benchmark. Nine Men's Morris is a non-trivial game, since it has about 56000 essentially different board positions, but still solvable be exhaustive search.

4000 patterns were randomly selected from the database and trained to a neural network using a feature based input coding. The starting network had 2110 weights and the topology 60-30-10-1. For simplicity the mean square error on a test set with 1000 patterns is taken as a measurement of performance. The allowed error on the training set was 0.005, where ENZO achieved 0.0107 on the test set compared to 0.149 for Pruning. Already 4 machines were sufficient to achieve the same time performance for both algorithms.

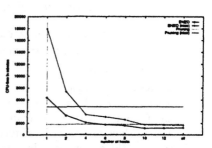

Figure 5: Absolute time requirement for ENZO compared to Pruning for strategy learning for the endgame of Nine Men's Morris.

Our experiments confirm the theoretical result that sequential evolution has about the same time complexity as Pruning, if $\lambda\sqrt{n} \approx n$, i.e., for large problems the time needed for Pruning grows faster compared to evolution. The parallel version based on PVM surpasses Pruning in absolute time requirement already for small to medium benchmarks using a standard workstation environment (with at least 5 machines). Furthermore, it is worth to remark that due to the design of both ENZO and PVM the development costs for the parallelization were very low, i.e., several weeks for design, implementation and testing were sufficient.

References

1. Christopher M. Bishop. *Neural Networks for Pattern Recognition.* Oxford Press, 1995.
2. Heinrich Braun and Thomas Ragg. ENZO – Evolution of Neural Networks, User Manual and Implementation Guide, ftp i11ftp.ira.uka.de, directory: pub/neuro/ENZO. Technical Report 21/96, Universität Karlsruhe, Fakultät für Informatik, 1996.
3. Heinrich Braun and Joachim Weisbrod. Evolving feedforward neural networks. *Proceedings of the International Conference on Artificial Neural Nets and Genetic Algorithms*, 1993.
4. A. Geist, A. Beguelin, J. Dongarra, W. Jiang, R. Manchek, and V. Sunderam. *PVM: Parallel Virtual Machine. A User's Guide and Tutorial for Networked Parallel Computing.* MIT Press, Cambridge, Mass. and London, England, 1994.
5. P.Sanders. Analysis of random polling dynamic load balancing. Technical Report 12/94, Fak. für Informatik, Universität Karlsruhe, 1994.
6. Thomas Ragg, Heinrich Braun, and Heiko Landsberg. A Comparative Study of Optimization Techniques. In *Evolutionary Computation and Machine Learning. Worksop at the International Conference on Machine Learning '96, Bari, Italy*, 1996.
7. M. Riedmiller and H. Braun. A Direct Adaptive Method for Faster Backpropagation Learning: The RPROP Algorithm. In *Proceedings of the ICNN*, 1993.
8. D. E. Rumelhart, G. E. Hinton, and R. Williams. Learning Internal Representations by Error Propagation. *Nature*, 1986.
9. Behrooz A. Shirazi, Ali R. Hurson, and Krishna M. Kavi. *Scheduling and load balancing in parallel and distributed systems.* IEEE Computer Soc. Press, Los Alamitos, Calif, 1995.

On the Synergistic Use of Parallel Exact Solvers and Heuristic/Stochastic Methods for Combinatorial Optimisation Problems: A Study Case for the TSP on the Meiko CS-2

Antonio d' Acierno and Salvatore Palma

IRSIP - CNR, Via P. Castellino 111, I - 80131, Napoli
emails: {antonio|salvio}@irsip.na.cnr.it

Abstract. This paper deals with the exact solutions of combinatorial optimisation problems. We describe a parallel implementation of a Branch and Bound algorithm; we complete such software with heuristic schemes and we show results with reference to the well known TSP problem. We evaluate the performance by introducing the concept of *implementation efficiency* that allows to estimate the communication overhead notwithstanding the over–search problem.

1 Introduction

Combinatorial Optimisation Problems (COPs) consist in the search of optima for functions of discrete variables; many of those belong to the class of NP-hard problems for which the search of the actual optimum could require prohibitive amount of computing time.

COP's can be approached via exact techniques such as the Branch and Bound (B&B) algorithm, i.e. a *divide-and-conquer* strategy. At each iteration step this algorithm selects (according to a *selection* rule) a sub-problem, computes its lower bound (by supposing, for example, that minimisation problems are considered) and, if this bound is better than the current best solution, the sub-problem is split (according to a *branching* rule) in two or more sub-problems or, if possible, solved; otherwise it is skipped. When the list of sub-problems becomes empty, the algorithm terminates. A B&B algorithm, clearly, can easily keep memory of the *Best Case Solution (BCS)* simply by evaluating, in the case of minimisation problem, the minimum among nodes bounds in the current list; this allows to define, at each instant, the quality of the running solution. Even using massively parallel computers, however, an exact technique requires, in the worst case, a time that grows exponentially with the input size. For this reason heuristic/stochastic approaches, that find sub-optimal solutions using well bound amounts of time and of computing power, are widely used. These methods typically converge to local minima and, in general, they *do not* provide any information about the distance from the actual optimum.

In literature there are many examples of application of parallel computers to solve optimisation problems. Parallel versions of exact and/or heuristic techniques have been in fact proposed to improve the speed and the efficiency of classical methods. All these examples show that the efforts have been focused on the possibility of improving each single approach.

In a previous paper [1] it has been described a fundamentally new approach to the exact solution of combinatorial optimisation problems on parallel computers, based on the synergistic use of exact and stochastic/heuristic techniques. The basic idea of the Multi-algorithmic Parallel Approach (MPA) [1] was the attempt to gather different techniques developed in literature for a certain problem with general purpose approaches in order to reduce the computational time needed to reach the *optimal* solution. Starting from the informal description of the B&B algorithm, in fact, it should be clear that the availability, during the search process, of sub-optimal good solutions should allow the B&B algorithm to cut branches, so lowering the computing time.

In this paper we further work out the original idea proposed in [1] by integrating the MPA with a parallel implementation of a classical B&B algorithm [2] proposed for the well known Traveling Salesman Problem (TSP); this algorithm uses, as bounding function, the $O(n^3)$ Hungarian solution for the assignment problem. The whole parallel algorithm has been realised using the *de facto* standard PVM [3] and has been tested on the Meiko CS-2.

2   The Parallel B&B Implementation

A B&B algorithm can be implemented in parallel according to two main strategies; the first one (*small-grained approach, SGA*) is based on the parallelisation of bounds computation by using, for example, SIMD machines. Using the SGA we have that the number of nodes split (*branches*) is exactly the same both in the sequential case and in the parallel one. Such an approach, on the other hand, is useless when the bound function is simple (say linear in the dimension of the problem). Alternatively and/or concurrently, we can use a *coarse-grained approach (CGA)*, where the search across the solutions tree is parallelised. This paper is focused on the latter strategy, owing to the parallel machine we used (Meiko CS-2). In this case, of course, the number of branches in the sequential case and in the parallel case are likely to be different, with the parallel implementation performing, on average, an higher number of branches. This phenomenon (*over-searching*) is, clearly, as higher as the sequential B&B is efficient and avoids useless paths in the search tree.

The B&B we choose can be implemented both using a *best in first out (BIFO)* strategy and a *last in first out (LIFO)* strategy. The BIFO strategy is, by definition, not bound in the memory complexity; moreover an actual BIFO parallel strategy (where at each iteration *each* processor selects the node from a *global* list) needs a lot of communication and sincronization so that it seems not well suited for the distributed memory MIMD parallel machine we are dealing with. For such reasons we used a mixed approach, that can be summarised as follows.

We used a *master-slave* structure. The master reads input data and branches the root problem; as there are nodes to be examined they are sent to slaves that can start working. The master branches sub-problems until a node has been sent to each slave. At this point it becomes idle and its list could be empty as well as could contain some nodes.

As soon as a slave finds a solution improving the current one, it sends a message to the master that update the current optimum, output this value and broadcast it to other slaves; this allows keeping the over–search as low as possible.

When a slave becomes idle since its list is empty, it sends a message to the master; if the master list is not empty, a node is sent back to the idle slave. Otherwise, the master broadcast a message to each slave asking for nodes; each slave with more than a node in its list send one of these (selected according its complexity) to the master that serves the idle son and will use other nodes for future requests. This strategy allows to have a well balanced load. The parallel algorithm of course terminates when each slave is idle and the master list is empty.

In our experiments, each slave acts according a LIFO strategy but it could act according a BIFO strategy. What is worth noting is that, in the latter case, we do not have an actual BIFO parallel B&B since, at each iteration, the nodes are selected having visibility just of the local list.

The parallel B&B described has been extended according to the idea suggested in [1]. Namely, we implemented two stochastic solvers (a simulated annealing and two-opt search) whose time duration is controlled via the number of iterations they perform. The slaves running such solvers are not served with nodes from the master; when they finish running heuristic solvers they ask for a node and become idle.

3   Results

Measuring the performance of a parallel B&B is not a simple task because of the over–search problem. Thus, we decided to distinguish between the *implementation efficiency* (that measures the communication overhead and the achieved load balancing) and the *global efficiency* (that gives an overall idea about the performance of the obtained implementation).

As concerns the *implementation efficiency*, this has been obtained by considering the time employed to solve the problem as a linear function of the number of branches. (This clearly introduces an approximation, since nodes at different levels in the search tree have different complexities as regards the bounding function). Figure 1 shows the difference between the linear approximation and the actual points for 10 random selected asymmetric problems of dimension 70 and 100, in the sequential case as well as in the parallel case with 12 slaves. The implementation efficiency, defined as the ratio between angular coefficients of approximating lines, is reported in tab. 1.

These results seem very interesting and are due to the following facts; *(i)* the custom version of PVM on the Meiko CS-2 introduces a negligible overhead, *(ii)* our choice of implementing an asynchronous parallel B&B allows to have a low number of communications and *(iii)* our load balancing strategy works very well (see tab. 1)

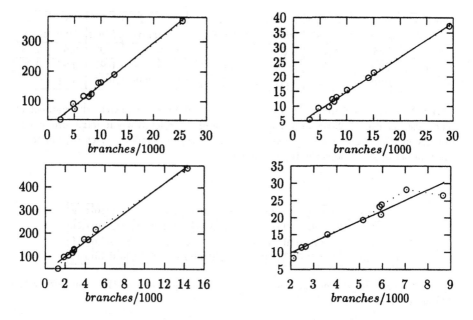

Fig. 1. The number of branches vs. the time for problems of dimension 70 (upper) and of dimension 100, in the sequential case (left) and in the parallel one (12 slaves).

Table 2 shows the global efficiency when 12 slaves are used, whose average value (0.87) while very interesting is lower than the average implementation efficiency for the same case (0.91), due to the over–search problem.

To give an idea of the whole performance of our algorithm, we show in figure 2 the results obtained on two classical problems from the TSPLIB archive. In these tests we simulate a *real* situation, i. e. we guess that X seconds (with, for example, $X = 1200$) can be used at most to find the solution (of course this solution, in many cases, does not represent the actual optimum) and we plot as a function of time the *Ratio* representing the quality of the solution (*Ratio=BCS/(current optimum)*, *Ratio*= 1 means problem solved). We show the performance of the parallel B&B, as well as the improvement obtained when the parallel B&B is integrated with a *simulated annealing* (SA) and with a *Two-Opt* heuristic (TO). It is worth noting in the latter case that we used 12 slaves for the parallel B&B; slave 12 (11) performs the SA (the TO) before becoming actually a slave in the parallel B&B.

4 Conclusions

We implemented a parallel B&B with a LIFO strategy, that can be simply extended to a local BIFO strategy, the actual BIFO strategy not seeming well suited for the hardware at hand. We showed how results can be improved by integrating into the parallel algorithm heuristic/stochastic methods; such an im-

problem	branches slave5	branches slave6	branches slave7	branches slave8
1	754	800	811	775
5	280	293	296	300
8	1081	1083	1135	1085

dim.	2 slaves	4 slaves	8 slaves	12 slaves
60	0.91	0.92	0.92	0.88
70	0.98	0.97	0.97	0.99
100	0.95	0.95	0.91	0.87

Table 1. Left: the number of branches performed by some slaves; the parallel implementation uses 8 slaves and it is run on random selected asymmetric problems of dimension 100. Right: the *implementation efficiency* for random selected problems as the number of slaves and the input size vary.

dimension	#1	#2	#3	#4	#5	#6	#7	#8	#9	#10	avg
60	0.74	0.73	0.80	1.06	1.68	3.57	0.56	0.75	0.84	0.84	1.16
70	0.79	0.58	0.63	1.04	0.69	0.82	0.49	0.82	0.87	0.90	0.76
100	0.68	0.78	0.51	0.55	0.78	0.49	0.71	1.52	0.65	0.46	0.71

Table 2. The global efficiency on random selected problem as the input size varies. The number of slaves is 12.

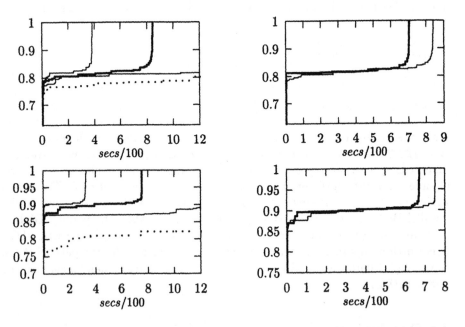

Fig. 2. The *Ratio* vs. the time using the parallel version of the branch and bound as the number of slaves varies (1, 6, 12, 24) using the parallel branch and bound (left). The same using 12 slaves and combining the parallel branch and bound with simulated annealing and two-opt (right). The problem are *swiss42* (upper) and hk48.

provement is likely to be greater if a local BIFO strategy is used; In fact, the BIFO strategy exploits in width the search tree so that the BCS is improved rapidly, while the current optimum not.

About the obtained results, we introduced the concept of *implementation efficiency* used to evaluate the communication overhead related to the exchange of global informations and to the load balancing mechanism. We obtained an interesting performance even if we do not have any miraculous solution. The average efficiency is less than 1, even if we have, in some cases, a super–efficiency. To have a parallel B&B that shows a steady super–efficiency it is necessary, in our opinion according to [4][5], to start from a sequential algorithm that does not optimise its search across the solution tree. As an examples of this common misunderstanding about the problem at hand, we cite [1]; the results presented, while still interesting, have been put into the perspective after an improvement of the used B&B.

To conclude, the problem related to the portability of the code has to be addressed. The PVM on the Meiko CS-2 does not need the presence of the *daemon*; since, in our implementation, each slaves have to ask for messages at each iteration, this allows to have negligible overhead as a consequence of the fact that the slave interrogates directly the communication hardware. To have good performance on classical platforms, where the slave should interrogate the daemon with a time–consuming operation, we can use the simple solution suggested in [1], where a catcher process is coupled with each slave. The work of the catcher is simply to wait for messages and to communicates with the slave via shared memory.

References

1. Bruno, G., d' Acierno , A.: The Multi–Algorithmic Approach to Optimisation Problems, High–Performance Computing and Networking, Lecture Notes in Computer Science, 919, Springer, Berlin, 1995.
2. Carpaneto, G., Toth, P.: Some New Branching and Bounding Criteria for the Asymmetric Traveling Salesman Problem, Management Science, 26 (7), 1980.
3. Geist, G. A., Sanderam, V. S.: Network-Based Concurrent Computing on the PVM System, Concurrency: Practice and Experience, 4 (1992), 293-311.
4. Lai, T. H., Sahni., Anomalies in Parallel Branch and Bound Algorithms, Comm. of ACM, 27, 6, 1984.
5. Lai, T.H., Sprague, A.: Performance of Parallel Branch and Bound Algorithms, IEEE Trans. on Computers, C34, 10, 1985.

Author Index

Springer-Verlag
and the Environment

We at Springer-Verlag firmly believe that an international science publisher has a special obligation to the environment, and our corporate policies consistently reflect this conviction.

We also expect our business partners – paper mills, printers, packaging manufacturers, etc. – to commit themselves to using environmentally friendly materials and production processes.

The paper in this book is made from low- or no-chlorine pulp and is acid free, in conformance with international standards for paper permanency.

Lecture Notes in Computer Science

For information about Vols. 1–1081

please contact your bookseller or Springer-Verlag